# Functional Analysis of the Human Genome

# The HUMAN MOLECULAR GENETICS series

Series Advisors

**D.N. Cooper,** *Charter Molecular Genetics Laboratory, Thrombosis Research Institute, University of London, UK*

**S.E. Humphries,** *Division of Cardiovascular Genetics, University College London Medical School, London, UK*

**T. Strachan,** *Department of Human Genetics, University of Newcastle-upon-Tyne, Newcastle-upon-Tyne, UK*

---

Human Gene Mutation
From Genotype to Phenotype
Functional Analysis of the Human Genome

*Forthcoming titles*

Molecular Genetics of Cancer
Environmental Mutagenesis
Human Genome Evolution

# Functional Analysis of the Human Genome

**Farzin Farzaneh**
*Department of Molecular Medicine, King's College School of Medicine and Dentistry, London, UK*

**David N. Cooper**
*Charter Molecular Genetics Laboratory, Thrombosis Research Institute, London, UK*

© BIOS Scientific Publishers Limited, 1995

First published 1995

All rights reserved. No part of this book may be reproduced or transmitted, in any form or by any means, without permission.

A CIP catalogue record for this book is available from the British Library.

ISBN 1 872748 46 5

**BIOS Scientific Publishers Ltd**
**9 Newtec Place, Magdalen Road, Oxford OX4 1RE, UK.**
**Tel. +44 (0)1865 726286. Fax +44 (0)1865 246823**

DISTRIBUTORS

Australia and New Zealand
   DA Information Services
   648 Whitehorse Road, Mitcham
   Victoria 3132

India
   Viva Books Private Limited
   4346/4C Ansari Road
   New Delhi 110002

Singapore and South East Asia
   Toppan Company (S) PTE Ltd
   38 Liu Fang Road, Jurong
   Singapore 2262

USA and Canada
   Books International Inc
   PO Box 605, Herndon,
   VA 22070

## TO ALI, BERYL AND NEIL

Typeset by Touchpaper, Abingdon, UK.
Printed by Biddles Ltd, Guildford, UK.

# Contents

|   |   |   |
|---|---|---|
| | Contributors | xi |
| | Abbreviations | xiii |
| | Preface | xvii |
| | Foreword. *V.A. McKusick* | xix |
| **1** | **Structure and function in the human genome.** *D.N. Cooper* | **1** |
| | Introduction | 1 |
| | Chromatin structure and transcription | 2 |
| |    Chromatin structure | 2 |
| |    Nucleosome positioning | 3 |
| |    Transcriptional domains | 4 |
| |    Chiasmata, recombination and recombination hotspots | 4 |
| |    Scaffold attachment regions | 5 |
| |    Centromeres | 6 |
| |    Telomeres | 6 |
| |    Origins of DNA replication | 7 |
| | Genes | 7 |
| |    Gene structure and organization | 7 |
| |    Pseudogenes | 8 |
| |    Functional organization of human genes | 9 |
| | Repetitive sequence elements | 11 |
| |    Tandem repeats | 12 |
| |    *Alu* sequences | 12 |
| |    LINE elements | 13 |
| |    Endogenous retroviral sequences | 14 |
| | Transcriptional regulation | 14 |
| |    Promoter elements | 14 |
| |    Enhancers | 15 |
| |    Negative regulatory elements | 16 |
| |    Locus control regions | 17 |
| |    Boundary elements | 19 |
| |    *Trans*-acting protein factors | 20 |
| |    Sequences involved in transcriptional termination | 21 |
| |    mRNA splicing and processing | 21 |
| |    Sequences involved in determining mRNA stability | 22 |
| |    Role of sequences in 5′ untranslated regions | 23 |
| | DNA methylation | 23 |
| |    Distribution of 5-methylcytosine | 23 |
| |    Replication of the methylation pattern and *de novo* methylation | 24 |
| |    Role of DNA methylation in the regulation of transcription | 25 |
| |    Role of DNA methylation in X-inactivation | 25 |
| |    Changes in DNA methylation during embryogenesis | 26 |
| |    DNA methylation and imprinting | 27 |
| | References | 28 |

## Contents

| | | |
|---|---|---|
| 2 | **Mapping the human genome.** *D.N. Cooper* | 43 |
| | Introduction | 43 |
| | Markers | 45 |
| |    Gene sequences | 45 |
| |    DNA polymorphisms | 46 |
| |    D-segments | 47 |
| |    Sequence-tagged sites | 47 |
| |    Inter-*Alu* PCR probes | 48 |
| |    Allele-specific oligonucleotides | 49 |
| | Cytogenetic mapping | 49 |
| |    Somatic cell hybrid analysis | 49 |
| |    Radiation hybrid mapping | 49 |
| |    Fluorescence *in situ* hybridization | 50 |
| |    *In situ* PCR | 50 |
| | High-resolution physical mapping | 51 |
| |    Yeast artificial chromosome cloning | 51 |
| |    Contig assembly | 52 |
| |    Pulsed-field gel electrophoresis and CpG island mapping | 52 |
| |    Chromosome jumping/linking libraries | 53 |
| |    DNA sequencing | 54 |
| |    Progress in physical mapping | 54 |
| | Genetic mapping | 54 |
| | Transcription map of the human genome | 57 |
| | Comparative gene mapping | 58 |
| | References | 59 |
| 3 | **Cloning the transcribed portion of the genome.** *P. Towner* | 69 |
| | Introduction | 69 |
| | Gene detection | 70 |
| | Preparation of target material | 71 |
| |    Isolation of total RNA | 71 |
| |    Isolation of mRNA | 72 |
| |    Preparation of cDNA | 72 |
| |    Selection of specific genes | 73 |
| | Library-based cDNA cloning strategies | 73 |
| |    Construction of a cDNA library | 74 |
| |    Screening cDNA libraries | 74 |
| |    Manipulation of identified cDNA sequences | 76 |
| | PCR-based isolation of genes from cDNA | 76 |
| |    Primer design | 77 |
| |    Mixed-pool or redundant oligonucleotide primers | 78 |
| |    Primary PCR reaction | 78 |
| |    Isolation of the 3' end of a cDNA | 79 |
| |    Isolation of the 5' end of a cDNA | 79 |
| |    Gene identification by differential display | 81 |
| | Expression systems | 81 |
| |    Expression using *E. coli* | 82 |
| |    Eukaryotic expression systems | 83 |
| | References | 83 |

| | | |
|---|---|---|
| **4** | **Retroviral insertional mutagenesis.** *F. Farzaneh, J. Gäken and S.-U. Gan* | **87** |
| | Introduction | 87 |
| | The retroviral life cycle | 88 |
| | Host range | 91 |
| | Replication-defective retroviral vectors | 92 |
| | Packaging cell lines | 93 |
| | Conditions required for efficient mutagenesis | 95 |
| | Mechanisms involved in retroviral insertional mutagenesis | 96 |
| | Mutation frequency | 98 |
| | Multiplicity of infection | 99 |
| | Mutant selection procedures | 99 |
| | Cloning of the sites of provirus integration | 99 |
| |     Construction of genomic libraries | 100 |
| |     PCR-mediated amplification | 100 |
| | Identification of the gene of interest | 100 |
| |     Identification of common sites of provirus integration | 101 |
| |     Library screening by nuclear run-on probes | 102 |
| | References | 103 |
| **5** | **Gene entrapment.** *H. von Melchner and H.E. Ruley* | **109** |
| | Introduction | 109 |
| | Gene trap vectors | 110 |
| | Cloning and analysis of flanking sequences | 116 |
| | Isolation and use of promoter-tagged sites | 116 |
| | Insertional mutagenesis in cultured cells | 118 |
| | Insertional mutagenesis in mice | 119 |
| | Identification of regulated genes | 121 |
| | References | 124 |
| **6** | **Gene transfer studies.** *D. Darling and M. Kuiper* | **131** |
| | Introduction | 131 |
| | What is transfection? | 131 |
| | What form should the DNA be in? | 132 |
| | Generalized requirements for eukaryotic gene transcription | 133 |
| |     Eukaryotic gene transcription | 133 |
| |     SV40-based plasmids | 134 |
| | Specialized eukaryotic host cells | 135 |
| | Specialized plasmids | 136 |
| |     Double insert plasmids | 136 |
| |     Inducible expression | 137 |
| |     Epstein–Barr virus-based plasmids | 138 |
| |     Shuttle vectors | 140 |
| |     Multifunctional plasmids | 140 |
| | Transfection procedures | 140 |
| |     Calcium phosphate co-precipitation | 141 |
| |     DEAE–dextran | 142 |
| |     Electroporation | 143 |
| |     Liposomes and lipid-based transfection | 143 |

| | |
|---|---|
| Adenovirus and poly-L-lysine-conjugated complexes | 144 |
| Alternative transfection procedures | 145 |
| Assays for new protein synthesis | 146 |
|    Dominant selectable marker genes | 146 |
|    Reporter genes | 147 |
| Analysis of cloned genes | 148 |
|    Identification of ligands for novel receptors | 148 |
|    Identification of transcription factors | 149 |
| References | 151 |

## 7 Foreign DNA integration and DNA methylation patterns. *W. Doerfler* — 155

| | |
|---|---|
| Introduction | 155 |
| The adenovirus system as a model | 156 |
|    Site selection in the integration of adenovirus DNA | 157 |
|    Modes of adenovirus DNA integration – a synopsis of data | 159 |
| On the mechanism of integrative recombination | 163 |
|    Insertion of foreign DNA by a versatile mechanism | 163 |
|    Studies on the mechanism of integrative recombination in a cell-free system | 165 |
| *De novo* DNA methylation of integrated foreign DNA | 167 |
|    *De novo* methylation of integrated foreign DNA: a cellular defence mechanism? | 167 |
|    Initiation of *de novo* methylation in mammalian cells is not predominantly dependent upon the nucleotide sequence of foreign DNA | 169 |
|    Methylation of triplet repeat amplifications in the human genome: manifestation of the cellular defence mechanism | 171 |
|    Alterations in patterns of cellular DNA methylation and gene expression as consequences of foreign DNA insertions into mammalian genomes | 172 |
| DNA methylation and gene activity | 173 |
|    A fully 5'-CG-3' but not a 5'-CCGG-3' methylated late FV3 promoter retains activity | 173 |
|    Topology of the promoter of RNA polymerase II- and III-transcribed genes is modified by the methylation of 5'-CG-3' dinucleotides | 174 |
|    Impact of 5'-CG-3' methylation on the activity of different eukaryotic promoters: a comparison | 175 |
| Uptake of foreign DNA through the gastrointestinal tract | 175 |
| A concept of oncogenesis – implications for gene therapy and research on transgenic organisms | 177 |
| References | 178 |

## 8 Transgenic animals in human gene analysis. *F. Theuring* — 185

| | |
|---|---|
| Introduction | 185 |
| Methodology | 187 |
| Transgenes to study gene regulation | 189 |
| Transgenes to study gene function | 193 |

|   |   |
|---|---|
| Functional analysis: gain-of-function | 194 |
| Functional analysis: loss-of-function | 197 |
| Conclusions | 199 |
| References | 200 |

## 9 Homologous recombination. *A. Mansouri* — 207

|   |   |
|---|---|
| Introduction | 207 |
| Embryonic stem cells | 208 |
| Principles of homologous recombination in mammalian cells | 209 |
|     Targeting vectors | 209 |
|     Promoterless constructs | 212 |
|     Positive–negative selection procedure | 212 |
|     Hit-and-run and in–out targeting strategies | 213 |
| Potential of homologous recombination in embryonic stem cells | 213 |
|     Developmental biology | 213 |
|     Animal models of human disease | 214 |
| Homologous recombination and gene therapy | 216 |
| Future perspectives: *Cre–LoxP* mediated gene targeting | 216 |
| References | 217 |

## 10 Complementation analysis. *A. Patel* — 221

|   |   |
|---|---|
| Introduction | 221 |
| Principles of somatic cell hybridization | 222 |
|     *De novo* and salvage pathways of nucleotide synthesis | 223 |
|     Purine nucleotide synthesis | 223 |
|     The *HPRT* gene | 224 |
|     *HPRT* variants | 224 |
|     Pyrimidine nucleotide synthesis | 225 |
|     Metabolic cooperation | 225 |
|     The HAT selection system | 226 |
|     Selection procedures for the isolation of hybrid cells | 226 |
| Identification of complementation groups and topological relationships | 227 |
|     Extinction and activation | 227 |
|     Assignment of complementation groups in clinical diseases | 228 |
|     Assignment of complementation groups in senescence | 229 |
|     Assignment of complementation groups in biochemical pathways | 230 |
|     Assignment of complementation groups in cytokine activity | 230 |
| Identification of the dominant/recessive nature of genetic lesions | 231 |
|     Chromosome segregation | 231 |
|     Dominant and recessive genetic changes involved in senescence | 232 |
|     Dominant and recessive nature of viral genes | 232 |
|     Dominant and recessive events in tumour progression | 232 |
|     Dominant nature of multi-drug resistance genes | 233 |
|     Dominant and recessive events involved in the immunological process | 233 |
|     Dominant and recessive developmentally regulated genes | 233 |
| Microcell fusion: principles and application to the chromosomal localization of genes | 234 |
|     Introduction to microcell fusion | 234 |

| | |
|---|---|
| General principles for microcell-mediated transfer | 234 |
| Pinpointing chromosomes involved in specific disease processes | 235 |
| Identification of tumour suppressor genes | 235 |
| Identification of genes involved in cellular senescence | 236 |
| References | 236 |

## 11 Antisense oligonucleotides: a survey of recent literature, possible mechanisms of action and therapeutic progress.

| | |
|---|---|
| *D. Pollock and J. Gäken* | 241 |
| Introduction | 241 |
| Some examples of antisense action in different systems | 242 |
| Targeting and design | 243 |
| Uptake of antisense oligonucleotides | 245 |
| Toxicity of antisense oligonucleotides | 246 |
| Modifications to the structure of antisense oligonucleotides | 246 |
| Possible mechanisms of action | 248 |
|    Steric inhibition | 248 |
|    RNase H-like cleavage of target RNA | 249 |
|    Triplex DNA formation | 251 |
| Double-stranded oligonucleotides | 253 |
| Circular oligonucleotides | 253 |
| Ribozymes | 253 |
| Non-specific cleavage of host RNA | 255 |
| Therapeutic applications | 258 |
| References | 260 |

**Index**     267

# Contributors

**Cooper, D.N.** Charter Molecular Genetics Laboratory, Thrombosis Research Institute, Manresa Road, London SW3 6LR, UK

**Darling, D.** Department of Molecular Medicine, Rayne Institute, King's College School of Medicine and Dentistry, 123 Coldharbour Lane, London SE5 9NU, UK

**Doerfler, W.** Institut für Genetik, Universität zu Köln, Weyertal 121, D-50931 Köln, Germany

**Farzaneh, F.** Department of Molecular Medicine, Rayne Institute, King's College School of Medicine and Dentistry, 123 Coldharbour Lane, London SE5 9NU, UK

**Gäken, J.** Department of Molecular Medicine, Rayne Institute, King's College School of Medicine and Dentistry, 123 Coldharbour Lane, London SE5 9NU, UK

**Gan, S.-U.** Department of Molecular Medicine, Rayne Institute, King's College School of Medicine and Dentistry, 123 Coldharbour Lane, London SE5 9NU, UK

**Kuiper, M.** Department of Molecular Medicine, Rayne Institute, King's College School of Medicine and Dentistry, 123 Coldharbour Lane, London SE5 9NU, UK

**Mansouri, A.** Max-Planck-Institute of Biophysical Chemistry, Department of Molecular Cell Biology, Am Fassberg, D-37077 Göttingen, Germany

**von Melchner, H.** Laboratory of Molecular Hematology, Department of Hematology, University of Frankfurt Medical School, Theodor-Stern Kai 7, D-60590 Frankfurt, Germany

**Patel, A.** Laboratoire de Biologie Cellulaire Haematopoietique, Université de Paris VII, Hôpital St Louis, Centre Hayem, 1 Avenue Claude Vellefaux, 75010 Paris, France

**Pollock, D.** Department of Molecular Medicine, Rayne Institute, King's College School of Medicine and Dentistry, 123 Coldharbour Lane, London SE5 9NU, UK

**Ruley, H.E.** Department of Microbiology and Immunology, Vanderbilt University School of Medicine, 1161 21st Avenue South, Nashville, TN 37232-2363, USA

**Theuring, F.** Schering AG Research Laboratories, Müllerstrasse 178, 13342 Berlin, Germany

**Towner, P.** Department of Molecular Medicine, Rayne Institute, King's College School of Medicine and Dentistry, 123 Coldharbour Lane, London SE5 9NU, UK

# Abbreviations

| | |
|---|---|
| Ad | adenovirus |
| AIDS | acquired immunodeficiency syndrome |
| ALS | amyotrophic lateral sclerosis |
| AP | activator protein |
| APRT | adenine phosphoribosyltransferase |
| ARBP | attachment region-binding protein |
| ARE | A–U rich element |
| AS | Angelman syndrome |
| ASO | allele-specific oligonucleotides |
| BES | $N,N$-bis(2-hydroxyethyl)-2-aminoethanesulphonic acid |
| Bluo-gal | 5-bromoindolyl-$\beta$-$o$-galactopyranoside |
| BPV | bovine papilloma virus |
| BrdU | bromodeoxyuridine |
| BWS | Beckwith–Wiedemann syndrome |
| CAT | chloramphenicol acetyltransferase |
| CEPH | Centre d'Étude du Polymorphisme Humain |
| CFTR | cystic fibrosis transmembrane conductance regulator |
| CHE | Chinese hamster embryo |
| CHO | Chinese hamster ovary |
| CMV | cytomegalovirus |
| CPE | cytoplasmic polyadenylation element |
| CPSF | cleavage-polyadenylation specificity factor |
| CREB | cAMP response element-binding factor |
| CstF | cleavage-stimulatory factor |
| DBD | DNA-binding domain |
| DEAE | diethylaminoethyl |
| DEPC | diethylpyrocarbonate |
| DHFR | dihydrofolate reductase |
| DZQ | 2,5 diaziridinyl-1,4-benzoquinone |
| EBNA | Epstein–Barr virus nuclear antigen |
| EBV | Epstein–Barr virus |
| ERE | oestrogen-responsive element |
| ES cells | embryonic stem cells |
| EST | expressed sequence tag |
| FHM | fat head minnow |
| FISH | fluorescence *in situ* hybridization |
| GM-CSF | granulocyte–macrophage colony-stimulating factor |
| HAT | hypoxanthine–aminopterin–thymidine |
| HEPES | $N$-(2-hydroxyethyl)piperazine-$N'$-(2-ethanesulphonic acid) |
| HGP | Human Genome Project |
| HIV | human immunodeficiency virus |
| HNF | hepatocyte nuclear factor |
| HMG | high mobility group protein |

| | |
|---|---|
| HPRT | hypoxanthine phosphoribosyltransferase |
| HSV | herpes simplex virus |
| HUGO | Human Genome Organization |
| IAP | intercisternal A particle |
| IFN | interferon |
| IMP | inosine monophosphate |
| IMPDH | inosine monophosphate dehydrogenase |
| IPTG | isopropyl thiogalactosidase |
| IRE | iron-responsive element |
| LCR | locus control region |
| LINEs | long interspersed repeat elements |
| LIF | leukaemia inhibitory factor |
| LTR | long terminal repeat |
| 5mC | 5-methylcytosine |
| MCS | multiple cloning site |
| MeP | methylphosphonate |
| MMLV | Moloney murine leukaemia virus |
| MMTV | murine mammary tumour virus |
| MSV | murine sarcoma virus |
| NER | nucleotide excision repair |
| NIH | National Institutes of Health |
| nt | nucleotides |
| ORF | open reading frame |
| PBS | phosphate-buffered saline |
| PCR | polymerase chain reaction |
| PEG | polyethylene glycol |
| PFGE | pulsed-field gel electrophoresis |
| PIR | glycoprotein processing inhibitor-resistant |
| PKC | protein kinase C |
| PMA | 4β-phorbol-12-myristate-13-acetate |
| PNET | primitive neuroectodermal tumour |
| PNS | positive–negative selection |
| PO | phosphodiester |
| polα | polymerase α |
| PRPP | 5-phosphoribosyl-1-pyrophosphate |
| PS | phosphorothioate |
| PTS | promoter-tagged site |
| PWS | Prader–Willi syndrome |
| RA | retinoic acid |
| RACE | rapid amplification of cDNA ends |
| RARβ | retinoic acid receptor-β |
| RCC | renal cell carcinoma |
| RFLP | restriction fragment length polymorphism |
| Ri | inner region |
| Ro | outer region |
| RSV | Rous sarcoma virus |
| SAR | scaffold-associated regions |
| SCID | severe combined immunodeficiency |
| SDS–PAGE | sodium dodecyl sulphate–polyacrylamide gel electrophoresis |

| | |
|---|---|
| SINEs | short interspersed repeat elements |
| snRNP | small ribonucleoprotein particle |
| SOD1 | superoxide dismutase |
| SSR | sample sequence repeats |
| STS | sequence-tagged site |
| $T_m$ | melting temperature |
| TCR | T-cell receptor |
| TdT | terminal deoxynucleotidyltransferase |
| TK | thymidine kinase |
| t-PA | tissue plasminogen activator |
| TRF | TTAGGG repeat-binding factor |
| ts | temperature sensitive |
| UTR | untranslated region |
| X-gal | 5-bromo-4-chloro-3-indolyl-β-D-galactoside |
| XGPRT | xanthine guanine phosphoribosyltransferase |
| XP | xeroderma pigmentosum |
| YAC | yeast artificial chromosome |

# Preface

Analysis of the structure and organization of the human genome is proceeding apace, bringing with it new insights into its function. This volume seeks to provide an overview of current knowledge of the relationship between structure and function in the human genome, together with a detailed description of some of the most important methodologies currently used for unravelling the function of genes and genomic structures.

The first two chapters provide an introduction to the relationship between structure and function in the human genome, focusing upon the role of specific DNA sequences in mediating particular cellular functions, and an up-to-date account of progress into the mapping of the human genome. The remaining chapters provide detailed accounts of some of the most important strategies aimed at the cloning and functional analysis of the genome. Topics covered include cDNA synthesis, gene transfer and expression strategies, retroviral insertional mutagenesis, gene entrapment, analysis of DNA methylation and imprinting, homologous recombination strategies for targeted gene replacement and inactivation, somatic cell hybridization and complementation studies, transgenic animals, including their use for the development of models for the study of human disease, and antisense oligonucleotides and ribozymes as tools for the experimental and therapeutic inhibition of expression of specific gene products. Together, these strategies promise to allow the functional analysis of the human genome to keep pace with, and complement, its structural analysis.

Farzin Farzaneh (*London*)
David N. Cooper (*London*)

# Acknowledgements

The editors gratefully acknowledge Jonathan Ray for his invaluable help in the planning and preparation of this volume. D.N.C. wishes to thank Vijay Kakkar for his continuing support and enthusiasm, Mike Chipperfield for providing information on the current status of the Genome Data Base, the Thrombosis Research Trust, the Welton Foundation, Charter plc., British–American Tobacco plc. and Sun-Life Assurance Co. Ltd for their much appreciated research funding, Jackie Vasper for keeping him fit and Margaret McLaughlin for her forebearance and understanding. F.F. wishes to acknowledge the Leukaemia Research Fund, Cancer Research Campaign, Medical Research Council and the Lewis Family Charitable Trust for the support of work in his laboratory, and Lindsay for her support and patience.

# Foreword

Genomics is the science of the structure, the organization and the function of the human genome. The Human Genome Project aims to identify all the functional genes and determine the nucleotide sequence of the entire genome. Although it is fundamentally a mapping project – the nucleotide sequence is the ultimate map – it was first formally proposed as a sequencing project. The Human Genome Project was initially conceived in 1985 by molecular geneticists. Many other scientists were under the impression that the proposed project would be an expensive exercise in mindless sequencing without much biology, and for that reason strenuously opposed it. Their criticisms have been largely silenced by the demonstrated usefulness of gene mapping as the basis for understanding normal function and particularly for elucidating the nature of specific diseases. Map-based gene discovery has become a (perhaps the) leading paradigm in contemporary biomedical research. Specialists in all branches of medicine are using the mapping strategy to study their most puzzling disorders.

It is particularly since 1980 that molecular genetic methods have come to the aid of gene mapping. Genes with unknown function have been identified. Determining their function has called for innovative approaches: conjectures concerning function made on the basis of structural similarity of the gene and gene product to previously known genes and gene products; analysis of gene function through the creation of directed mutations (thus mimicking the 'experiments of nature' on which the medical geneticist depends for analysis of gene function); inhibition of gene expression with antisense oligonucleotides; gene transfer techniques, especially those creating transgenic mice; and others.

There are many types of genome map. One of the most useful is the cDNA map (also known as the exon or functional map) which locates all the genes that are functional as indicated by their transcription into messenger RNA. The cDNA map is a useful beginning for the functional analysis of the human genome and a good basis for complete sequencing. This is the approach that has been taken by the group of Craig Venter and by others. The efficient mapping of their so-called ESTs (expressed sequence tags) will greatly facilitate the genome project.

When the Human Genome Project has achieved its goal, around the year 2005 by present prediction, what will it provide? It will provide a source book for study by biologists and clinicians for a long time thereafter. This is because the function will still not be known for all 60 000–70 000 genes (or whatever the correct number is). Furthermore, the variation in all these genes in the 6 billion people who will probably inhabit the Earth at the time the project is completed,

## Foreword

will also not be known. And finally, the relationship between that variation in genomic structure and variation in function will remain to be established. This collection of reviews on the functional analysis of the human genome provides a good background for an important aspect of genomics and for determining the function of the many genes uncovered by the Human Genome Project.

Victor A. McKusick (*Baltimore, MD*)

# 1

# Structure and function in the human genome

David N. Cooper

## 1.1 Introduction

The haploid human genome comprises some $3.2 \times 10^9$ base pairs (bp) of DNA. The bulk of the genome comprises DNA sequence of varying degrees of repetitivity, whilst about 10% of the genome represents single copy sequence containing perhaps 70 000 genes (Figure 1.1). The repetitive portion contains a mixture of DNA sequence elements which may have an architectural or regulatory role or may be merely 'junk DNA'. (It is as well here to remember Sydney Brenner's admonition that junk is not synonymous with rubbish – the latter is thrown out, the former is stored in the attic!) Our formidable task is to dissect the structure of the human genome in its entirety in order to discern the

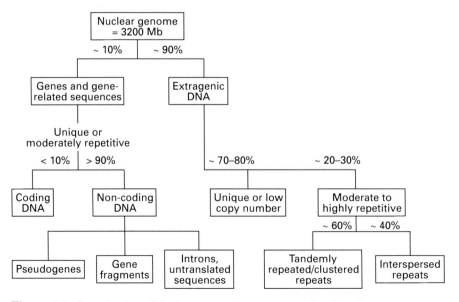

**Figure 1.1.** Organization of the human nuclear genome (after Strachan, 1992).

function of its individual component parts. An adequate knowledge of the structure of the human genome is a prerequisite for understanding its function.

Structure can be studied at many levels from the cytogenetic (chromosome banding) to the linear DNA sequence of the chromosome. In between, the structure of chromatin and the nature and distribution of different types of DNA sequence element provide intermediate levels of organization. The structure of the genome may soon be understood, at least on one level, in terms of its linear DNA sequence (see Chapter 2). However, this sequence will also contain other levels of encoded information concealed within sequence motif distributions and periodicities. The task ahead is to discern function from structure; this may involve the analysis of the distribution of long tracts of repetitive DNA or of single bases in a regulatory motif.

With the elucidation of the genetic code some 30 years ago, the amino acid sequence of a protein could be predicted from the DNA sequence of a gene. The sequencing of long stretches of chromosomal DNA is now providing us with new codes to break; much more challenging than the original genetic code. Within the non-coding portion of the genome are sequences that determine chromosome architecture, mediate both chromatin condensation and decondensation, organize transcriptionally active domains or silence them, sequences that promote or repress gene transcription, sequences that interact with a myriad of DNA-binding proteins and which in turn allow these proteins to interact with each other to regulate gene activity, sequences that potentiate the basic processes of DNA replication, recombination, repair, etc. The list is enormous and we are clearly on the threshold of a new genetic revolution which will require new techniques of data analysis rather than merely of data collection. It is the task of this chapter to attempt to relate the structure of the human genome to its function as well as providing both an introduction and overview to the rest of the volume.

## 1.2 Chromatin structure and transcription

### 1.2.1 Chromatin structure

Eukaryotic chromosomes contain DNA in a highly coiled and condensed form. DNA is first packaged by the histones of the nucleosome (see Section 1.2.2). Chains of nucleosomes comprise the 10 nm fibre and this is coiled to form the 30 nm fibre which is in turn further coiled to form chromatin (reviewed by Kornberg and Lorch, 1992; Wolffe, 1992; Paranjape et al., 1994). The highest degree of condensation is found in transcriptionally inactive regions. To allow transcriptional activation, chromatin must be uncoiled; a process which occurs in at least three stages: the unfolding of large chromosomal domains (25–100 kb), the remodelling of the chromatin structure of gene regulatory regions and the alteration of nucleosome structure in transcribed regions. The unfolding of the higher order structure of chromatin may be initiated by sequences known as locus control regions (LCRs; see below). Unfolding then reveals binding sites for activator proteins which, once bound, alter nucleosome positioning and reveal further binding sites for activator proteins. Decondensed chromatin then provides an accessible template for assembly of the transcription initiation complex.

### 1.2.2 Nucleosome positioning

DNA is organized and packaged into chromatin by nucleosomes. A nucleosome comprises a histone tetramer $(H3/H4)_2$ and two histone H2A/H2B dimers wrapped around two turns (160 bp) of DNA and stabilized by a single linker histone H1 (van Holde, 1993). The assembly of promoter elements into nucleosomes brings about the repression of basal transcription (Grunstein, 1990; Felsenfeld, 1992; Paranjape *et al.*, 1994).

Most DNA sequences are readily incorporated into nucleosomes. The exceptions are poly(dA)·poly(dT) tracts and Z-DNA (Wolffe, 1992) on account of their rigidity or unusual structure. Incorporation of DNA into nucleosomes may be strongly influenced by DNA sequence. Wada-Kiyama and Kiyama (1994) have suggested that an unusual structure (base slippage or partial triplex structure) formed by a polypurine·polypyrimidine sequence allows bending of the DNA, which in turn influences nucleosome positioning.

Clearly, in order that the chromatin may be transcribed, the regulatory sequences of a gene promoter must be accessible to incoming transcription factors. In this context, nucleosomes appear to provide more than just packaging. Indeed, histone–DNA interactions within the nucleosome are critically important in determining the accessibility of the DNA to these factors (reviewed by Wolffe, 1994). Regulatory sequences may be in contact with specific histones within the nucleosome so that modification [e.g. acetylation (Lee *et al.*, 1993; Sommerville *et al.*, 1993; Turner, 1993) or dissociation (Jackson, 1990; Clark and Felsenfeld, 1992)] of the histones rather than displacement or complete removal of the nucleosome (Felsenfeld, 1992) will provide access to *trans*-acting factors. The incorporation of high mobility group proteins, HMG14 and HMG17, can also facilitate transcription factor binding by disrupting nucleosome structure (Crippa *et al.*, 1993; Paranjape *et al.*, 1994). The affinity with which a transcription factor interacts with DNA in the nucleosome appears to depend upon the exact location of the recognition sequence within the nucleosome ('translational positioning'; Li and Wrange, 1993). Nucleosome positioning may in turn be influenced by other transcription factors (e.g. hepatocyte nuclear factor 3, HNF3; McPherson *et al.*, 1993).

The folding of DNA as a result of nucleosome assembly and packing may serve to bring regulatory elements, and any proteins bound to them, into closer proximity in order to maximize the transcriptional competence of the polymerase complex. One example of this process is the juxtaposition of the oestrogen-responsive enhancer and the liver-specific regulatory elements of the *Xenopus* vitellogenin B1 gene promoter, normally located some 180 bp apart; this results in a 100-fold enhancement of transcription (Schild *et al.*, 1993). Nucleosome positioning therefore both provides a three-dimensional template for the assembly of the transcriptional initiation complex and serves to prevent the formation of inhibitory DNA–histone interactions.

### 1.2.3 Transcriptional domains

The idea that chromatin structure and chromosomal organization can influence gene expression is hardly new but its most recent form, the concept of transcriptional domains whose boundaries are determined by chromatin accessibility, has considerable explanatory value (reviewed by Sippel et al., 1992; Dillon and Grosveld, 1994). Actively transcribed genes are more accessible to DNase I than are inactive genes (Weintraub and Groudine, 1976; Wood and Felsenfeld, 1982) and the DNA of the former is known to adopt a more extended conformation (Björkroth et al., 1988).

Transgenic assays have provided a different kind of evidence for transcriptional domains: whereas the integration of minimal promoter/gene combinations often yield only a low level of expression depending upon the chromosomal location of the transgene ('position effects'), the transfer of much longer sequences, putatively corresponding to chromosomal domains, invariably results in a high level of expression and a normal regulatory capacity (Bonifer et al., 1990; Schedl et al., 1993; Strauss et al., 1993). These findings suggest that the functional elements controlling a gene's expression may often be located at a very considerable distance from the gene and that the gene as a functional unit may be much larger than the extent of the coding region. Another example of this phenomenon is provided by the LCR approximately 60 kb upstream of the human β-globin (*HBB*) gene (see Section 1.5.4). The LCR is associated with an altered chromatin conformation around the β-globin gene region spanning some 150 kb (Forrester et al., 1990).

The transcription of human genes is therefore concentrated into units and these units appear to be clustered in groups within nuclear domains in which the transcriptional machinery is concentrated (Jackson et al., 1993a). Structural loops arising from the attachment of specific regions of chromatin to the nuclear matrix appear to coincide with these functional domains. Indeed, scaffold-associated regions (SARs; see Section 1.2.5) often mark the boundaries of transcriptionally active chromatin domains (Phi-Van and Strätling, 1988; Levy-Wilson and Fortier, 1989; McKnight et al., 1992).

### 1.2.4 Chiasmata, recombination and recombination hotspots

Chiasmata represent the cytological evidence for recombination. At meiosis, each pair of homologous chromosomes has at least one chiasma and the number of chiasma per pair is proportional to the size of the chromosome. Chiasma frequency is a function both of distance from the centromere (Laurie and Hulten, 1985) and chromosome band 'flavour': T bands exhibit a sixfold higher chiasma frequency than G bands (Holmquist, 1992).

Recombination is considerably higher in human females than in males, as evidenced by the average distance between two markers in the female genetic map being 85% longer than in males (see Chapter 2). Recombination also tends to increase towards the telomeres (the distal 15% of chromosomes contains 40% of the chiasmata) and intrachromosomal homologous recombination may be enhanced by transcription in mammalian cells (Thomas and Rothstein, 1989; Nickoloff, 1992)

Obligatory recombination occurs during male meiosis within the pseudoautosomal region, a 2.6 Mb stretch of homologous sequence at the tip of the short arms of the X and Y chromosomes (Rouyer *et al.*, 1986; Brown, 1988; Petit *et al.*, 1988; Ellis and Goodfellow, 1989).

A number of different DNA sequences appear to be recombinational hotspots in the genomes of mice and men. $(CAGA)_6$ and $(CAGG)_{7-9}$ represent hotspots of recombination in the murine MHC gene cluster (Steinmetz, 1987). Other sequences thought to promote recombinational instability are alphoid repeats (see below; Heartline *et al.*, 1988), Z-DNA ('left-handed DNA'; Wahls *et al.*, 1990a) and minisatellite sequences (see below; Chandley and Mitchell, 1988; Wahls *et al.*, 1990b). Recombinational breakpoints have also been found to be associated with topoisomerase I cleavage sites in the rat genome (Bullock *et al.*, 1985). The majority of these cleavage sites contain the sequences CTT and GTT. It may therefore be that the process of non-homologous recombination is mediated by topoisomerase I.

### *1.2.5 Scaffold attachment regions*

Chromatin is attached to the nuclear matrix or scaffold at specific sites, SARs. The organization of chromatin with respect to the nuclear scaffold is thought to determine chromosome architecture in terms of its functional domains and may therefore influence gene activity (reviewed by Laemmli *et al.*, 1992). These regions do not appear to share extensive sequence homology but often comprise 200 bp of AT-rich DNA [e.g. the sequence AATATTTTT in the murine immunoglobulin κ gene locus (Cockerill and Garrard 1986) or the artificial sequence, AATATATTT (Mielke *et al.*, 1990)]. A number of proteins are known to bind to SARs including the attachment region-binding protein, ARBP (von Kries *et al.*, 1991) and histone H1 (Izaurralde *et al.*, 1989).

SARs appear to be preferentially associated with topoisomerase II cleavage sites (Cockerill and Garrard, 1986; Mielke *et al.*, 1990; reviewed by Laemmli *et al.*, 1992) and share sequence homology with binding sites for homeobox proteins (Boulikas, 1992). Topoisomerase II plays a role in the segregation of daughter chromosomes after DNA replication and also in chromosome condensation; it binds preferentially to SARs (Adachi *et al.*, 1989). Vertebrate topoisomerase II cleavage sites also occur in association with SARs and manifest a consensus sequence, A/GNT/CNNCNNGT/CNGG/TTNT/CNT/C (Spitzner and Muller, 1988).

SARs also appear to be preferentially associated with enhancer-type elements. Indeed, SARs stimulate heterologous gene expression in reporter gene experiments (Stief *et al.*, 1989; Xu *et al.*, 1989; Phi-Van *et al.*, 1990; Klehr *et al.*, 1991). SAR regions from the chicken lysozyme gene locus have been shown to confer position-independent and tissue-specific regulation on the whey acid protein transgene in murine mammary tissue (McKnight *et al.*, 1992), indicating that SARs can promote the establishment of independent genetic domains. A *cis*-acting regulatory element 3' to the human $^{A}\gamma$-globin (*HBG1*) gene, known to be associated with the nuclear matrix, has been shown to bind specifically to an A–T rich binding protein (SATB1) that binds to SARs (Cunningham *et al.*, 1994).

## 1.2.6 Centromeres

The centromere is essential for normal disjunction of the chromosomes following cellular division at meiosis and mitosis (reviewed by Rattner and Lin, 1988). Centromeric DNA consists of arrays of tandemly repeated DNA sequences which have undergone an homogenization process by repeated genetic exchanges (Jackson *et al.*, 1993b; Warburton *et al.*, 1993). The basic 171 bp repeat of alphoid satellite DNA comprises the bulk of centromeric DNA and contains a 17 bp binding site for the centromere-specific protein CENP-B (Willard, 1990). Using a combination of oligonucleotide primer extension and immunocytochemistry, Mitchell *et al.* (1992) showed that the alphoid repeats were closely associated with the kinetochore (the structural element on the chromosome that binds to the mitotic spindle). Further, the presence of $(AATGG)_n$ $(CCATT)_n$ repeats in the centromeric region suggests that stem–loop structures might form which could serve as specific recognition sites for kinetochore function (Catasti *et al.*, 1994).

## 1.2.7 Telomeres

Telomeres allow the end of the chromosomal DNA to be replicated completely without the loss of bases at the termini (reviewed by Blackburn, 1991, 1994; Gilson *et al.*, 1993). They are the sites at which the pairing of homologous chromosomes is initiated and in humans contain long arrays (averaging about 10–15 kb) of minisatellite DNA (see Section 1.4.1) comprising tandem hexanucleotide repeats, most frequently TTAGGG (Brown, 1989). Other sequences (e.g. TTGGGG, TGAGGG) are also known (Allshire *et al.*, 1989; Brown, 1989). These sequences and their variants are tandemly repeated but are non-randomly distributed and are polymorphic in terms of their location (Brown *et al.*, 1990).

The simple sequence of telomeres is synthesized by the ribonucleoprotein polymerase, telomerase (Blackburn, 1992), and is thought to protect the ends of chromosomes from degradation during DNA replication. Telomere length decreases with age or the number of cell divisions. Two proteins that bind to these repeats, TTAGGG repeat-binding factor (TRF) and a second factor related to the ribonucleoproteins, have been described (Zhong *et al.*, 1992; Ishikawa *et al.*, 1993).

*1.2.8 Origins of DNA replication*

Chromosomal DNA replication has long been thought to initiate at specific points (origins) and to proceed outward bidirectionally. Although a number of putative origins of replication have been identified in mammalian species (reviewed by De Pamphilis 1993; Coverley and Laskey, 1994), data are sparse. Perhaps the best characterized is that found in human between the δ- and β-globin genes on chromosome 11 (Kitsberg *et al.*, 1993a). This replication origin is bi-directional and is functional regardless of the transcriptional state of the β-globin gene. Computer analysis of six putative origins of replication (including one in the human c-*myc* oncogene) has identified a fairly redundant consensus sequence, A/TAA/TTTA/G/TA/G/TA/TA/TA/TA/G/TA/C/TA/T GA/TA/C/TA/CAA/TTT (Dobbs *et al.*, 1994).

The units of DNA replication in human range in size from 50 to 600 kb and are often clustered (Hand, 1978). The gene-rich R bands replicate early in S phase whilst the G bands replicate late. Housekeeping genes invariably replicate early whereas tissue-specific genes can be early or late replicating (some replicate earlier when transcriptionally active; Goldman *et al.*, 1984; Hatton *et al.*, 1988). Non-transcribed genes on the X chromosome also replicate late (Torchia *et al.*, 1994).

## 1.3 Genes

*1.3.1 Gene structure and organization*

The coding portion of the human genome, roughly 5% of the total DNA complement, probably contains between 50 000 and 100 000 different gene sequences. Thus the smallest human chromosome, 21, may well contain as many as 2000 genes. It is now known that most genes in higher organisms are not contiguous but rather are a complex mosaic of coding (exon) and non-coding (intron) sequences. Each individual gene differs not only with respect to its DNA sequence specifying the amino acid sequence of the protein it encodes, but also with respect to its structure. A few human genes are devoid of introns [e.g. thrombomodulin (*THBD*) which spans 3.7 kb; Shirai *et al.*, 1988] whilst others possess a considerable number [e.g. 52 in the von Willebrand factor (*VWF*) gene (length 175 kb; Mancuso *et al.*, 1989) and 79 in the dystrophin (*DMD*) gene (length 2400 kb; Den Dunnen *et al.*, 1992; Roberts *et al.*, 1992)].

Exons in human genes rarely exceed 800 bp in length (Hawkins, 1988). However, exceptionally long internal exons of 3106 bp and 7572 bp are present in the human factor VIII (*F8*) and apolipoprotein B (*APOB*) genes respectively. An 837 bp 5′ exon is present in the human cytochrome P450 (*CYP4A11*) gene whilst 1802 bp and 2703 bp 3′ exons are present in the human factor VII (*F7*) and Duchenne muscular dystrophy (*DMD*) genes respectively. Introns vary enormously in size in eukaryotes from 30 bp or less to tens of kilobases.

Well over 1000 genes have now been mapped to within single chromosome bands. Some 80% map to the G–C rich R bands whilst 20% map to G bands (Bickmore and Sumner, 1989; Craig and Bickmore, 1993). A similar distribution is apparent for CpG islands: 86% are located in R bands (Larsen *et al.*, 1992; Craig and Bickmore, 1994). 'Housekeeping' genes are strictly confined to the R bands, together with about half of the tissue-specific genes (Holmquist, 1992), whereas the remainder of the tissue-specific genes are present in the G bands. Holmquist (1992) has recognized four different types of R band; T bands, which are often found at telomeres, represent a subset or 'flavour' of R bands with the highest G–C content and contain between 58 and 68% of R band genes (Ikemura and Wada, 1991; Mouchiroud *et al.*, 1991; Saccone *et al.*, 1992; Craig and Bickmore, 1993) and the majority of CpG islands (Craig and Bickmore, 1994).

Chromosomes 13 and 18 appear to possess a relatively low gene density and chromosome 19 a relatively high density as evidenced by the chromosomal assignment of some 320 cDNAs derived from a brain cDNA library (Polymeropoulos *et al.*, 1993). These observations are consistent with those first made by McKusick (1991). It should be noted that, except for chromosome 21, the smallest chromosome, only chromosomes 13 and 18 are associated with trisomies that are compatible with live birth.

Several examples are known of human genes being located within the introns of other genes. Thus the *F8A* gene is located within intron 22 of the factor VIII (*F8*) gene and is transcribed in the opposite orientation to that of the *F8* gene (Levinson *et al.*, 1990). Similarly, the oligodendrocyte-myelin glycoprotein (*OMG*) gene is located within an intron of the neurofibromatosis type 1 (*NF1*) gene (Viskochil *et al.*, 1991).

Two genes encoding *erbA* homologues, *ear*-1 and *ear*-7, which are located at the same locus on human chromosome 17, possess overlapping exons but are transcribed from opposite DNA strands (Miyajima *et al.*, 1989). Another example of this phenomenon of exon sharing is provided by the rat locus encoding gonadotropin-releasing hormone; a second gene, *SH*, is encoded by the other strand (Adelman *et al.*, 1987). It would appear likely that significantly long antisense overlapping open reading frames also exist in the cDNA of other human genes (Merino *et al.*, 1994). Perhaps both DNA strands possessed coding capacity at some earlier stage in evolution and we may still observe the remnants of this ancient organization.

### 1.3.2 Pseudogenes

Pseudogenes are DNA sequences which are closely related to functional genes but which are incapable of encoding a protein product on account of the presence of deletions, insertions and nonsense mutations which abolish the reading frame (reviewed by Wilde, 1985). Some human pseudogenes are transcribed (e.g. Nguyen *et al.*, 1991; Takahashi *et al.*, 1992; Bristow *et al.*, 1993) but these transcripts are not translated. There are two major types of pseudogene: the first arises through the duplication and subsequent inactivation of a gene. This type of pseudogene retains the exon/intron organization of the parental gene and are often closely linked to the parental gene. Examples include the pseudogenes in the α- and β-globin clusters (e.g. Cheng *et al.*, 1988). The second type of pseudogene contains only the exons of the parental gene, usually possesses a poly(A) tail at the 3' end and is dispersed randomly in the genome. These *processed* genes are thought to have originated as mRNAs which have then become integrated into the genome by retrotransposition (i.e. the reverse transcription of the mRNA and the integration of the resulting cDNA). One example of a retrotranscribed pseudogene is that related to the Pi class glutathione S-transferase gene (*GSTPP*; Board *et al.*, 1992).

Pseudogenes are relatively common in the genome (McAlpine *et al.*, 1993; several hundred are known) and may be especially prevalent in multigene families (e.g. β-globin, actin, HLA, interferons, snRNAs, keratins, T-cell receptors, immunoglobulin gene clusters). However, single copy genes may also have multiple pseudogenes (e.g. prohibitin, *PHB*, 4 pseudogenes; argininosuccinate synthetase, *ASS*, 14 pseudogenes).

### 1.3.3 Functional organization of human genes

Well before the organization of the human genome is known in its entirety, some trends governing the distribution of genes are already becoming apparent (McKusick, 1986; Strachan, 1992; Schinzel *et al.*, 1993).

(i) Genes which encode the same product [e.g. ribosomal RNA (*RNR*), histones (*H1F2, H1F3, H1F4, H2A, H2B, H3F2, H4F2*), *HLA*, homeobox proteins (*HOXA, B, C* and *D*), immmunoglobulins (*IGK, IGL, IGH*)] are often clustered. However, these clusters are usually distributed between several different chromosomes (e.g. *RNR*, chromosomes 13, 14, 15, 21 and 22; histones, chromosomes 1, 6 and 12; *HOX*, chromosomes 7, 17, 12 and 2). Thus multigene families have evolved by duplication and divergence but the duplicated copies may no longer be syntenic (i.e. linkage on the same chromosome conserved) owing to the translocation of discrete chromosomal regions during evolution. The likelihood of synteny may well be related to the time since duplication.

(ii) Genes which encode tissue-specific protein isoforms or isoenzymes are sometimes clustered [e.g. pancreatic (*AMY2A, AMY2B*) and salivary (*AMY1A*) amylase genes on chromosome 1p21] but sometimes not [e.g. cardiac (*ACTC*), skeletal muscle (*ACTA1*), smooth muscle, aorta (*ACTA2*) and smooth muscle, enteric (*ACTA3*) α-actin genes which are located on chromosomes 15, 1, 10 and 2, respectively]. Again, time since the duplication event is likely to be an important factor in determining synteny.

(iii) Genes encoding isozymes specific for different subcellular compartments are usually not syntenic [e.g. soluble/extracellular and mitochondrial forms of superoxide dismutase (*SOD1/SOD3, SOD2* on chromosomes 21/4 and 6 respectively), aconitase (*ACO1, ACO2* on chromosomes 9 and 22) and thymidine kinase (*TK1, TK2* on chromosomes 17 and 16)]. This must reflect the different evolutionary origins of the nuclear and mitochondrial genomes.

(iv) Genes encoding enzymes catalysing successive steps in a particular metabolic pathway are usually not syntenic [e.g. the five enzymes of the urea cycle (*ARG1, ASL, ASS, CPS1, OTC*) are encoded by genes on chromosomes 6, 7, 9, 2 and X; four enzymes involved in galactose metabolism (*GALE, GALK1, GALK2, GALT*) are encoded by genes on chromosomes 1, 17, 15 and 9]. However, there are exceptions to this rule: four genes encoding enzymes of the glycolytic pathway (*TPI, GAPD, ENO2, LDHB*) are located on the short arm of chromosome 12 in the region p13–p12. The *GDH* and *PGD* genes encoding enzymes of the phosphogluconate pathway are encoded by linked genes on chromosome 1. The reasons for this syntenic organization, when it occurs, are not known.

(v) Genes encoding different subunits of a heteromeric protein are often not syntenic [e.g. the genes encoding α- (*HBA*, chromosome 16) and β-globin (*HBB*, chromosome 11), lactate dehydrogenase A (*LDHA*, chromosome 11) and B (*LDHB*, chromosome 12), factor XIII subunits a (*F13A*, chromosome 6) and b (*F13B*, chromosome 1), the immunoglobulin light chains (*IGK, IGL*, chromosomes 2 and 22) and heavy chains (*IGH*, chromosome 14)]. However, several cases of synteny are known [e.g. the genes encoding the three chains of fibrinogen (*FGA, FGB, FGG*, all closely linked on chromosome 4), the α- and β-chains of C4b-binding protein (*C4BPA* and *C4BPB*, closely linked on chromosome 1q32), the complement component 1Q α- and β-chains (*C1QA, C1QB*, linked on chromosome 1p) and the platelet membrane glycoproteins IIb and IIIa (*GP2B, GP3A*, closely linked on chromosome 17q21–q22)]. The various subunits of the T-cell antigen receptor (TCR) are intriguing: the α- and δ-subunits are encoded by genes on chromosome 14, the β- and γ-subunit genes are on chromosome 7, whereas the ε-subunit gene lies on chromosome 11. Synteny probably implies a common evolutionary origin.

(vi) Clustering of genes of similar function and common evolutionary origin is common [e.g. the genes encoding blood coagulation factors VII (*F7*) and X (*F10*) on chromosome 13, the γ-crystallin (*CRYG*) gene cluster on chromosome 2q33 and the four alcohol dehydrogenase (*ADH*) genes on chromosome 4q22]. However, there is no chromosomal clustering of genes with respect to the structure/function of particular organs or subcellular organelles (e.g. mitochondria). Most intriguing is the clustering of cytokines (including several haematopoietic growth factors) and their receptors on the long arm of chromosome 5; the genes encoding granulocyte–macrophage colony-stimulating factor (*GMCSF*), macrophage colony-stimulating factor (*CSF2*), the CSF1 receptor (*CSF1R*, colony-stimulating factor-1 receptor, also known as c-*fms*), interleukins 3, 4, 5, 9, 12B and 13 (*IL-3, IL-4, IL-5, IL-9, IL-12B, IL-13*), platelet-derived growth factor receptor-β (*PDGFRB*), acidic fibroblast growth factor (*FGF1*) and fibroblast growth factor receptor 4 (*FGFR4*).

The genes encoding the IL-3 receptor α-chain (*IL3RA*) and the GM-CSF receptor α-chain (*CSF2RA*) both map to the pseudoautosomal region of the sex chromosomes.

(vii) Genes encoding ligands and their associated receptors are sometimes syntenic [e.g. the genes encoding transferrin (*TF*) and its receptor (*TFRC*) are both located on chromosome 3q whilst the genes encoding apolipoprotein E (*APOE*) and the low density lipoprotein receptor (*LDLR*) are both located on chromosome 19]. However, not surprisingly, this is far from always being the case [e.g. insulin (*INS*, chromosome 11) and insulin receptor (*INSR*, chromosome 19); epidermal growth factor (*EGF*, chromosome 4), epidermal growth factor receptor (EGFR, chromosome 7); growth hormone (*GH1, GH2*, chromosome 17) and growth hormone receptor (*GHR*, chromosome 5); interferons α, β, γ and ω1 (*IFNA, IFNB, IFNG, IFNW1*, chromosomes 9, 8 and 9, 12, 9), interferon receptors α/β and γ (*IFNAR, IFNGR1*, chromosome 6)].

(viii) The linear order of a family of related genes can reflect the order in which they become activated during development [e.g. the *HBE* (embryonic), *HBG2, HBG1* (fetal), *HBD, HBB* (postnatal) genes of the human β-globin cluster]. The expression of these genes is controlled by an upstream LCR (see Section 1.5.4) and correct gene order is required for the normal temporal pattern of developmental expression (Hanscombe *et al.*, 1991). Another example of this phenomenon is provided by the *HOX* genes which are organized chromosomally according to their order of expression. Intriguingly, the Hox proteins encoded vary in their affinity for their target DNA sequences and these affinities also correlate with linear order (Pellerin *et al.*, 1994).

## 1.4 Repetitive sequence elements

Repetitive DNA comprises the bulk (>90%) of the human genome. A large number of different types of repetitive sequence element are found in the human genome (reviewed by Jelinek and Schmid, 1982; Hardman, 1986; Vogt, 1990) and their analysis may go some way toward explaining the patterns of chromosome bands noted after the staining of metaphase chromosomes. Three main categories are now recognized:

(i) a highly repetitive class including sequence families with more than $10^5$ copies per haploid genome;
(ii) a middle repetitive sequence class ($10^2$–$10^5$ copies); and
(iii) a low repetitive class whose members possess between 2 and 100 copies per haploid genome.

The different classes of highly repetitive satellite DNA (tandem repeats) are reviewed here together with *Alu*I repeats and LINE (L1) elements, the two most abundant and best characterized of the interspersed middle repetitive sequence families in the human genome.

## 1.4.1 Tandem repeats

Tandemly repetitive DNA comprises satellite DNA, minisatellite DNA and microsatellite DNA. Satellite DNA comprises the majority of heterochromatin and is clustered in tandem arrays of up to several megabases (Mb) in length. A number of different families [e.g. simple sequence (5–25 bp repeats), alphoid (171 bp repeat), $Sau$3A (~68 bp repeat)] have been identified (reviewed by Vogt, 1990) which are largely confined to the heterochromatic C bands at the centromeres of chromosomes 1, 9, 16 and Y.

The hypervariable minisatellite sequences (about $10^4$ copies/genome) share a core consensus sequence (GGTGGGCAGARG, where R = purine) which is reminiscent of the *Escherichia coli* Chi element known to be a signal for generalized recombination (Jeffreys, 1987). These minisatellites exhibit substantial copy number variability in terms of the number of constituent repeat units and are often telomeric in location (see Section 1.2.7). Another family of telomeric minisatellite sequences consists of a hexanucleotide repeat unit (TTAGGG).

Microsatellite DNA families are simple sequence repeats, the most common being $(A)_n/(T)_n$, $(CA)_n/(TG)_n$ and $(CT)_n/(AG)_n$ types (Vogt *et al.*, 1990; Williamson *et al.*, 1991; Beckmann and Weber, 1992). Mini- and microsatellites account for between 0.2 and 0.5% of the genome, respectively, and are widely scattered on many chromosomes. Their high copy number variability and association with a considerable number of different genes has meant that they provide a very valuable source of highly informative markers for the indirect diagnosis of human genetic disease.

## 1.4.2 Alu sequences

The *Alu*I family of short interspersed repeated elements (SINEs) is present in all primates. Up to 900 000 copies are thought to exist in the human genome (some 5% of the total DNA complement) with an average spacing of 4 kb (Hwu *et al.*, 1986). Most occur in non-coding DNA but some are known to be located in untranslated regions of mRNAs (Makalowski *et al.*, 1994) and even within coding regions (Margalit *et al.*, 1994).

*Alu* repeats share a recognizable consensus sequence but the extent of homology to this consensus varies from 72 to 99% (Kariya *et al.*, 1987; Batzer *et al.*, 1990). Human *Alu* sequences are approximately 300 bp in length, are polyadenylated and consist of two related sequences each between 120 and 150 bp long separated by an A-rich region. *Alu* sequences appear to be degenerate forms of 7SL RNA that have been reverse transcribed and integrated into the genome (Ullu and Tschudi, 1984). Several reports of transcription of *Alu* sequences either by RNA polymerase II or III have appeared (Sinnett *et al.*, 1992; Maraia *et al.*, 1993). *Alu* sequences contain an internal RNA polymerase III promoter (Jelinek and Schmid, 1982) and a negative regulatory element which can serve to play a role in the positive or negative regulation of transcription from a variety of promoters *in vitro* (Saffer and Thurston, 1989; Brini *et al.*, 1993). *In vivo*, however, the transcription of these sequences appears to be effectively silenced by DNA methylation (Liu *et al.*, 1994) and/or nucleosome positioning (Englander *et al.*, 1993). *Alu* sequences are almost comply methylated in somatic tissues but are hypomethylated in sperm (Hellmann-Blumberg *et al.*, 1993).

*Alu* repeats are concentrated in R bands in metaphase chromosomes (Korenberg and Rykowski, 1988) whereas these repeats are under-represented in other regions (e.g. centric heterochromatin; Moyzis *et al.*, 1989). R bands are G–C rich, replicate their DNA early in S phase and condense late in mitotic prophase. R bands also contain the bulk of active gene sequences. One consequence of the non-random distribution pattern of *Alu* sequences in the genome is that procedures which screen for these repeats in genomic DNA clones (see Chapter 2) will preferentially locate gene sequences.

There appear to be at least four different types of *Alu* sequence which belong to two distinct subfamilies (Britten *et al.*, 1988; Jurka and Smith, 1988). Some types of *Alu* sequence are human-specific (Batzer *et al.*, 1990; Batzer and Deininger, 1991). These appear to be derived from a number of different but closely related master copies or 'source genes' (Matera *et al.*, 1990a) . Whilst the vast majority of human *Alu* sequences appear to be transcriptionally inert, one of the transpositionally competent human-specific subfamilies is also transcriptionally active (Matera *et al.*, 1990b; Sinnett *et al.*, 1992).

### 1.4.3 LINE elements

Some $10^5$ copies of long interspersed repeat elements (LINEs) are present in the human genome (Hwu *et al.*, 1986) and account for perhaps 2–3% of the total DNA complement (reviewed by Skowronski and Singer, 1986; Singer *et al.*, 1993). Human LINE elements vary in size from as little as 60 bp up to 6–7 kb; about 95% are truncated at their 5' ends but they mostly appear to contain the same 3' sequences as well as a poly(A) tail of variable length. Each individual LINE element differs from the consensus sequence (Scott *et al.*, 1987) by about 13% although many exhibit internal deletions and rearrangements. The majority of human LINE elements appear to have been generated within the last 30 million years (Scott *et al.*, 1987).

LINE elements are confined to the G/Q (Giemsa/Quinacrine) bands of the euchromatin (Korenberg and Rykowski, 1988). G/Q bands are A–T rich, replicate their DNA late during the DNA synthetic period, condense early during mitosis and are relatively poor in expressed genes.

LINEs probably represent processed pseudogene-like copies of reverse transcripts which have been re-integrated into the genome (Hattori *et al.*, 1986). A full-length LINE element possesses two open reading frames, ORF1 (1 kb) and ORF2 (4 kb); the latter possesses reverse transcriptase activity (Mathias *et al.*, 1991) which may serve to mediate the retrotransposition of LINE elements.

LINE element transcripts have only been found in undifferentiated teratocarcinoma cells (Skowronski and Singer, 1985; Skowronski *et al.*, 1988) suggesting that they may be expressed early on in mammalian development. A promoter at the 5' end appears to be responsible for the specific expression of LINE elements in teratocarcinoma cells (Swergold, 1990). However, only a small subset of all LINE elements is capable of being transcribed.

### 1.4.4 Endogenous retroviral sequences

In excess of 10% of the human genome comprises integrated copies of RNA molecules including retroviruses, retroviral-like DNAs, retroposons and retrotranscripts (reviewed by Cohen and Larsson, 1988; Amariglio and Rechavi, 1993; McDonald, 1993). This represents more than 500 000 separate integration events. Non-viral retroposons include the *Alu* sequences and LINE elements discussed above. A number of endogenous retroviral or retroviral-like sequence families have been identified and characterized. These include HERV-K (Ono *et al.*, 1987), RTVL-1 (Maeda and Kim, 1990), RTVL-H (Wilkinson *et al.*, 1993), MaLR (Smit, 1993) and the transposon-like human repeat element (THE1; Deka *et al.*, 1988; Fields *et al.*, 1992; Hakim *et al.*, 1994). Retroviral sequences have sometimes become integrated into human genes (e.g. the endogenous retrovirus, HRES-1 lies within the coding sequence of the transaldolase gene; Banki *et al.*, 1994) and once transposed, they may even be recruited to play a role in the transcriptional regulation of a gene [e.g. as in the human salivary amylase (*AMY1C*) gene; Ting *et al.*, 1992].

## 1.5 Transcriptional regulation

### 1.5.1 Promoter elements

The archetypal gene contains promoter elements upstream (5') of the transcriptional initiation site (or cap site, consensus, CTTYTG, Y = T, C) at +1 which may itself be some distance upstream of the AUG translational initiation codon (consensus, GCCRCCATGG, R = A, G; Kozak, 1987, 1991). The cap site is usually preceded by upstream (5') *cis*-acting regulatory elements of defined sequence, for example TATAAA (25–30 bp 5' to the cap site) and CCAAT (~90 bp 5' to the cap site) which play a role in constitutive gene expression. Housekeeping genes usually lack TATAAA boxes and instead contain G–C rich sequences which may contain binding sites for the transcription factor Sp1 (consensus sequence, GGGCGG 'GC box').

In some genes, transcription may initiate from more than one promoter. Thus, for example, at least six alternative promoters are present 5' to the *DMD* gene; one of these promoters is 500 kb upstream of the others (Nishio *et al.*, 1994). Such instances of multiple promoters are not uncommon and presumably serve to provide the potential for the production of several tissue-specific or developmental stage-specific isoforms of a given protein.

The primary control of gene expression is usually exerted at the level of transcription. A large number of *cis*-acting DNA sequence motifs have been identified which represent binding sites for DNA-binding proteins (*trans*-acting factors; Johnson and McKnight, 1989) required to confer appropriate regulation upon the genes bearing them (reviewed by Latchman, 1990; Freemont *et al.*, 1991; Faisst and Meyer, 1992). The removal of these *cis*-acting elements abolishes the gene's specific pattern of expression whereas their combination with a reporter gene [e.g. chloramphenicol acetyltransferase (CAT) or luciferase; Rosenthal, 1987] confers this pattern of expression upon the heterologous gene *in vitro*. Together, the *cis*-acting sequences bind a host of different basal transcription factors (e.g. TFIID, TFIIA, TFIIB, TFIIE, etc.) that must be brought into close proximity to allow correct assembly of the transcriptional initiation complex (Buratowski, 1994), a prerequisite for RNA polymerase II to initiate transcription. The multiprotein transcriptional initiation complex is assembled on the TATAAA box in genes that possess it and at the initiator (Inr) element (consensus, CTCANTCT) in promoters that lack it (Martinez *et al.*, 1994). The transcriptional initiation complex possesses low basal activity which is then increased by transcription factors binding in *cis*. Bound transcription factors may interact directly with the transcriptional initiation complex or they may exert their influence through the action of intermediary factors and co-factors (reviewed by Clark and Doherty, 1993).

Some *cis*-acting regulatory motifs bind tissue-specific transcription factors [e.g. GATA1 (erythroid-specific; A/TGATAA/G), HNF1 (liver-enriched; GGTTAATNATTAAC), MyoD (muscle-specific, CANNTG)]. Other elements (e.g. octamer; ATTTGCAT) bind numerous different factors in different tissues.

Responsive elements, as the name suggests, are able to confer transcriptional responsiveness to various external trigger stimuli such as hormones, growth factors, etc. Together, these upstream DNA sequences are involved in controlling regulation and induction and conferring tissue-specificity (reviewed by Maniatis *et al.*, 1987; Mitchell and Tjian, 1989). Elements conferring responsiveness to serum (SRE), cAMP (CRE), heat shock (HSE), oestrogen (ERE), glucocorticoids (GRE), thyroid hormone (TRE), retinoic acid (RARE), vitamin D (VDRE), insulin and heavy metals (MRE) have been described among others (Locker, 1993). These elements often occur upstream of the cap site but may also sometimes occur downstream of the gene (e.g. the SRE and CRE sequences of the murine *JunB* gene; Perez-Albuerne *et al.*, 1993).

Some genes may be regulated both spatially and temporally (e.g. the β-globin cluster); 'haemoglobin switching' appears to involve the LCR upstream of the β-globin gene cluster (see Section 1.5.4).

### 1.5.2 Enhancers

Enhancers are DNA sequences that are present 5' or 3' to a gene (or within an exon or an intron) and which are capable of activating the transcription of the gene in a tissue-specific fashion, independently of their orientation and distance from the correct initiation site (Müller *et al.*, 1988; Wasylyk, 1988). Enhancer elements have a modular structure, consisting of patchworks of *cis*-acting sequences that serve as binding sites for different transcription factors

(Dynan, 1989). They function by acting as templates for the assembly of multiprotein complexes on the promoter. Thompson and McKnight (1992) have likened enhancer–protein interactions to a three-dimensional jigsaw puzzle:

> "the arrangement of *cis* regulatory motifs forms the puzzle template. Specific regulatory proteins, by forming contours of appropriate fit for both DNA template and neighbouring proteins, constitute the puzzle pieces".

Mechanistic models involving the looping out of DNA between the enhancer and the transcriptional initiation complex have been proposed in order to explain how enhancers manage to influence the activity of their target promoters at considerable distances (Ptashne, 1988).

Enhancers required for normal tissue-specific expression have been found approximately 3.5 kb upstream of the human α-fetoprotein (*AFP*) gene (Watanabe *et al.*, 1987) and as far as 11 kb upstream of the rat ornithine transcarbamylase (*OAT*) gene (Murakami *et al.*, 1990).

### 1.5.3 Negative regulatory elements

The negative regulation of transcription is now so well documented (reviewed by Levine and Manley, 1989; Jackson, 1991; Clark and Doherty, 1993; Herschbach and Johnson, 1993) that it is likely that most genes are subject to their inhibitory influence. Indeed, every gene promoter region possesses its own unique combination of positive and negative regulatory elements which serves to determine its temporal and spatial pattern of expression. These elements permit the binding of a specific set of DNA-binding proteins which are thereby brought into sufficient proximity to allow their interaction both with each other and with the RNA polymerase in order to influence transcription either positively or negatively. Interestingly, some genes encode both transcriptional activators and repressors, a feat made possible either by alternative splicing or by the alternative use of translational initiation codons.

Negatively acting factors can act indirectly and essentially passively or alternatively directly and actively (Clark and Doherty, 1993; Herschbach and Johnson, 1993). Factors which act passively are termed *repressors*. Interference with the binding of a transcriptional activator by binding or partially binding to the activator's binding site would represent a passive mechanism. An example of this is the down-regulation of activator protein 1 (AP-1) activity by cAMP response element-binding factor (CREB) which shares AP-1's DNA binding specificity (Masquilier and Sassone-Corsi, 1992). Transcriptional repression by the sequestration of positively acting transcription factors is also a passive mechanism (reviewed by Clark and Doherty, 1993; Herschbach and Johnson, 1993): either a negatively acting factor forms a dimeric complex which fails to bind DNA but sequesters positive transcription factors in an inactive form or the dimeric complex binds DNA but lacks a domain required for transcription activation thereby both competing for occupation of *cis* elements and sequestering activators in a transcriptionally non-functional complex. Finally, negatively acting proteins need not bind directly to DNA but may regulate transcription by masking the activation domains of positively acting factors (e.g. the retinoblastoma gene product, Rb; Weintraub *et al.*, 1992).

Active mechanisms of transcriptional repression involve elements known as *silencers*. Silencers often influence gene expression negatively in a position- and orientation-independent fashion in much the same way as enhancers stimulate gene expression. They are modular in structure (reviewed by Clark and Doherty, 1993) and directly inhibit transcriptional initiation either by disrupting the transcriptional initiation complex or by interfering with its catalytic activity.

Negative regulation therefore serves to prevent the expression of a gene in an inappropriate tissue or at an inappropriate time or at an inappropriate level. They also potentiate the down-regulation of the expression of a gene following its transient induced response.

## 1.5.4 Locus control regions

*Cis*-acting elements that exert their effects on the expression of downstream genes over great distances have been described in the human α-globin (*HBA*), β-globin (*HBB*) and red/green cone pigment (*RCP, GCP*) genes; these are the LCRs. The *HBB* LCR is essential for the high level, tissue-specific expression of the *HBB* gene in transgenic mice; expression is dependent upon copy number but independent of the site of integration in the genome.

Removal of an LCR has dramatic consequences for the expression of downstream genes. Deletions between 9 and 60 kb upstream of the human *HBB* gene cause β-thalassaemia, a disease which is normally associated with lesions within the *HBB* gene itself (reviewed by Cooper and Krawczak, 1993; Dillon and Grosveld, 1993). Since the *HBB* gene was intact, the implication was that the removal of sequences far upstream of the *HBB* gene resulted in suppression of its transcriptional activity. The deleted material spanned a region containing four erythroid cell-specific DNase I-hypersensitive sites some 60 kb upstream of the *HBB* gene. These sites are present prior to transcriptional activation of the globin genes and are developmentally stable. Grosveld *et al.* (1987) have shown that DNA containing the four erythroid-specific hypersensitive sites is capable of directing a high level of position-independent *HBB* gene expression *in vitro*. Three of the hypersensitive sites confer position-independent expression on the *HBB* gene but maximal activation requires the entire LCR.

The above-mentioned deletions were associated with an inactive chromatin conformation as far as 100 kb 3′ to the *HBB* gene (Kioussis *et al.*, 1983; Forrester *et al.*, 1990). The LCR may therefore serve to initiate the process of chromatin decondensation necessary for transcriptional activation to occur. In one deletion chromosome, the β-globin locus has been shown to replicate late, in contrast to the early replication exhibited by the wild-type chromosome. This suggests that the LCR might contain an origin of replication. Since the effect on chromatin conformation appears to be unidirectional and that on replication timing bidirectional, it appears likely that these properties of the LCR are separable and that the β-globin LCR may contain a variety of different regulatory elements.

The location of the LCR has now been further narrowed down to a 0.8 kb DNA fragment, between 10.2 and 11 kb 5' to the *HBE1* gene (itself 40 kb upstream of the *HBB* gene), and is believed to contain an enhancer capable of directing the expression of downstream sequences in an erythroid-specific fashion (Tuan *et al.*, 1989). A 36 bp core sequence around the second hypersensitive site has been shown to enhance *HBB* gene expression and contains two binding sites for AP-1 and one for the erythroid-specific transcription factor, NFE-2 (Chang *et al.*, 1992). Experiments in transgenic mice show the LCR to be both necessary and sufficient for developmental regulation of the human β-globin gene cluster (Morley *et al.*, 1992). Indeed, *HBB* gene expression is abolished by insertional inactivation of the LCR (Kim *et al.*, 1992a). This LCR therefore serves both to organize the β-globin gene cluster into an active chromatin domain and to enhance the transcription of individual globin genes. Haemoglobin gene switching is now thought to come about through competition between the different globin genes for access to the LCR. Thus, whilst the transcription of the *HBE1* gene is enhanced by the closely linked LCR in the embryo, *HBE1* gene expression is suppressed in the fetus through the activation of a silencer element. Expression of the γ-globin (*HBG*) genes then becomes dominant.

A similar situation also pertains in the α-globin gene cluster. Again, deletions far upstream of the *HBA1* and *HBA2* genes affect their expression and give rise to α-thalassaemia (reviewed by Cooper and Krawczak, 1993). These deletions exhibit an area of overlap between 30 and 50 kb upstream of the *HBA* genes which contains several DNase I-hypersensitive sites (two erythroid-specific) and is capable of directing the high level expression of an *HBA* gene both in stably transfected mouse erythroleukaemia cells and when integrated into the genomes of transgenic mice (Higgs *et al.*, 1990; Vyas *et al.*, 1992). However, only one hypersensitive site has a significant effect on *HBA* gene expression and this influence does not exhibit copy number dependence (Sharpe *et al.*, 1993). Since expression levels are also not developmentally stable (Sharpe *et al.*, 1992), it may be that the *HBA* LCR is not exactly analogous to the *HBB* LCR. The regulatory element has been further localized to a 350 bp stretch and the presence of functional binding sites for the transcription factors GATA-1 and AP-1 confirmed by DNase footprinting and gel retardation assays (Jarman *et al.*, 1991).

Using transgenic mice, Wang *et al.* (1992) have provided evidence for the existence of an LCR between 3.1 and 3.7 kb 5' to the red and green cone pigment (*RCP/GCP*) genes; deletion mutagenesis studies and evolutionary conservation data pinpointed a 200 bp region which probably represents the core element. However, it is not yet known if this LCR can confer cone-specific expression on a heterologous promoter or if LCR activity is affected by the site of integration.

The region 3' to the human Cα gene at the immunoglobulin heavy chain locus contains a cluster of DNase I-hypersensitive sites which are tissue-specific and cell stage-specific and can up-regulate c-*myc* gene transcription, consistent with an LCR function (Madisen and Groudine, 1994). An LCR-like sequence 3' to the human *CD2* gene has been reported which is capable of conferring high level, tissue-specific and position-independent expression of a linked minigene in transgenic mice (Greaves *et al.*, 1989). Ganss *et al.* (1994) have reported a cell-specific enhancer 12 kb upstream of the murine tyrosinase gene which confers copy number-dependent expression but not position independence.

## 1.5.5 Boundary elements

The position and orientation independence of enhancers and LCRs clearly raises the potential problem of the inappropriate activation of promoters from neighbouring genes. Sequences which constrain the activity of enhancers have been reported in mice and *Drosophila* (Kellum and Schedl, 1991; reviewed by Eissenberg and Elgin, 1991). Termed 'boundary elements' or 'insulators', these sequences serve to insulate a gene from the effects of either enhancer or suppressor elements emanating from the surrounding chromatin. This insulator function appears to work either by blocking interactions with other sequences past the boundary element and/or by limiting the influence of an enhancer to the locality of its target gene(s). Boundary elements therefore serve as functional barriers when inserted between an enhancer and its downstream reporter gene. However, bracketing an enhancer–gene combination with boundary elements serves to confine the enhancer activity to that gene's promoter such that its expression is maintained or even increased. SARs (see Section 1.2.5) may act as boundary elements as evidenced by their ability to establish transcriptional domains around transgenes and act as buffers to shield these transgenes from position effects (McKnight *et al.*, 1992).

A boundary element has been described 5' to the LCR of the chicken β-globin gene cluster (hypersensitive site 4); it serves to insulate genes 5' to it from the influence of the LCR, a function which it also manifests in both human and *Drosophila* cells (Chung *et al.*, 1993). DNA around the DNase I-hypersensitive site 5 of the human β-globin gene appears to function in a similar fashion (Li and Stamatoyannopoulos, 1994). Another example of a boundary element in humans is to be found upstream of the coagulation factor X (*F10*) gene (Miao *et al.*, 1992). This gene lies 2.8 kb downstream of the gene (*F7*) encoding the homologous clotting protein, factor VII. Three positive regulatory elements (*FXP3*, *FXP2* and *FXP1*) have been characterized upstream of the *F10* gene. *FXP1* and *FXP2* act in an orientation- and position-independent fashion, whilst *FXP1* and *FXP3* are responsible for directing liver-specific gene expression. Miao *et al.* (1992) also identified a repressor element just upstream of the *FXP3* sequence and suggested that its presence is required in order to prevent transcriptional activation of the *F7* gene by the *F10* gene enhancers.

## 1.5.6 Trans-*acting protein factors*

As we have seen, the transcriptional activation of eukaryotic genes is made possible by the interaction of *trans*-acting protein factors with *cis*-acting DNA sequence motifs including enhancers. These transcription factors typically contain a sequence-specific DNA-binding domain, a multimerization domain which allows the formation of either homomultimers or heteromultimers, and a transcriptional activation domain. These domains can be combined in a modular fashion to generate an array of different transcription factors (Tjian and Maniatis, 1994).

It is the *cis*-acting DNA sequences within a gene promoter that allow transcription factors to be brought into close proximity so that they may either interact with each other in the transcriptional initiation complex or combine together in an enhancer complex. No one enhancer-binding protein can act on its own, rather it must act in concert with other enhancer-binding proteins. For example, one factor may induce a bend in the DNA thereby promoting the interaction of two already bound proteins with each other. Once the enhancer complex is assembled, it must be able to interact either directly or indirectly with the basal transcription apparatus via its activation domain (Ptashne, 1988; Ptashne and Gann, 1990: Pugh and Tjian, 1990).

Transcription factors can be grouped in families of related proteins whose relatedness extends to homology in their DNA-binding domains and therefore an ability to bind to related DNA sequences (reviewed by Pabo and Sauer, 1992). Specific transcription factors can then bind to more than one DNA sequence. Sometimes, a single DNA sequence motif may be bound by more than one transcription factor. DNA-binding domains fall into one of four main groups defined by homologous amino acid sequences that give rise to a particular structure capable of binding DNA: homeodomain, zinc finger, leucine zipper and helix–loop–helix (Pabo and Sauer, 1992). These domains usually bind the negatively charged DNA molecule through basic amino acid residues. Activation domains of transcription factors also come in four types: rich in either glutamine, proline, serine/threonine or acidic amino acids. Examples of these types are Sp1, CTF/NF-1, Pit-1 and GATA-1 respectively (reviewed by Ptashne, 1988; Ptashne and Gann, 1990).

The study of DNA–protein interactions has been enormously facilitated by the use of two techniques: gel retardation analysis (also termed band or mobility shift assays; Dent and Latchman, 1993) and DNase I footprinting (Lakin, 1993). The former technique is extremely useful for searching for DNA-binding proteins in crude nuclear extracts, whereas the latter method provides information as to the precise location of the binding site on the DNA sequence under study.

## 1.5.7 Sequences involved in transcriptional termination

Downstream (3′) of the termination codon, other sequences have been implicated in the termination of transcription and the addition of a poly(A) tail, thought to play a role in determining mRNA stability. These include the polyadenylation motif, AAUAAA, usually 20–30 nucleotides 5' to the site of poly(A) addition and the G/T cluster immediately 3' to the end of the mRNA which appears to play a crucial role in 3' end formation (Birnstiel *et al.*, 1985; Wahle and Keller, 1992; Manley and Proudfoot, 1994). The AAUAAA motif is thought to interact with at least two RNA-binding proteins, cleavage-polyadenylation specificity factor (CPSF) and cleavage-stimulatory factor (CstF), which are thought to promote assembly of the polyadenylation complex (Wahle, 1992; Manley and Proudfoot, 1994). A further U-rich element, upstream of the AAUAAA motif (the 'cytoplasmic poladenylation element'; CPE) is also involved. The G/T cluster (consensus sequence, YGUGUUYY) downstream of the AAUAAA signal may also be involved in promoting efficient 3' end formation (McLaughlan *et al.*, 1985).

A wide variety of RNA-binding proteins are now known which play an important role in the post-translational regulation of gene expression (reviewed by Burd and Dreyfuss, 1994).

### 1.5.8 mRNA splicing and processing

One of the characteristics of eukaryotic genes which distinguishes them from their prokaryotic counterparts is the production of large pre-mRNAs which contain intervening non-coding sequences (introns) that are removed by a highly accurate cleavage/ligation reaction known as splicing before the mRNA is transported to the cytoplasm for translation (reviewed by Green, 1986; Padgett *et al.*, 1986). Splicing not only permits the removal of introns from the primary transcript but also allows the generation of different mRNAs from the same gene by alternative splicing, an important mechanism for tissue-specific or developmental regulation of gene expression and a very economical means of generating biological diversity (Nadal-Ginard *et al.*, 1987; Norton, 1994). Alternative splicing may be regulated by variation in the intracellular levels of antagonistic splicing factors (Caceres *et al.*, 1994).

Whilst few eukaryotic genes are completely devoid of introns, intron number may vary from as few as one to as many as 79 in the case of the gene (*DMD*) encoding the Duchenne muscular dystrophy protein, dystrophin (Den Dunnen *et al.*, 1989; Roberts *et al.*, 1992). Although introns may in some cases be very large, vertebrate exons internal to a gene rarely exceed 800 bp in length (Hawkins, 1988).

The splicing of a eukaryotic mRNA appears to occur as a two-stage process. In the case of a simple two exon gene, the pre-mRNA is first cleaved at the 5' (donor) splice site to generate two splicing intermediates, an exon-containing RNA species and a lariat RNA species containing the second exon plus intervening intron. Cleavage at the 3' (acceptor) splice site and ligation of the exons then occurs resulting in the excision of the intervening intron in the form of a lariat. Splicing efficiency is critically dependent upon the accuracy of cleavage and rejoining. This accuracy appears to be determined, at least in part, by the virtually invariant GT and AG dinucleotides present at the 5' and 3' exon/intron junctions respectively. However, more extensive consensus sequences spanning the 5' and 3' splice junctions are evident (Mount, 1982; Padgett et al., 1986).

A further conserved sequence element, the 'branch-point', has been identified in the introns of eukaryotic genes at a site some 18–40 bp upstream of the 3' splice site (Green, 1986). Although this sequence appears to play a role in forming a branch with the 5' terminus of the intron, it exhibits a rather weak consensus sequence [ $Y_{81}NY_{100}T_{87}R_{81}A_{100}Y_{94}$ (R = purine); Krainer and Maniatis, 1988]. The sequence UACUAAC appears to be the most efficient branch site for mammalian mRNA splicing both *in vitro* and *in vivo* (Zhuang et al., 1989). Whereas both the length and location of the pyrimidine tract may be important determinants of branch-point and acceptor splice site utilization, the 3' acceptor splice site itself appears to possess little specificity and may serve merely as the first AG dinucleotide downstream of the branch-point/pyrimidine tract.

Splicing occurs within the *spliceosome*, a complex assembly of small ribonucleoprotein particles (snRNPs) composed of a variety of snRNAs and associated proteins (reviewed in detail by Guthrie and Patterson, 1988; Lührmann et al., 1990). The pre-mRNA is folded in such a way that splice sites are optimally aligned for cleavage and ligation. In this process, the snRNAs play a vital role. Our current understanding of mRNA splicing suggests that the formation of the 5' splice site complex is contingent upon the prior formation of the 3' splice site complex (Robberson et al., 1990; Talerico and Berget, 1990; Smith et al., 1993).

### 1.5.9 Sequences involved in determining mRNA stability

The regulation of gene expression may occur at any one of a number of different stages in the pathway from gene activation to synthesis of the protein product and its post-translational processing and export. This may operate at the level of gene activation, initiation of transcription, or instead occur post-transcriptionally (e.g. in the rate of addition of the 5' 7-methylguanosine triphosphate cap, the control of mRNA modification, polyadenylation, splicing, transport, stability, translation, post-translational modification, etc.; reviewed by Atwater et al., 1990; Hentze, 1991).

DNA sequences downstream of the translational termination codons of many genes specify motifs that fulfil various functions at the RNA level including polyadenylation and cleavage as well as the stability of the mRNA. For example, the 3' untranslated regions (UTRs) of many oncogene, cytokine and lymphokine mRNAs contain A–U rich elements (AREs) which are thought to promote poly(A) removal and mRNA destabilization (reviewed by Ross, 1988; Hargrove and Schmidt, 1989; Peltz and Jacobson, 1992; Sachs, 1993; Decker and Parker, 1994). The AREs of these transcripts often possess an AUUUA motif or a U-rich element which are capable of binding a number of cytoplasmic proteins that serve to accelerate the turnover of the mRNA transcript. The ARE motif appears to mediate mRNA decay by stimulating deadenylation of the poly(A) tail.

## 1.5.10 Role of sequences in 5′ untranslated regions

The presence of the 5′ UTR (the sequence lying between the transcriptional initiation site and the translational start codon) is often essential for the normal expression of a gene. Sequences in the 5′ UTRs of various genes are thought to play a role in controlling the translation of the encoded mRNA (reviewed by Sachs, 1993; Melefors and Hentze, 1993). Perhaps the best characterized post-transcriptional control mechanism involving the 5′ UTR is that of the iron-responsive element (IRE). The IRE is found in the 5′ UTRs of several human genes [e.g. ferritin (*FTH1*), transferrin (*TF*), erythroid 5-aminolevulinate acid synthase (*ALAS*); Cox and Adrian, 1993; Bhasker *et al.*, 1993] and is capable of adopting a stem–loop structure that interacts with a cytosolic RNA-binding protein thereby inhibiting mRNA translation. The post-transcriptional regulation of several other human genes [e.g. transforming growth factor-β1 (*TGFB1*; Kim *et al.*, 1992b) and basic fibroblast growth factor (*FGF2*; Prats *et al.*, 1992)] also appears to involve regulatory elements that modulate the efficiency of translation.

# 1.6 DNA methylation

## 1.6.1 Distribution of 5-methylcytosine

5-Methylcytosine (5mC) is the most common form of DNA modification in eukaryotic genomes. Soon after DNA synthesis is complete, target cytosines are modified by a DNA methyltransferase using S-adenosylmethionine as methyl donor. In humans, between 70 and 90% of 5mC occurs in CpG dinucleotides, the majority of which appear to be methylated (reviewed by Cooper, 1983). Spatially, the distribution of CpG also appears to be non-random in the human genome; about 1% of the genome consists of stretches very rich in CpG which together account for roughly 15% of all CpG dinucleotides (Bird, 1986). In contrast to most of the scattered CpG dinucleotides, these 'CpG islands' represent unmethylated domains and comprise approximately 50% of all non-methylated CpGs in the genome (Bird *et al.*, 1985). CpG islands occur, on average, every 100 kb in the murine genome (Brown and Bird, 1986) and often correspond to gene coding regions both in the mouse (Lindsay and Bird, 1987) and in human (Gardiner *et al.*, 1990; Larsen *et al.*, 1992). Not all vertebrate genes, however, possess these unmethylated CpG islands (Bird, 1986; Gardiner-Garden and Frommer, 1987) and many are partially or even heavily methylated (reviewed by Cooper, 1983).

The target of DNA methylation, the CpG dinucleotide, is dramatically under-represented in vertebrate genomes, occurring at between 20 and 25% of the frequency predicted from observed mononucleotide frequencies (cf. 37% expected within gene coding regions). It has become clear that this 'CpG suppression' and the level of DNA methylation are intimately related. This is thought to be due to the propensity of cytosine, methylated at the 5 position, to undergo deamination to form thymidine (Cooper and Krawczak, 1993).

## 1.6.2 Replication of the methylation pattern and de novo methylation

DNA methylation patterns associated with a variety of different types of human DNA sequence have been shown to be indistinguishable between individuals with different genetic backgrounds (Behn-Krappa et al., 1991). This implies the existence of a very efficient replicative mechanism. By contrast, DNA methylation patterns exhibited by genes in somatic cells are often tissue-specific but these patterns are still inherited clonally in semi-conservative fashion (reviewed by Razin, 1984). Maintenance of the somatic DNA methylation pattern is achieved by virtue of the methyltransferase being capable of recognizing asymmetrically methylated CpG dinucleotides and converting the symmetrically placed cytosines on the daughter strand to 5mC, thus restoring symmetry and reproducing the parental pattern of DNA methylation. Although this enzymatic process appears to possess high fidelity (Smith et al., 1992), there is some evidence to suggest that partially methylated sites (sites only methylated in a specific proportion of cells) may, under some circumstances, be stably maintained in mammalian tissues/cell lines (Turker et al., 1989).

Both demethylation and *de novo* methylation of specific sites have been reported (Trasler et al., 1990; reviewed by Doerfler et al., 1990) suggesting that the control of DNA methylation pattern formation could be rather complex. The efficiency of *de novo* methylation (which initially creates hemimethylated DNA in which only one DNA strand is modified) is considerably lower than maintenance methylation (Hare and Taylor, 1988; Pfeifer et al., 1990); patterns of hemimethylated DNA have sometimes been observed to be stable over several cell generations (Saluz et al., 1986; Toth et al., 1989) and may serve to prevent chromatin expression (Deobagkar et al., 1990). The means by which CpG islands are protected from germline DNA methylation are unknown, but some DNA sequences still appear able to avoid *de novo* methylation (Kolsto et al., 1986). Data are now emerging which suggest that Sp1 sites (GC boxes) confer protection against methylation upon CpG islands (Brandeis et al., 1994; Macleod et al., 1994). Demethylation during differentiation, simply the replacement of 5mC by cytosine, may occur (at least in some cases) by an active and enzymatic mechanism (Vairapandi and Duker, 1993) rather than merely by a passive failure to methylate the hemimethylated site (Razin et al., 1986).

## 1.6.3 Role of DNA methylation in the regulation of transcription

The possible role(s) of DNA methylation in the formation of active/inactive chromatin and in transcriptional regulation is an exciting area of research. It has long been clear that non-transcribed genes are more heavily methylated than transcribed genes but it has often not been possible to determine whether methylation has been a cause or a consequence of transcriptional inactivity. Where DNA methylation does inhibit transcription, it appears to do so either by interfering directly with the interaction between transcription factors and their binding sites (Watt and Molloy, 1988; Iguchi-Ariga and Schaffner, 1989; Jost *et al.*, 1991; Levine *et al.*, 1991; Muiznieks and Doerfler, 1994) and/or indirectly by promoter inactivation through the binding of methylated CpG-specific binding proteins (Antequera *et al.*, 1989; Meehan *et al.*, 1989). In the latter case, repression appears to be dependent upon CpG dinucleotide density, methylation density and the strength of the natural gene promoter (Boyes and Bird, 1991, 1992; Hsieh, 1994). Methylation may exert its effects on protein binding by inducing local distortions in the structure of duplex DNA (Hodges-Garcia and Hagerman, 1992). Protein binding is not however always affected by DNA methylation of the DNA binding motif in the promoter (Harrington *et al.*, 1988; Höller *et al.*, 1988; Weih *et al.*, 1991; Muiznieks and Doerfler, 1994) The interested reader is referred to reviews on this subject for further information (Lewis and Bird, 1991; Bird, 1992).

Active roles for DNA methylation have been postulated in DNA replication and repair as well as in transcriptional regulation (Adams, 1990). Whatever the actual function(s) of DNA methylation, this modification is clearly important since the mutational inactivation of the murine DNA methyltransferase gene causes abnormal development and embryonic lethality (Li *et al.*, 1992). More indirectly, its legacy of deamination-type mutations represents a significant contribution to the incidence of human genetic disease (Cooper and Krawczak, 1993).

### 1.6.4 Role of DNA methylation in X-inactivation

X-inactivation provides the means by which the mammalian cell achieves dosage compensation for its X-linked genes; in females, one of the two X chromosomes is inactivated in each cell. X-inactivation may involve DNA methylation since CpG island methylation commonly occurs in genes present on the inactive X chromosome in females (reviewed by Cedar and Razin, 1990). Although genes on the inactive X chromosome tend to be more heavily methylated than those on the active X (Monk, 1986; Grant and Chapman, 1988), a strong correlation between DNA methylation and transcriptional inactivation has really only been found for CpG islands as opposed to gene coding regions, and for 'housekeeping' genes as opposed to tissue-specific genes. Although the onset of DNA methylation of some X-linked genes coincides closely with X-inactivation (Singer-Sam *et al.*, 1990; Grant *et al.*, 1992), the transcriptional silencing of the hypoxanthine phosphoribosyltransferase (*HPRT*) gene appears to occur prior to methylation of its 5′ CpG island (Lock *et al.*, 1987; Monk, 1990; Singer-Sam *et al.*, 1990). Interestingly, patterns of DNA methylation differ between housekeeping genes: the CpG island associated with the 5′ end of the X-linked human phosphoglycerate kinase (*PGK1*) gene is completely unmethylated on the active X but almost completely methylated on the inactive X (Pfeifer *et al.*, 1990). The 5′ CpG island of the human *HPRT* gene exhibits a different pattern: the active allele is again almost completely unmethylated whereas the inactive allele is heavily methylated except for a region containing four GC boxes which has been shown to interact with DNA-binding proteins only on the active allele (Hornstra and Yang, 1994). These proteins may serve to protect the DNA from methylation before they are themselves displaced following inactivation. Since there is no general or consistent correlation between DNA methylation and the binding sites of transcription factors, it may be that DNA methylation plays a role in stabilizing transcriptionally repressed chromatin rather than in the inactivation process *per se*.

X-inactivation appears to be a special case of genomic imprinting (see Section 1.6.6; reviewed by Lyon, 1993; Pfeifer and Tilghman, 1994). However, it appears to be a complex process that may involve multiple mechanisms and which entails the spreading of inactivation from an 'inactivation centre'. Considerable evidence for the existence of an inactivation centre is now available (reviewed by Migeon, 1994). Within this region is a gene (*XIST*) which represents the only known locus to be transcribed solely from the inactive X chromosome (Migeon, 1994) and which is therefore a prime candidate for a role in initiating X-inactivation. Although the *XIST* mRNA is an abundant species in the nucleus, it does not appear to be translated. Some X-linked loci escape inactivation (e.g. genes in the pseudoautosomal region) but it is not yet clear how this is brought about.

### 1.6.5 Changes in DNA methylation during embryogenesis

Some variation in the methylation status of homologous sites on different alleles is apparent; this variation is tissue-specific and the pattern is reproducible after transmission through the germline (Silva and White, 1988). By contrast, methylation patterns of many genes are altered during early embryonic development in a cell type-specific manner (reviewed by Cedar and Razin, 1990; Ghazi *et al.*, 1992; Brandeis *et al.*, 1993). As far as bulk genomic DNA is concerned, murine oocyte DNA is relatively unmethylated, sperm is heavily methylated, whilst the methylation status of the pre-implantation embryo is intermediate (Monk *et al.*, 1987; Sanford *et al.*, 1987; Howlett and Reik, 1991). Studies of individual CpG sites in specific genes have demonstrated that many CpGs are methylated in both sperm and oocyte but that these sites become demethylated prior to the 16-cell morula stage and remain unmodified through blastulation (Trasler *et al.*, 1990; Shemer *et al.*, 1990; Kafri *et al.*, 1992). The methylation patterns of the parental alleles are therefore erased during embryogenesis. Following implantation and just prior to gastrulation, extensive *de novo* methylation occurs, thereby resetting the methylation pattern of the embryonic genes (Monk *et al.*, 1987; Shemer *et al.*, 1990, 1991a; Howlett and Reik, 1991; Kafri *et al.*, 1992). CpG islands are however protected from this wave of remethylation (Kolsto *et al.*, 1986; Frank *et al.*, 1991; Shemer *et al.*, 1991b). The adult tissue-specific patterns are eventually established by demethylation of tissue-specific genes in the cell types in which they are expressed (reviewed by Brandeis *et al.*, 1993).

### 1.6.6 DNA methylation and imprinting

Monk (1988) loosely defined genomic imprinting as the

> "differential modification of the maternal and paternal contributions to the zygote, resulting in the differential expression of parental alleles during development and in the adult".

This differential modification appears to be essential for normal development, since parthenogenetic embryos (whether diploid paternal or diploid maternal) do not survive to term: in diploid maternal embryos, fetal development is normal but development of the extraembryonic membranes is abnormal. In diploid paternal embryos, it is the other way around (reviewed by Monk, 1988). Clearly, maternal and paternal chromosomes must differ epigenetically and in such a way that different developmental programmes are followed.

In principle, imprinting must be established either before or early in gametogenesis, must be stable enough to be retained in somatic cells and must be capable of being erased in the germline at every generation prior to the fresh establishment of the imprinting pattern. The idea that imprinting might involve DNA methylation was proposed by Monk *et al.* (1987) on the grounds that this post-synthetic modification was heritable, reversible and known to play a role in the control of gene expression. This view soon received solid support from the results of studies on transgenic mice (reviewed by Surani *et al.*, 1988; Hall, 1990; Reik, 1992); in most cases, transgenes inherited from the father were less methylated than transgenes inherited from the mother. Further, the expression of the transgene correlated with its methylation status; the less methylated paternally inherited transgenes exhibited a higher level of expression. Direct evidence that imprinting involves DNA methylation came from the analysis of three imprinted genes in transgenic mice deficient in DNA methyltransferase (Li *et al.*, 1993). The normally silent paternal *H19* allele was activated whereas the normally active paternal *Igf-2* allele and the normally active maternal *Igf-2r* allele were repressed, demonstrating that an alteration in methylation results in altered patterns of expression. The regulation of the process of imprinting both spatially and temporally remains unclear however (reviewed by Barlow, 1994; Efstratiadis, 1994; Razin and Cedar, 1994).

Assuming that imprinting is critical for normal development, then it may reasonably be supposed that a failure to modify the maternal and paternal chromosomes differentially would give rise to a clinically recognizable phenotype. Imprinting is now known to be involved in both inherited and somatic disease phenotypes in humans (reviewed by Peterson and Sapienza, 1993). Indeed, several possible examples of chromosome disorders are explicable in terms of imprinting. In Beckwith–Wiedemann syndrome (BWS), sporadic cases are associated with trisomy of 11p15.5 and the duplicated region is invariably of paternal origin. By contrast, the familial form of the disease, which maps to the same chromosomal location, is transmitted three times more frequently by mothers than by fathers. Observations such as these originally led

Koufos et al. (1989) to propose that the paternally transmitted allele is always inactivated by imprinting whilst children are only affected if the mutant allele is inherited from their mother. This postulate is now supported by data from many other BWS families, although the reduced fecundity of affected males also contributes to the excess of transmitting females. Perhaps the clearest example of imprinting is provided by the Angelman (AS) and Prader–Willi syndromes (PWS), two inherited disorders often associated with cytogenetically detectable (and indistinguishable) deletions of chromosome 15q11–q13; in cases of AS, the deletion is exclusively maternal in origin whereas, in the great majority of cases of PWS, the deletion originates from the father.

Ten imprinted loci are known in the mouse (six repressed on the maternal chromosome, two repressed on the paternal chromosome, two uncharacterized) and at least five in human [*IGF2* (11p15.5), *ZNF127* (15q11–q12), *SNRPN* (15q11–q12), *XIST* and the human homologue of the murine gene *H19*; Surani 1994]. Extrapolating from searches for differential methylation imprints between parental alleles, performed by means of a genomic scanning technique (Hatada et al., 1993; Hayashizaki et al., 1994), we may surmise that there may be as many as 100 imprinted loci in mammalian genomes. Whereas most genes replicate synchronously during S phase, all known imprinted genes in mice and human are embedded within domains of 1–2 Mb that replicate asynchronously; the paternal allele usually replicates before the maternal allele during each cell cycle (Kitsberg et al., 1993b; Knoll et al., 1994). This suggests that imprinting may also be a property of chromatin domains. There has been much discussion of possible imprinting control regions on the autosomes analogous to the putative X-inactivation centre, *XIST*. Some evidence is now beginning to emerge for such a control region in the vicinity of the *SNRPN* gene (Sutcliffe et al., 1994).

## Acknowledgements

The author wishes to thank Drs Paula Hallam, David Millar and Paul Towner for their helpful comments on the manuscript.

## References

**Adachi Y, Käs E, Laemmli UK.** (1989) Preferential, cooperative binding of DNA topoisomerase II to scaffold-associated regions. *EMBO J.* **8:** 3997–4006.
**Adams RLP.** (1990) DNA methylation. The effect of minor bases on DNA–protein interactions. *Biochem. J.* **265:** 309–320.
**Adelman JP, Bond CT, Douglass J, Herbert E.** (1987) Two mammalian genes transcribed from opposite strands of the same DNA locus. *Science* **235:** 1514–1517.
**Allshire RC, Dempster M, Hastie ND.** (1989) Human telomeres contain at least three types of G rich repeat distributed non-randomly. *Nucleic Acids Res.* **17:** 4611–4627.
**Amariglio N, Rechavi G.** (1993) Insertional mutagenesis by transposable elements in the mammalian genome. *Environ. Mol. Mutagen.* **21:** 212–218.
**Antequera F, Macleod D, Bird AP.** (1989) Specific protection of methylated CpGs in mammalian nuclei. *Cell* **58:** 509–517.

Atwater JA, Wisdom R, Verma IM. (1990) Regulated mRNA stability. *Annu. Rev. Genet.* **24**: 519–541.
Banki K, Halladay D, Perl A. (1994) Cloning and expression of the human gene for transaldolase. A novel highly repetitive element constitutes an integral part of the coding sequence. *J. Biol. Chem.* **269**: 2847–2851.
Barlow DP. (1994) Imprinting: a gamete's point of view. *Trends Genet.* **10**: 194–199.
Batzer MA, Deininger PL. (1991) A human-specific subfamily of *Alu* sequences. *Genomics* **9**: 481–487.
Batzer MA, Kilroy GE, Richard PE, Shaikh TH, Desselle TD, Hoppens CL, Deininger PL. (1990) Structure and variability of recently inserted Alu family members. *Nucleic Acids Res.* **18**: 6793–6798.
Beckmann JS, Weber JL. (1992) Survey of human and rat microsatellites. *Genomics* **12**: 627–631.
Behn-Krappa A, Hölker I, Sandaradura de Silva U, Doerfler W. (1991) Patterns of DNA methylation are indistinguishable in different individuals over a wide range of human DNA sequences. *Genomics* **11**: 1–7.
Bhasker CR, Burgiel G, Neupert B, Emery-Goodman A, Kuhn LC, May BK. (1993) The putative iron-responsive element in the human erythroid 5-aminolevulinate synthase mRNA mediates translational control. *J. Biol. Chem.* **268**: 12699–12705.
Bickmore WA, Sumner AT. (1989) Mammalian chromosome banding – an expression of genome organization. *Trends Genet.* **5**: 144–148.
Bird AP. (1986) CpG-rich islands and the function of DNA methylation. *Nature* **321**: 209–213.
Bird AP. (1992) The essentials of DNA methylation. *Cell* **70**: 5–8.
Bird AP., Taggart M, Frommer M, Miller OJ, Macleod D. (1985) A fraction of the mouse genome that is derived from islands of nonmethylated CpG-rich DNA. *Cell* **40**: 91–99.
Birnstiel ML, Busslinger M, Strub K. (1985) Transcription termination and 3′ processing: the end is in sight. *Cell* **41**: 349–359.
Björkroth B, Ericsson C, Lamb MM, Daneholt B. (1988) Structure of the chromatin axis during transcription. *Chromosoma* **96**: 333–340.
Blackburn EH. (1991) Structure and function of telomeres. *Nature* **350**: 569–573.
Blackburn EH. (1992) Telomerases. *Annu. Rev. Biochem.* **61**: 113–129.
Blackburn EH. (1994) Telomeres: no end in sight. *Cell* **77**: 621–623.
Board PG, Coggan M, Woodcock DM. (1992) The human Pi class glutathione transferase sequence at 12q13–q14 is a reverse-transcribed pseudogene. *Genomics* **14**: 470–473.
Bonifer C, Vidal M, Grosveld F, Sippel AE. (1990) Tissue-specific and position independent expression of the complete gene domain for chicken lysozyme in transgenic mice. *EMBO J.* **9**: 2843–2848.
Boulikas T. (1992) Homeotic protein binding sites, origins of replication, and nuclear matrix anchorage sites share the ATTA and ATTTA motifs. *J. Cell. Biochem.* **50**: 111–116.
Boyes J, Bird A. (1991) DNA methylation inhibits transcription indirectly via a methyl-CpG binding protein. *Cell* **64**: 1123–1134.
Boyes J, Bird A. (1992) Repression of genes by DNA methylation depends on CpG density and promoter strength: evidence for involvement of a methyl-CpG binding protein. *EMBO J.* **11**: 327–333.
Brandeis M, Ariel M, Cedar H. (1993) Dynamics of DNA methylation during development. *BioEssays* **15**: 709–713.
Brandeis M, Frank D, Keshet I, Siegfried Z, Mendelsohn M, Nemes A, Temper V, Razin A, Cedar H. (1994) Sp1 elements protect a CpG island from *de novo* methylation. *Nature* **371**: 435–438.
Brini AT, Lee GM, Kinet J-P. (1993) Involvement of *Alu* sequences in the cell-specific regulation of transcription of the γ chain of Fc and T cell receptors. *J. Biol. Chem.* **268**: 1355–1361.
Bristow J, Gitelman SE, Tee MK, Staels B, Miller WL. (1993) Abundant adrenal-specific transcription of the human P450c21A 'pseudogene'. *J. Biol. Chem.* **268**: 12919–12924.
Britten RJ, Baron WF, Stout DB, Davidson EH. (1988) Sources and evolution of human *Alu* repeated sequences. *Proc. Natl Acad. Sci. USA* **85**: 4770–4774.

Brown WRA. (1988) A physical map of the human pseudoautosomal region. *EMBO J.* **7:** 2377–2385.
Brown WRA. (1989) Molecular cloning of human telomeres in yeast. *Nature* **338:** 774–776.
Brown WRA, Bird AP. (1986) Long-range restriction site mapping of mammalian genomic DNA. *Nature* **322:** 477–481.
Brown WRA, MacKinnon PJ, Villasanté A, Spurr N, Buckle VJ, Dobson MJ. (1990) Structure and polymorphism of human telomere-associated DNA. *Cell* **63:** 119–132.
Bullock P, Champoux JJ, Botchan M. (1985) Association of crossover points with topoisomerase I cleavage sites: a model for nonhomologous recombination. *Science* **230:** 954–957.
Buratowski S. (1994) The basics of basal transcription by RNA polymerase II. *Cell* **77:** 1–3.
Burd CG, Dreyfuss G. (1994) Conserved structures and diversity of functions of RNA-binding proteins. *Science* **265:** 615–621.
Caceres JF, Stamm S, Helfman DM, Krainer AR. (1994) Regulation of alternative splicing *in vivo* by overexpression of antagonistic splicing factors. *Science* **265:** 1706–1710.
Catasti P, Gupta G, Garcia AE, Ratliff R, Hong L, Yau P, Moyzis RK, Bradbury EM. (1994) Unusual structures of the tandem repetitive DNA sequences located at human centromeres. *Biochemistry* **33:** 3819–3830.
Cedar H, Razin A. (1990) DNA methylation and development. *Biochim. Biophys. Acta* **1049:** 1–8.
Chandley AC, Mitchell AR. (1988) Hypervariable minisatellite regions are sites for crossing over at meiosis in man. *Cytogenet. Cell Genet.* **48:** 152–155.
Chang JC, Liu D, Kan YW. (1992) A 36-base-pair core sequence of locus control region enhances retrovirally transferred human β-globin gene expression. *Proc. Natl Acad. Sci. USA* **89:** 3107–3110.
Cheng JF, Krane DE, Hardison RC. (1988) Nucleotide sequence and expression of rabbit globin genes ζ1, ζ2, and ζ3. Pseudogenes generated by block duplications are transcriptionally competent. *J. Biol. Chem.* **263:** 9981–9993.
Chung JH, Whiteley M, Felsenfeld G. (1993) A 5' element of the chicken β-globin domain serves as an insulator in human erythroid cells and protects against position effect in *Drosophila*. *Cell* **74:** 505–514.
Clark DJ, Felsenfeld G. (1992) A nucleosome core is transferred out of the path of a transcribing polymerase. *Cell* **71:** 11–22.
Clark AR, Doherty K. (1993) Negative regulation of transcription in eukaryotes. *Biochem. J.* **296:** 521–541.
Cockerill PN, Garrard WT. (1986) Chromosomal loop anchorage of the kappa immunoglobulin gene occurs next to the enhancer in a region containing topoisomerase II sites. *Cell* **44:** 273–282.
Cohen M, Larsson E. (1988) Human endogenous retroviruses. *BioEssays* **9:** 191–196.
Cooper DN. (1983) Eukaryotic DNA methylation. *Hum. Genet.* **64:** 315–333.
Cooper DN, Krawczak M. (1993) *Human Gene Mutation*. BIOS Scientific Publishers, Oxford.
Coverley D, Laskey RA. (1994) Regulation of eukaryotic DNA replication. *Annu. Rev. Biochem.* **63:** 745–776.
Cox LA, Adrian GS. (1993) Posttranscriptional regulation of chimeric human transferrin genes by iron. *Biochemistry* **32:** 4738–4745.
Craig JM, Bickmore WA. (1993) Chromosome bands – flavours to savour. *BioEssays* **15:** 349–354.
Craig JM, Bickmore WA. (1994) The distribution of CpG islands in mammalian chromosomes. *Nature Genetics* **7:** 376–379.
Crippa MP, Trieschmann L, Alfonso PJ, Wolffe AP, Bustin M. (1993) Deposition of chromosomal protein HMG-17 during replication affects the nucleosomal ladder and transcriptional potential of nascent chromatin. *EMBO J.* **12:** 3855–3864.
Cunningham JM, Purucker ME, Jane SM, Safer B, Vanin EF, Ney PA, Lowrey CH, Nienhuis AW. (1994) The regulatory element 3' to the $^{A}\gamma$-globin gene binds to the nuclear matrix and interacts with special A–T-rich binding protein 1 (SATB1), an SAR/MAR-associating region DNA binding protein. *Blood* **84:** 1298–1308.

**De Pamphilis ML.** (1993) Eukaryotic DNA replication – anatomy of an origin. *Annu. Rev. Biochem.* **62:** 29–63.

**Decker CJ, Parker R.** (1994) Mechanisms of mRNA degradation in eukaryotes. *Trends Biochem. Sci.* **19:** 336–340.

**Deka N, Willard CR, Wong E, Schmid CW.** (1988) Human transposon-like elements insert at a preferred target site: evidence for a retrovirally mediated process. *Nucleic Acids Res.* **16:** 1143–1151.

**Den Dunnen JT, Grootscholten PM, Dauwerse JG, Walker AP, Monaco AP, Butler R, Anand R, Coffey AJ, Bentley DR, Steensma HY, van Ommen GJB.** (1992) Reconstruction of the 2.4 Mb human DMD gene by homologous YAC recombination. *Hum. Mol. Genet.* **1:** 19–28.

**Dent CL, Latchman DS.** (1993) The DNA mobility shift assay. In: *Transcription Factors: a Practical Approach* (ed. DS Latchman). IRL Press, Oxford, pp. 1–26.

**Deobagkar DD, Liebler M, Graessmann M, Graessmann A.** (1990) Hemimethylation of DNA prevents chromatin expression. *Proc. Natl Acad. Sci. USA* **87:** 1691–1695.

**Dillon N, Grosveld F. (1993)** Transcriptional regulation of multigene loci: multilevel control. *Trends Genet.* **9:** 134–137.

**Dillon N, Grosveld F.** (1994) Chromatin domains as potential units of eukaryotic gene function. *Curr. Opin. Genet. Dev.* **4:** 260–264.

**Dobbs DL, Shaiu W-L, Benbow RM.** (1994) Modular sequence elements associated with origin regions in eukaryotic chromosomal DNA. *Nucleic Acids Res.* **22:** 2479–2489.

**Doerfler W, Toth M, Kochanek S, Achten S, Freisem-Rabien U, Behn-Krappa A, Orend G.** (1990) Eukaryotic DNA methylation: facts and problems. *FEBS Lett.* **268:** 329–333.

**Dynan WS.** (1989) Modularity in promoters and enhancers. *Cell* **58:** 1–4.

**Efstratiadis A.** (1994) Parental imprinting of autosomal mammalian genes. *Curr. Opin. Genet. Dev.* **4:** 265–280.

**Eissenberg JC, Elgin SCR.** (1991) Boundary functions in the control of gene expression. *Trends Genet.* **7:** 335–340.

**Ellis N, Goodfellow PN.** (1989) The mammalian pseudoautosomal region. *Trends Genet.* **5:** 406–409.

**Englander EW, Wolffe AP, Howard BH.** (1993) Nucleosome interactions with a human *Alu* element. *J. Biol. Chem.* **268:** 19565–19573.

**Faisst S, Meyer S.** (1992) Compilation of vertebrate-encoded transcription factors. *Nucleic Acids Res.* **20:** 3–26.

**Felsenfeld G. (1992)** Chromatin as an essential part of the transcriptional mechanism. *Nature* **355:** 219–224.

**Fields CA, Grady DL, Moyzis RK. (1992)** The human THE-LTR(O) and MstII interspersed repeats are subfamilies of a single widely distributed highly variable repeat family. *Genomics* **13:** 431–436.

**Forrester WC, Epner E, Driscoll MC, Enver T, Brice M, Papayannopoulou T, Groudine M.** (1990) A deletion of the human β-globin locus activation region causes a major alteration in chromatin structure and replication across the entire β-globin locus. *Genes Dev.* **4:** 1637–1649.

**Frank D, Keshet I, Shani M, Levine A, Razin A, Cedar H.** (1991) Demethylation of CpG islands in embryonic cells. *Nature* **351:** 239–241.

**Freemont PS, Lane AN, Sanderson MR.** (1991) Structural aspects of protein–DNA recognition. *Biochem. J.* **278:** 1–23.

**Ganss R, Montoliu L, Monaghan AP, Schütz G.** (1994) A cell-specific enhancer far upstream of the mouse tyrosinase gene confers high level and copy number-related expression in transgeneic mice. *EMBO J.* **13:** 3083–3093.

**Gardiner K, Horisberger M, Kraus J, Tantravahi U, Korenberg J, Rao V, Reddy S, Patterson D.** (1990) Analysis of human chromosome 21: correlation of physical and cytogenetic maps; gene and CpG island distributions. *EMBO J.* **9:** 25–34.

**Gardiner-Garden M, Frommer M.** (1987) CpG islands in vertebrate genomes. *J. Mol. Biol.* **196:** 261–282.

**Ghazi H, Gonzales FA, Jones PA.** (1992) Methylation of CpG island-containing genes in human sperm, fetal and adult tissues. *Gene* **114**: 203–210.

**Gilson E, Laroche T, Gasser SM.** (1993) Telomeres and the functional architecture of the nucleus. *Trends Cell Biol.* **3**: 128–134.

**Goldman MA, Holmquist GP, Gray MC, Caston LA, Nag A.** (1984) Replication timing of genes and middle repetitive sequences. *Science* **224**: 686–692.

**Grant SG, Chapman VM.** (1988) Mechanisms of X-chromosome regulation. *Annu. Rev. Genet.* **22**: 199–233.

**Grant M, Zuccotti M, Monk M.** (1992) Methylation of CpG sites of two X-linked genes coincides with X-inactivation in the female mouse embryo but not in the germ line. *Nature Genetics* **2**: 161–166.

**Greaves DR, Wilson FD, Lang G, Kioussis D.** (1989) Human CD2 3′-flanking sequences confer high-level, T cell-specific, position-independent gene expression in transgenic mice. *Cell* **56**: 979–986.

**Green MR.** (1986) Pre-mRNA splicing. *Annu. Rev. Genet.* **20**: 671–708.

**Grosveld F, van Assendelft GB, Greaves DR, Kollias G.** (1987) Position-independent high-level expression of the human β-globin gene in transgenic mice. *Cell* **51**: 975–985.

**Grunstein M.** (1990) Histone function in transcription. *Annu. Rev. Cell Biol.* **6**: 643–678.

**Guthrie C, Patterson B.** (1988) Spliceosomal snRNAs. *Annu. Rev. Genet.* **22**: 387–419.

**Hakim I, Amariglio N, Grossman Z, Simoni-Brok F, Ohno S, Rechavi G.** (1994) The genome of the THE I human transposable repetitive elements is composed of a basic motif homologous to an ancestral immunoglobulin gene sequence. *Proc. Natl Acad. Sci. USA* **91**: 7967–7969.

**Hall JG.** (1990) Genomic imprinting: review and relevance to human diseases. *Am. J. Hum. Genet.* **46**: 857–873.

**Hand R.** (1978) Eukaryotic DNA: organization of the genome for replication. *Cell* **15**: 317–325.

**Hanscombe O, Whyatt D, Fraser P, Yannoutsos N, Greaves D, Dillon N, Grosveld F.** (1991) Importance of globin gene order for correct developmental expression. *Genes Dev.* **5**: 1387–1394.

**Hardman N.** (1986) Structure and function of repetitive DNA in eukaryotes. *Biochem. J.* **234**: 1–11.

**Hare JT, Taylor JH.** (1988) Hemi-methylation dictates strand selection in repair of G/T and A/C mismatches in SV40. *Gene* **74**: 159–161.

**Hargrove JL, Schmidt FH.** (1989) The role of mRNA and protein stability in gene expression. *FASEB J.* **3**: 2360–2370.

**Harrington MA, Jones PA, Imagawa M, Karin M.** (1988) Cytosine methylation does not affect binding of transcription factor Sp1. *Proc. Natl Acad. Sci. USA* **85**: 2066–2070.

**Hatada I, Suguma T, Mukai T.** (1993) A new imprinted gene cloned by a methylation-sensitive genome scanning method. *Nucleic Acids Res.* **21**: 5577–5582.

**Hatton KS, Dhar V, Brown EH, Iqbal MA, Stuart S, Didano VT, Schildkraut CL.** (1988) Replication program of active and inactive multigene families in mammalian cells. *Mol. Cell. Biol.* **8**: 2149–2158.

**Hattori M, Kuhara S, Takenaka O, Sakaki Y.** (1986) L1 family of repetitive DNA sequences in primates may be derived from a sequence encoding a reverse-transcriptase-related protein. *Nature* **321**: 625–628.

**Hawkins JD.** (1988) A survey on intron and exon lengths. *Nucleic Acids Res.* **16**: 9893–9908.

**Hayashizaki Y, Shibata H, Hirotsune S, Sugino H, Okazaki Y, Sasaki N, Hirose K, Imoto H, Okuizumi H, Muramatsu M** (1994) Identification of an imprinted U2af binding protein related sequence on mouse chromosome 11 using the RLGS method. *Nature Genetics* **6**: 33–39.

**Heartlein MW, Knoll JHM, Latt SA.** (1988) Chromosome instability associated with human alphoid DNA transfected into the Chinese hamster genome. *Mol. Cell. Biol.* **8**: 3611–3618.

**Hellmann-Blumberg U, McCarthy Hintz MF, Gatewood JM, Schmid CW.** (1993) Developmental differences in methylation of human *Alu* repeats. *Mol. Cell. Biol.* **13**: 4523–4530.

**Hentze MW.** (1991) Determinants and regulation of cytoplasmic mRNA stability in eukaryotic cells. *Biochim. Biophys. Acta* **1090**: 281–292.

**Herschbach BM, Johnson AD.** (1993) Transcriptional repression in eukaryotes. *Annu. Rev. Cell Biol.* **9:** 479–509.

**Higgs DR, Wood WG, Jarman AP, Sharpe J, Lida J, Pretorius IM, Ayyub H.** (1990) A major positive regulatory region located far upstream of the human α-globin gene locus. *Genes Dev.* **4:** 1588–1593.

**Hodges-Garcia Y, Hagerman PJ.** (1992) Cytosine methylation can induce local distortions in the structure of duplex DNA. *Biochemistry* **31:** 7595–7599.

**van Holde K.** (1993) The omnipotent nucleosome. *Nature* **362:** 111–112.

**Höller M, Westin G, Jiricny J, Schaffner W.** (1988) Sp1 transcription factor binds DNA and activates transcription even when the binding site is CpG methylated. *Genes Dev.* **2:** 1127–1135.

**Holmquist GP.** (1992) Chromosome bands, their chromatin flavors and their functional features. *Am. J. Hum. Genet.* **51:** 17–37.

**Hornstra IK, Yang TP.** (1994) High-resolution methylation analysis of the human hypoxanthine phosphoribosyltransferase gene 5′ region on the active and inactive X chromosomes: correlation with binding sites for transcription factors. *Mol. Cell. Biol.* **14:** 1419–1430.

**Howlett SK, Reik W.** (1991) Methylation levels of maternal and paternal genomes during preimplantation development. *Development* **113:** 119–127.

**Hsieh C-L.** (1994) Dependence of transcriptional repression on CpG methylation density. *Mol. Cell. Biol.* **14:** 5487–5494.

**Hwu HR, Roberts JW, Davidson EH, Britten RJ.** (1986) Insertion and/or deletion of many repeated DNA sequences in human and higher ape evolution. *Proc. Natl Acad. Sci. USA* **83:** 3875–3879.

**Iguchi-Ariga SMM, Schaffner W.** (1989) CpG methylation of the cAMP-responsive enhancer/promoter sequence TGACGTCA abolishes specific factor binding as well as transcriptional activation. *Genes Dev.* **3:** 612–619.

**Ikemura T, Wada K.** (1991) Evident diversity of codon usage patterns of human genes with respect to chromosome banding patterns and chromosome numbers; relation between nucleotide sequence data and cytogenetic data. *Nucleic Acids Res.* **19:** 4333–4339.

**Ishikawa F, Matunis MJ, Dreyfuss G, Cech TR.** (1993) Nuclear proteins that bind the pre-mRNA 3′ splice site sequence r(UUAG/G) and the human telomeric DNA sequence d(TTAGGG)$_n$. *Mol. Cell. Biol.* **13:** 4301–4310.

**Izaurralde E, Kas E, Laemmli UK.** (1989) Highly preferential nucleation of histone H1 assembly on scaffold-associated regions. *J. Mol. Biol.* **210:** 573–585.

**Jackson V.** (1990) *In vivo* studies on the dynamics of histone–DNA interaction: evidence for nucleosome dissolution during replication and transcription and a low level of dissolution independent of both. *Biochemistry* **29:** 719–731.

**Jackson ME.** (1991) Negative regulation of eukaryotic transcription. *J. Cell Sci.* **100:** 1–7.

**Jackson DA, Hassan AB, Errington RJ, Cook PR.** (1993a) Visualization of focal sites of transcription within human nuclei. *EMBO J.* **12:** 1059–1065.

**Jackson MS, Slijepcevic P, Ponder BA.** (1993b) The organisation of repetitive sequences in the pericentromeric region of human chromosome 10. *Nucleic Acids Res.* **21:** 5865–5874.

**Jarman AP, Wood WG, Sharpe JA, Gourdon G, Ayyub H, Higgs DR.** (1991) Characterization of the major regulatory element upstream of the human α-globin gene cluster. *Mol. Cell. Biol.* **11:** 4679–4689.

**Jeffreys A.** (1987) Highly variable minisatellites and DNA fingerprints. *Biochem. Soc. Trans.* **15:** 309–317.

**Jelinek WR, Schmid CW.** (1982) Repetitive sequences in eukaryotic DNA and their expression. *Annu. Rev. Biochem.* **51:** 813–844.

**Johnson PF, McKnight SL.** (1989) Eukaryotic transcriptional regulatory proteins. *Annu. Rev. Biochem.* **58:** 799–839.

**Jost J-P, Saluz H-P, Pawlak A.** (1991) Estradiol down regulates the binding activity of an avian vitellogenin gene repressor (MDBP-2) and triggers a gradual demethylation of the mCpG pair of its DNA binding site. *Nucleic Acids Res.* **19:** 5771–5775.

**Jurka J, Smith T.** (1988) A fundamental division in the *Alu* family of repeated sequences. *Proc. Natl Acad. Sci. USA* **85:** 4775–4778.

**Kafri T, Ariel M, Brandeis M, Shemer R, Urven L, McCarrey J, Cedar H, Razin A.** (1992) Developmental pattern of gene-specific DNA methylation in the mouse embryo. *Genes Dev.* **6:** 704–714.

**Kariya Y, Kato K, Hayashizaki Y, Himeno S, Tarui S, Matsubara K.** (1987) Revision of consensus sequence of human *Alu* repeats – a review. *Gene* **53:** 1–10.

**Kellum R, Schedl P.** (1991) A position–effect assay for boundaries of higher order chromosomal domains. *Cell* **64:** 941–950.

**Kim CG, Epner EM, Forrester WC, Groudine M.** (1992a) Inactivation of the human β-globin gene by targeted insertion into the β-globin locus control region. *Genes Dev.* **6:** 928–938.

**Kim SJ, Park K, Koeller D, Kim KY, Wakefield LM, Sporn MB, Roberts AB.** (1992b) Post-transcriptional regulation of the human transforming growth factor-beta 1 gene. *J. Biol. Chem.* **267:** 13702–13707.

**Kioussis D, Vanin E, deLange T, Flavell RA, Grosveld FG.** (1983) β-globin gene inactivation by DNA translocation in γ-β-thalassaemia. *Nature* **306:** 662–666.

**Kitsberg D, Selig S, Keshet I, Cedar H.** (1993a) Replication structure of the human β-globin gene domain. *Nature* **366:** 588–590.

**Kitsberg D, Selig S, Brandeis M, Simon I, Keshet I, Driscoll DJ, Nicholls RD, Cedar H.** (1993b) Allele-specific replication timing of imprinted gene regions. *Nature* **364:** 459–463.

**Klehr D, Maass K, Bode J.** (1991) Scaffold-attached regions from the human interferon β domain can be used to enhance the stable expression of genes under the control of various promoters. *Biochemistry* **30:** 1264–1270.

**Knoll JHM, Cheng S-D, Lalande M.** (1994) Allele specificity of DNA replication timing in the Angelman/Prader–Willi syndrome imprinted chromosomal region. *Nature Genetics.* **6:** 41–46.

**Kolsto AB, Kallias G, Giguere V, Isobe KI, Prydz H, Grosveld F.** (1986) The maintenance of methylation-free islands in transgenic mice. *Nucleic Acids Res.* **14:** 9667–9677.

**Korenberg JR, Rykowski MC.** (1988) Human genome organization: *Alu*, Lines, and the molecular structure of metaphase chromosome bands. *Cell* **53:** 391–400.

**Kornberg RD, Lorch Y.** (1992) Chromatin structure and transcription. *Annu. Rev. Cell Biol.* **8:** 563–587.

**Koufos A, Grundy P, Morgan K, Aleck KA, Hadro T, Lampkin BC, Kalbakji A, Cavenee WK.** (1989) Familial Wiedemann–Beckwith syndrome and a second Wilms' tumor locus both map to 11p15.5. *Am. J. Hum. Genet.* **44:** 711–719.

**Kozak M.** (1987) An analysis of 5'-noncoding sequences from 699 vertebrate messenger RNAs. *Nucleic Acids Res.* **15:** 8125–8148.

**Kozak M.** (1991) An analysis of vertebrate mRNA sequences: intimations of translational control. *J. Cell Biol.* **115:** 887–903.

**Krainer AR, Maniatis T.** (1988) RNA splicing. In: *Transcription and Splicing* (eds BD Hames, DM Glover). IRL Press, Oxford, pp. 131–206.

**von Kries JP, Buhrmeister H, Strätling WH.** (1991) Matrix/scaffold attachment region binding protein: identification, purification and mode of binding. *Cell* **64:** 123–135.

**Laemmli UK, Käs E, Poljak L, Adachi Y.** (1992) Scaffold-associated regions: *cis*-acting determinants of chromatin structural loops and functional domains. *Curr. Opin. Genet. Dev.* **2:** 275–285.

**Lakin ND.** (1993) Determination of DNA sequences that bind transcription factors by DNA footprinting. In: *Transcription Factors: a Practical Approach* (ed. DS Latchman). IRL Press, Oxford, pp. 27–47.

**Larsen F, Gundersen G, Lopez R, Prydz H.** (1992) CpG islands as gene markers in the human genome. *Genomics* **13:** 1095–1107.

**Latchman DS.** (1990) Eukaryotic transcription factors. *Biochem. J.* **270:** 281–289.

**Laurie DA, Hulten MA.** (1985) Further studies on chiasma distribution and interference in the human male. *Ann. Hum. Genet.* **49:** 203–214.

**Lee DY, Hayes JJ, Pruss D, Wolffe AP.** (1993) A positive role for histone acetylation in transcription factor binding to nucleosomal DNA. *Cell* **72:** 73–84.

**Levine M, Manley JL.** (1989) Transcriptional repression of eukaryotic promoters. *Cell* **59**: 405–408.

**Levine A, Cantoni GL, Razin A.** (1991) Inhibition of promoter activity by methylation: possible involvement of protein mediators. *Proc. Natl Acad. Sci. USA* **88**: 6515–6518.

**Levinson B, Kenwrick S, Lakich D, Hammonds G, Gitschier J.** (1990) A transcribed gene in an intron of the factor VIII gene. *Genomics* **7**: 1–11.

**Levy-Wilson B, Fortier C.** (1989) The limits of the DNase I-sensitive domain of the human apolipoprotein gene coincide with the locations of chromosomal anchorage loops and define the 5' and 3' boundaries of the gene. *J. Biol. Chem.* **264**: 21196–21204.

**Lewis J, Bird A.** (1991) DNA methylation and chromatin structure. *FEBS Lett.* **285**: 155–159.

**Li Q, Stamatoyannopoulos G.** (1994) Hypersensitive site 5 of the human β locus control region functions as a chromatin insulator. *Blood* **84**: 1399–1401.

**Li Q, Wrange O.** (1993) Translational positioning of a nucleosomal GRE modulates glucocorticoid receptor affinity. *Genes Dev.* **7**: 2471–2482.

**Li E, Bestor TH, Jaenisch R.** (1992) Targeted mutation of the DNA methyltransferase gene results in embryonic lethality. *Cell* **69**: 915–926.

**Li E, Beard C, Jaenisch R.** (1993) Role for DNA methylation in genomic imprinting. *Nature* **366**: 362–365.

**Lindsay S, Bird AP.** (1987) Use of restriction enzymes to select potential gene sequences in mammalian DNA. *Nature* **327**: 336–338.

**Liu W-M, Maraia RJ, Rubin CM, Schmid CW.** (1994) *Alu* transcripts: cytoplasmic localisation and regulation by DNA methylation. *Nucleic Acids Res.* **22**: 1087–1095.

**Lock LF, Takagi N, Martin GR.** (1987) Methylation of the Hprt gene on the inactive X occurs after chromosome inactivation. *Cell* **48**: 39–46.

**Locker J.** (1993) Transcription controls: *cis*-elements and trans-factors. In: *Gene Transcription: a Practical Approach* (eds BD Hames, SJ Higgins). IRL Press, Oxford, pp. 321–345.

**Lührmann R, Kastner B, Bach M.** (1990) Structure of spliceosomal snRNPs and their role in pre-mRNA splicing. *Biochim. Biophys. Acta* **1087**: 265–292.

**Lyon MF.** (1993) Epigenetic inheritance in mammals. *Trends Genet.* **9**: 123–128.

**Macleod D, Charlton J, Mullins J, Bird AP.** (1994) Sp1 sites in the mouse *aprt* gene promoter are required to prevent methylation of the CpG island. *Genes Dev.* **8**: 2282–2292.

**Madisen L, Groudine M.** (1994) Identification of a locus control region in the immunoglobulin heavy-chain locus that deregulates c-*myc* expression in plasmacytoma and Burkitt's lymphoma cells. *Genes Dev.* **8**: 2212–2226.

**Maeda N, Kim HS.** (1990) Three independent insertions of retrovirus-like sequences in the haptoglobin gene cluster of primates. *Genomics* **8**: 671–683.

**Makalowski W, Mitchell GA, Labuda D.** (1994) *Alu* sequences in the coding regions of mRNA: a source of protein variability. *Trends Genet.* **10**: 188–193.

**Mancuso DJ, Tuley EA, Westfield LA, Worrall NK, Shelton-Inloes BB, Sorace JM, Alevy YG, Sadler JE.** (1989) Structure of the gene for human von Willebrand factor. *J. Biol. Chem.* **264**: 19514–19527.

**Maniatis T, Goodbourn S, Fischer JA.** (1987) Regulation of inducible and tissue-specific gene expression. *Science* **236**: 1237–1244.

**Manley JL, Proudfoot NJ.** (1994) RNA 3' ends: formation and function – meeting review. *Genes Dev.* **8**: 259–264.

**Maraia RJ, Driscoll CT, Bilyeu T, Hsu K, Darlington GJ.** (1993) Multiple dispersed loci produce small cytoplasmic Alu RNA. *Mol. Cell Biol.* **13**: 4233–4241.

**Margalit H, Nadir E, Ben-Sasson SA.** (1994) A complete *Alu* element within the coding sequence of a central gene. *Cell* **78**: 173–174.

**Martinez E, Chiang C-M, Ge H, Roeder RG.** (1994) TATA-binding protein-associated factor(s) in TFIID function through the initiator to direct basal transcription from a TATA-less class II promoter. *EMBO J.* **13**: 3115–3126.

**Masquilier D, Sassone-Corsi P.** (1992) Transcriptional cross-talk: nuclear factors CREM and CREB bind to AP-1 sites and inhibit activation by Jun. *J. Biol. Chem.* **267**: 22460–22466.

Matera AG, Hellmann U, Hintz MF, Schmid CW. (1990a) Recently transposed Alu repeats result from multiple source genes. *Nucleic Acids Res.* **18:** 6019–6023.

Matera AG, Hellmann U, Schmid CW. (1990b) A transpositionally and transcriptionally competent *Alu* subfamily. *Mol. Cell. Biol.* **10:** 5424–5432.

Mathias SL, Scott AF, Kazazian HH, Boeke JD, Gabriel A. (1991) Reverse transcriptase encoded by a human transposable element. *Science* **254:** 1808–1810.

McAlpine PJ, Shows TB, Povey S, Carritt B, Pericak-Vance MA, Boucheix C, Anderson WA, White JA. (1993) The 1993 catalog of approved genes and report of the nomenclature committee. In: *Human Gene Mapping; a Compendium* (eds AJ Cuticchia, PL Pearson). Johns Hopkins University Press, Baltimore, MD, pp. 6–124.

McDonald JF. (1993) Evolution and consequences of transposable elements. *Curr. Opin. Genet. Dev.* **3:** 855–864.

McKnight RA, Shamay A, Sankaran L, Wall RJ, Hennighausen L. (1992) Matrix-attachment regions can impart position-independent regulation of a tissue-specific gene in transgenic mice. *Proc. Natl Acad. Sci. USA* **89:** 6943–6947.

McKusick VA. (1986) The morbid anatomy of the human genome. *Medicine* **65:** 1–33.

McKusick VA. (1991) Current trends in mapping human genes. *FASEB J.* **5:** 12–20.

McLaughlan J, Gaffney D, Whitton JL, Clements JB. (1985) The consensus sequence YGTGTTYY located downstream from the AATAAA signal is required for efficient formation of mRNA 3' termini. *Nucleic Acids Res.* **13:** 1347–1368.

McPherson CE, Shim EY, Friedman DS, Zaret KS. (1993) An active tissue-specific enhancer and bound transcription factors existing in a precisely positioned array. *Cell* **75:** 387–398.

Meehan RR, Lewis JD, McKay S, Kleiner EL, Bird AP. (1989) Identification of a mammalian protein that binds specifically to DNA containing methylated CpGs. *Cell* **58:** 499–507.

Melefors Ö, Hentze MW. (1993) Translational regulation by mRNA/protein interactions in eukaryotic cells: ferritin and beyond. *BioEssays* **15:** 85–90.

Merino E, Balbas P, Puente JL, Bolivar F. (1994) Antisense overlapping open reading frames in genes from bacteria to humans. *Nucleic Acids Res.* **22:** 1903–1908.

Miao CH, Leytus SP, Chung DW, Davie EW. (1992) Liver-specific expression of the gene coding for human factor X, a blood coagulation factor. *J. Biol. Chem.* **267:** 7395–7401.

Mielke C, Kohwi Y, Kohwi-Shigematsu T, Bode J. (1990) Hierarchical binding of DNA fragments derived from scaffold-attached regions: correlation of properties *in vitro* and function *in vivo*. *Biochemistry* **29:** 7475–7485.

Migeon BR. (1994) X-chromosome inactivation: molecular mechanisms and genetic consequences. *Trends Genet.* **10:** 230–235.

Mitchell PJ, Tjian R. (1989) Transcriptional regulation in mammalian cells by sequence-specific DNA binding proteins. *Science* **245:** 371–378.

Mitchell A, Jeppesen P, Hanratty D, Gosden J. (1992) The organization of repetitive DNA sequences on human chromosomes with respect to the kinetochore analysed using a combination of oligonucleotide primers and CREST anticentromere serum. *Chromosoma* **101:** 333–341.

Miyajima N, Horiuchi R, Shibuya Y, Fukushige S, Matsubara K, Toyoshima K, Yamamoto T. (1989) Two erbA homologs encoding proteins with different T3 binding capacities are transcribed from opposite DNA strands of the same genetic locus. *Cell* **57:** 31–39.

Monk M. (1986) Methylation and the X-chromosome. *BioEssays* **4:** 204–208.

Monk M. (1988) Genomic imprinting. *Genes Dev.* **2:** 921–925.

Monk M. (1990) Changes in DNA methylation during mouse embryonic development in relation to X-chromosome activity and imprinting. *Phil. Trans. R. Soc. Lond. Biol.* **326:** 299–312.

Monk M, Boubelik M, Lehnert S. (1987) Temporal and regional changes in DNA methylation in the embryonic, extraembryonic and germ cell lineages during mouse embryo development. *Development* **99:** 371–382.

Morley BJ, Abbott CA, Sharpe JA, Lida J, Chan-Thomas PS, Wood WG. (1992) A single β-globin locus control region element (5' hypersensitive site 2) is sufficient for developmental regulation of human globin genes in transgenic mice. *Mol. Cell. Biol.* **12**: 2057–2066.

**Mount SM.** (1982) A catalogue of splice junction sequences. *Nucleic Acids Res.* **10**: 459–472.

**Mouchiroud D, D'Onofrio G, Aissani B, Macaya G, Gautier C, Bernardi G.** (1991) The distribution of genes in the human genome. *Gene* **100**: 181–187.

**Moyzis RK, Torney DC, Meyne J, Buckingham JM, Wu J-R, Burks C, Sirotkin KM, Goad WB.** (1989) The distribution of interspersed repetitive DNA sequences in the human geneome. *Genomics* **4**: 273–289.

**Muiznieks I, Doerfler W.** (1994) The impact of 5'-CG-3' methylation on the activity of different eukaryotic promoters: a comparative study. *FEBS Lett.* **344**: 251–254.

**Müller MM, Gerster T, Schaffner W.** (1988) Enhancer sequences and the regulation of gene transcription. *Eur. J. Biochem.* **176**: 485–495.

**Murakami T, Nishiyori A, Takiguchi M, Mori M.** (1990) Promoter and 11 kb upstream enhancer elements responsible for hepatoma cell-specific expression of the rat ornithine transcarbamylase gene. *Mol. Cell. Biol.* **10**: 1180–1191.

**Nadal-Ginard B, Gallego ME, Andreadis A.** (1987) Alternative splicing: mechanistic and biological implications of generating multiple proteins from a single gene. In: *Genetic Engineering. Principles and Methods*, Vol. 9 (ed. JK Setlow). Plenum Press, New York, pp. 42–51.

**Nguyen T, Sunahara R, Marchese A, Van Tol HH, Seeman P, O'Dowd BF.** (1991) Transcription of a human dopamine D5 pseudogene. *Biochem. Biophys. Res. Commun.* **181**: 16–21.

**Nickoloff JA.** (1992) Transcription enhances intrachromosomal homologous recombination in mammalian cells. *Mol. Cell. Biol.* **12**: 5311–5318.

**Nishio H, Takeshima Y, Narita N, Yanagawa H, Suzuki Y, Ishikawa Y, Ishikawa Y, Minami R, Nakamura H, Matsuo M.** (1994) Identification of a novel first exon in the human dystrophin gene and of a new promoter located more than 500 kb upstream of the nearest known promoter. *J. Clin. Invest.* **94**: 1037–1042.

**Norton PA.** (1994) Alternative pre-mRNA splicing: factors involved in splice site selection. *J. Cell Sci.* **107**: 1–7.

**Ono M, Kawakami M, Takezawa T.** (1987) A novel human nonviral retroposon derived from an endogenous retrovirus. *Nucleic Acids Res.* **15**: 8725–8737.

**Pabo CO, Sauer RT.** (1992) Transcription factors: structural families and principles of DNA recognition. *Annu. Rev. Biochem.* **61**: 1053–1095.

**Padgett RA, Grabowski PJ, Konarska MM, Seiler S, Sharp PA.** (1986) Splicing of messenger RNA precursors. *Annu. Rev. Biochem.* **55**: 1119–1150.

**Paranjape SM, Kamakaka RT, Kadonaga JT.** (1994) Role of chromatin structure in the regulation of transcription by RNA polymerase II. *Annu. Rev. Biochem.* **63**: 265–297.

**Pellerin I, Schnabel C, Catron KM, Abate C.** (1994) Hox proteins have different affinities for a consensus DNA site that correlate with the positions of their genes on the hox cluster. *Mol. Cell. Biol.* **14**: 4532–4545.

**Peltz SW, Jacobson A.** (1992) mRNA stability: in *trans*-it. *Curr. Opin. Cell. Biol.* **4**: 979–983.

**Perez-Albuerne ED, Schatteman G, Sanders LK, Nathans D.** (1993) Transcriptional regulatory elements downstream of the JunB gene. *Proc. Natl Acad. Sci. USA* **90**: 11960–11964.

**Peterson K, Sapienza C.** (1993) Imprinting the genome: imprinted genes, imprinting genes, and a hypothesis for their interaction. *Annu. Rev. Genet.* **27**: 7–31.

**Petit C, Levilliers J, Weissenbach J.** (1988) Physical mapping of the human pseudoautosomal region; comparison with genetic linkage map. *EMBO J.* **7**: 2369–2376.

**Pfeifer K, Tilghman SM.** (1994) Allele-specific gene expression in mammals: the curious case of the imprinted RNAs. *Genes Dev.* **8**: 1867–1874.

**Pfeifer GP, Steigerwald SD, Hansen RS, Gartler SM, Riggs AD.** (1990) Polymerase chain reaction-aided genomic sequencing of an X chromosome-linked CpG island: methylation patterns suggest clonal inheritance, CpG site autonomy and an explanation of activity state stability. *Proc. Natl Acad. Sci. USA* **87**: 8252–8256.

**Phi-Van L, Strätling WH.** (1988) The matrix attachment regions of the chicken lysozyme gene co-map with the boundaries of the chromatin domain. *EMBO J.* **7**: 655–664.

**Phi-Van L, von Kries JP, Ostertag W, Strätling WH.** (1990) The chicken lysozyme 5' matrix attachment region increases transcription from a heterologous promoter in heterologous cells and dampens position effects on the expression of transfected genes. *Mol. Cell. Biol.* **10:** 2302–2307.

**Polymeropoulos MH, Xiao H, Sikela JM, Adams M, Venter JC, Merril CR.** (1993) Chromosomal distribution of 320 genes from a brain cDNA library. *Nature Genetics* **4:** 381–386.

**Prats AC, Vagner S, Prats H, Amalric F.** (1992) *Cis*-acting elements involved in the alternative translation initiation process of human basic fibroblast growth factor mRNA. *Mol. Cell. Biol.* **12:** 4796–4805.

**Ptashne M.** (1988) How eucaryotic transcriptional activators work. *Nature* **335:** 683–689.

**Ptashne M, Gann AAF.** (1990) Activators and targets. *Nature* **346:** 329–331.

**Pugh BF, Tjian R.** (1990) Mechanism of activation by Sp1: evidence for coactivators. *Cell* **61:** 1187–1197.

**Rattner JB, Lin C-C.** (1988) The organization of the centromere and centromeric heterochromatin. In: *Heterochromatin; Molecular and Structural Aspects* (eds. RS Verma). Cambridge University Press, Cambridge, pp. 203–227.

**Razin A.** (1984) DNA methylation patterns: formation and biological functions. In: *DNA Methylation; Biochemistry and Biological Significance* (eds A Razin, H Cedar, AD Riggs). Springer, New York.

**Razin A, Cedar H.** (1994) DNA methylation and genomic imprinting. *Cell* **77:** 473–476.

**Razin A, Szyf M, Kafri T, Roll M, Giloh H, Scarpa S, Carotti D, Cantoni GL.** (1986) Replacement of 5-methylcytosine by cytosine: a possible mechanism for transient DNA demethylation during differentiation. *Proc. Natl Acad. Sci. USA* **83:** 2827–2831.

**Reik W. (1992)** Genome imprinting. In: *Transgenic Animals* (eds F Grosveld, G Kollias). Academic Press, London, pp. 99–126.

**Robberson BL, Cote GJ, Berget SM.** (1990) Exon definition may facilitate splice site selection in RNAs with multiple exons. *Mol. Cell. Biol.* **10:** 84–94.

**Roberts RG, Coffey AJ, Bobrow M, Bentley DR.** (1992) Determination of the exon structure of the distal portion of the dystrophin gene by vectorette PCR. *Genomics* **13:** 942–950.

**Rosenthal N.** (1987) Identification of regulatory elements of cloned genes with functional assays. *Methods Enzymol.* **152:** 704–720.

**Ross J.** (1988) Messenger RNA turnover in eukaryotic cells. *Mol. Biol. Med.* **5:** 1–14.

**Rouyer F, Simmler MC, Johnson C, Vergnaud G, Cooke HJ, Weiessenbach J.** (1986) A gradient of sex linkage in the pseudoautosomal region of the human sex chromosomes. *Nature* **319:** 291–295.

**Saccone S, De Sario A, Della Valle G, Bernardi G.** (1992) The highest gene concentrations in the human genome are in telomeric bands of metaphase chromosomes. *Proc. Natl Acad. Sci. USA* **89:** 4913–4917.

**Sachs AB.** (1993) Messenger RNA degradation in eukaryotes. *Cell* **74:** 413–421.

**Saffer JD, Thurston SJ.** (1989) A negative regulatory element with properties similar to those of enhancers is contained within an *Alu* sequence. *Mol. Cell. Biol.* **9:** 355–364.

**Saluz HP, Jiricny J, Jost JP.** (1986) Genomic sequencing reveals a positive correlation between the kinetics of strand-specific DNA demethylation of the overlapping estradiol/glucocorticoid-receptor binding sites and the rate of avian vitellogenin mRNA synthesis. *Proc. Natl Acad. Sci. USA* **83:** 7167–7171.

**Sanford JP, Clark HJ, Chapman VM, Rossant J.** (1987) Differences in DNA methylation during oogenesis and spermatogenesis and their persistence during early embryogenesis in the mouse. *Genes Dev.* **1:** 1039–1046.

**Schedl A, Montollu L, Kelsey G, Schütz G.** (1993) A yeast artificial chromosome covering the tyrosinase gene confers copy number-dependent expression in transgenic mice. *Nature* **362:** 258–261.

**Schild C, Claret F-X, Wahli W, Wolffe AP.** (1993) A nucleosome-dependent static loop potentiates estrogen-regulated transcription from the *Xenopus* vitellogenin B1 promoter *in vitro*. *EMBO J.* **12:** 423–433.

Schinzel A, McKusick VA, Francomano C, Pearson PL. (1993) Report of the committee for clinical disorders and chromosomal aberrations. In: *Human Gene Mapping; a Compendium* (eds AJ Cuticchia, PL Pearson). Johns Hopkins University Press, Baltimore, MD, pp. 735–772.

Scott AF, Schmeckpeper BJ, Abdelrazik M, Comey CT, O'Hara B, Rossiter JP, Cooley T, Heath P, Smith KD, Margolet L. (1987) Origin of the human L1 elements: proposed progenitor genes deduced from a consensus DNA sequence. *Genomics* **1**: 113–125.

Sharpe JA, Chan-Thomas PS, Lida J, Ayyub H, Woods WG, Higgs DR. (1992) Analysis of the human α-globin upstream regulatory element (HS-40) in transgenic mice. *EMBO J.* **11**: 4565–4572.

Sharpe JA, Wells DJ, Whitelaw E, Vyas P, Higgs DR, Wood WG. (1993) Analysis of the human α-globin gene cluster in transgenic mice. *Proc. Natl. Acad. Sci. USA* **90**: 11262–11266.

Shemer R, Walsh A, Eisenberg S, Breslow JL, Razin A. (1990) Tissue-specific expression and methylation of the human apolipoprotein AI gene. *J. Biol. Chem.* **265**: 1010–1015.

Shemer R, Kafri T, O'Connell A, Eisenberg S, Breslow JL, Razin A. (1991a) Methylation changes in the ApoAI gene during embryonic development of the mouse. *Proc. Natl Acad. Sci. USA* **88**: 11300–11304.

Shemer R, Eisenberg S, Breslow JL, Razin A. (1991b) Methylation patterns of the ApoAI-CIII–AIV gene cluster in adult and embryonic tissue suggest dynamic changes in methylation during development. *J. Biol. Chem.* **266**: 23676–23681.

Shirai T, Shiojiri S, Ito H, Yamamoto S, Kusumoto H, Deyashika Y, Maruyama I, Suzuki K. (1988) Gene structure of human thrombomodulin, a cofactor for thrombin-catalysed activation of protein C. *J. Biochem.* **103**: 281–285.

Silva AJ, White R. (1988) Inheritance of allelic blueprints for methylation patterns. *Cell* **54**: 145–152.

Singer MF, Krek V, McMillan JP, Swergold GD, Thayer RE. (1993) LINE-1: a human transposable element. *Gene* **135**: 183–188.

Singer-Sam J, Grant M, LeBon JM, Okuyama K, Chapman V, Monk M, Riggs AD. (1990) Use of a *Hpa*II-polymerase chain reaction assay to study DNA methylation in the *Pgk*-1 CpG island of mouse embryos at the time of X-chromosome inactivation. *Mol. Cell. Biol.* **10**: 4987–4989.

Sinnett D, Richer C, Deragon J-M, Labuda D. (1992) *Alu* RNA transcripts in human embryonal carcinoma cells. *J. Mol. Biol.* **226**: 689–706.

Sippel AE, Saueressig H, Winter D, Grewal T, Faust N, Hecht A, Bonifer C. (1992) The regulatory domain organization of eukaryotic genomes: implications for stable gene transfer. In: *Transgenic Animals* (eds F Grosveld, G Kollias). Academic Press, London, pp. 1–26.

Skowronski J, Singer MF. (1986) The abundant LINE-1 family of repeated DNA sequences in mammals: genes and pseudogenes. *Cold Spring Harbor Symp. Quant. Biol.* **51**: 457–464.

Skowronski J, Fanning TG, Singer MF. (1988) Unit-length LINE-1 transcripts in human teratocarcinoma cells. *Mol. Cell. Biol.* **8**: 1385–1397.

Smit AF. (1993) Identification of a new, abundant superfamily of mammalian LTR-transposons. *Nucleic Acids Res.* **21**: 1863–1872.

Smith SS, Kaplan BE, Sowers LC, Newman EM. (1992) Mechanism of human methyl-directed DNA methyltransferase and the fidelity of cytosine methylation. *Proc. Natl Acad. Sci. USA* **89**: 4744–4748.

Smith CWJ, Chu TT, Nadal-Ginard B. (1993) Scanning and competition between Ags are involved in 3' splice site selection in mammalian introns. *Mol. Cell. Biol.* **13**: 4939–4952.

Sommerville J, Baird J, Turner BM. (1993) Histone H4 acetylation and transcription in amphibian chromatin. *J. Cell. Biol.* **120**: 277–290.

Spitzner JR, Muller MT. (1988) A consensus sequence for cleavage by vertebrate DNA topoisomerase II. *Nucleic Acids Res.* **16**: 5533–5538.

Steinmetz M, Uematsu Y, Lindahl KF. (1987) Hotspots of homologous recombination in mammalian genomes. *Trends Genet.* **3**: 7–10.

Stief A, Winter DM, Strätling WH, Sippel AE. (1989) A nuclear DNA attachment element mediates elevated and position-independent gene activity. *Nature* **341**: 343–346.

**Strachan T.** (1992) *The Human Genome.* BIOS Scientific Publishers, Oxford.
**Strauss WM, Dausman J, Beard C, Johnson C, Lawrence JB, Jaenisch R.** (1993) Germline transmission of a yeast artificial chromosome spanning the murine α1(I) collagen locus. *Science* **259**: 1904–1907.
**Surani MA, Reik W, Allen ND.** (1988) Transgenes as molecular probes for genomic imprinting. *Trends Genet.* **4**: 59–62.
**Surani MA.** (1994) Genomic imprinting: control of gene expression by epigenetic inheritance. *Curr. Opin. Cell Biol.* **6**: 390–395.
**Sutcliffe JS, Nakao M, Christian S, Örstavik KH, Tommerup N, Ledbetter DH, Beaudet AL.** (1994) Deletions of a differentially methylated CpG island at the SNRPN gene define a putative imprinting control region. *Nature Genetics* **8**: 52–56.
**Swergold GD.** (1990) Identification, characterization and cell specificity of a human LINE-1 promoter. *Mol. Cell. Biol.* **10**: 6718–6729.
**Takahashi N, Nagai Y, Ueno S, Saeki Y, Yanagihara T.** (1992) Human peripheral blood lymphocytes express D5 dopamine receptor gene and transcribe the two pseudogenes. *FEBS Lett.* **314**: 23–25.
**Talerico M, Berget SM.** (1990) Effect of 5' splice site mutations on splicing of the preceding intron. *Mol. Cell. Biol.* **10**: 6299–6305.
**Thomas BJ, Rothstein R.** (1989) Elevated recombination rates in transcriptionally active DNA. *Cell* **56**: 619–630.
**Thompson CC, McKnight SL.** (1992) Anatomy of an enhancer. *Trends Genet.* **8**: 232–236.
**Ting CN, Rosenberg MP, Snow CM, Samuelson LC, Meisler MH.** (1992) Endogenous retroviral sequences are required for tissue-specific expression of a human salivary amylase gene. *Genes Dev.* **6**: 1457–1465.
**Tjian R, Maniatis T.** (1994) Transcriptional activation: a complex puzzle with few easy pieces. *Cell* **77**: 5–8.
**Torchia BS, Call LM, Migeon BR.** (1994) DNA replication analysis of FMR1, XIST and factor 8C loci by FISH shows nontranscribed X-linked genes replicate late. *Am. J. Hum. Genet.* **55**: 96–104.
**Toth M, Lichtenberg U, Doerfler W.** (1989) Genomic sequencing reveals a 5-methylcytosine-free domain in active promoters and the spreading of preimposed methylation patterns. *Proc. Natl Acad. Sci. USA* **86**: 3728–3732.
**Trasler JM, Hake LE, Johnson PA, Alcivar AA, Millette CF, Hecht NB.** (1990) DNA methylation and demethylation events during meiotic prophase in the mouse testis. *Mol. Cell. Biol.* **10**: 1828–1834.
**Tuan DYH, Solomon WB, London IM, Lee DP.** (1989) An erythroid-specific, developmental-stage-independent enhancer far upstream of the human "β-like globin" genes. *Proc. Natl Acad. Sci. USA* **86**: 2554–2558.
**Turker MS, Swisshelm K, Smith AC, Martin GM.** (1989) A partial methylation profile for a CpG site is stably maintained in mammalian tissues and cultured cell lines. *J. Biol. Chem.* **264**: 11632–11636.
**Turner BM.** (1993) Decoding the nucleosome. *Cell* **75**: 5–8.
**Ullu E, Tschudi C.** (1984) *Alu* sequences are processed 7SL RNA genes. *Nature* **312**: 171–172.
**Vairapandi M, Duker NJ.** (1993) Enzymic removal of 5-methylcytosine from DNA by a human DNA-glycosylase. *Nucleic Acids Res.* **21**: 5323–5327.
**Viskochil D, Cawthon R, O'Connell P, Xu G, Stevens J, Culver M, Carey J, White R.** (1991) The gene encoding the oligodendrocyte-myelin glycoprotein is embedded within the neurofibromatosis type I gene. *Mol. Cell. Biol.* **11**: 906–912.
**Vogt P.** (1990) Potential genetic functions of tandem repeated DNA sequence blocks in the human genome are based on a highly conserved "chromatin folding code". *Hum. Genet.* **84**: 301–336.
**Vyas P, Vickers MA, Simmons DL, Ayyub H, Craddock CF, Higgs DR.** (1992) *Cis*-acting sequences regulating expression of the human α-globin cluster lie within constitutively open chromatin. *Cell* **69**: 781–793.
**Wada-Kiyama Y, Kiyama R.** (1994) Periodicity of DNA bend sites in human ε-globin gene region. *J. Biol. Chem.* **269**: 22238–22244.

**Wahle E.** (1992) The end of the message: 3′-end processing leading to polyadenylated messenger RNA. *BioEssays* **14:** 113–118.

**Wahle E, Keller W.** (1992) The biochemistry of 3′-end cleavage and polyadenylation of messenger RNA precursors. *Annu. Rev. Biochem.* **61:** 419–440.

**Wahls WP, Wallace LJ, Moore PD.** (1990a) The Z-DNA motif d(TG)30 promotes reception of information during gene conversion events while stimulating homologous recombination in human cells in culture. *Mol. Cell. Biol.* **10:** 785–793.

**Wahls WP, Wallace LJ, Moore PD.** (1990b) Hypervariable minisatellite DNA is a hotspot for homologous recombination in human cells. *Cell* **60:** 95–103.

**Wang Y, Macke JP, Merbs SL, Zack DJ, Klaunberg B, Bennett J, Gearhart J, Nathans J.** (1992) A locus control region adjacent to the human red and green visual pigment genes. *Neuron* **9:** 429–440.

**Warburton PE, Waye JS, Willard HF.** (1993) Nonrandom localization of recombination events in human alpha satellite repeat unit variants: implications for higher order structural characteristics within centromeric heterochromatin. *Mol. Cell. Biol.* **13:** 6520–6529.

**Wasylyk B.** (1988) Enhancers and transcription factors in the control of gene expression. *Biochim. Biophys. Acta* **951:** 17–35.

**Watanabe K, Saito A, Tamaoki T.** (1987) Cell-specific enhancer activity in a far upstream region of the human α-fetoprotein gene. *J. Biol. Chem.* **262:** 4812–4818.

**Watt F, Molloy PL.** (1988) Cytosine methylation prevents binding to DNA of a HeLa cell transcription factor required for optimal expression of the adenovirus major late promoter. *Genes Dev.* **2:** 1136–1143.

**Weih F, Nitsch D, Reik A, Schütz G, Becker PB.** (1991) Analysis of CpG methylation and genomic footprinting at the tyrosine aminotransferase gene: DNA methylation alone is not sufficient to prevent protein binding *in vivo*. *EMBO J.* **10:** 2559–2567.

**Weintraub H, Groudine M.** (1976) Chromosomal subunits in active genes have an altered conformation. *Science* **193:** 848–856.

**Weintraub SJ, Prater CA, Dean DC.** (1992) Retinoblastoma protein switches the E2F site from positive to negative element. *Nature* **358:** 259–261.

**Wilde CD.** (1985) Pseudogenes. *CRC Crit. Rev. Biochem.* **19:** 323–352.

**Wilkinson DA, Goodchild NL, Saxton TM, Wood S, Mager DL.** (1993) Evidence for a functional subclass of the RTVL-H family of human endogenous retrovirus-like sequences. *J. Virol.* **67:** 2981–2989.

**Willard HF.** (1990) Centromeres of mammalian chromosomes. *Trends Genet.* **6:** 410–416.

**Williamson R, Bowcock A, Kidd K, Pearson P, Schmidtke J, Ceverha P, Chipperfield M, Cooper DN, Coutelle C, Hewitt J, Klinger K, Langley K, Weber J, Beckman J, Tolley M, Maidak B.** (1991) Report of the DNA Committee and catalogues of cloned and mapped genes, markers formatted for PCR and DNA polymorphisms. *Cytogenet. Cell Genet.* **58:** 1190–1832.

**Wolffe A.** (1992) *Chromatin. Structure and Function.* Academic Press, London.

**Wolffe A.** (1994) Nucleosome positioning and modification: chromatin structures that potentiate transcription. *Trends Biochem. Sci.* **19:** 240–244.

**Wood WI, Felsenfeld G.** (1982) Chromatin structure of the chicken β-globin gene region: sensitivity to DNase I, micrococcal nuclease and DNase II. *J. Biol. Chem.* **257:** 7730–7736.

**Xu M, Hammer RE, Blasquez VC, Jones SL, Garrard WT.** (1989) Immunoglobulin kappa gene expression after stable integration. II. Role of the intronic MAR and enhancer in transgenic mice. *J. Biol. Chem.* **264:** 21190–21195.

**Zhong Z, Shiue L, Kaplan S, de Lange T.** (1992) A mammalian factor that binds telomeric TTAGGG repeats *in vitro*. *Mol. Cell. Biol.* **12:** 4834–4843.

**Zhuang Y, Goldstein AM, Weiner AM.** (1989) UACUAAC is the preferred branch site for mammalian mRNA splicing. *Proc. Natl Acad. Sci. USA* **86:** 2752–2756.

# 2

# Mapping the human genome

David N. Cooper

## 2.1 Introduction

A detailed map of all chromosomes, with the positions of genes accurately located, is a prerequisite for the eventual in-depth functional analysis of the human genome. Such a map would contain genetic, physical and cytogenetic data and should be accessible at many different levels of resolution from chromosome bands to single base pairs (Stephens *et al.*, 1990; Pearson and Söll, 1991). Together with data on chromosome structural features (e.g. telomeres, centromeres, origins of replication, etc.; see Chapter 1), pseudogenes, repetitive sequences and regulatory elements, it should be possible to relate the structure of the genome to both its function and evolution.

The last 15 years have seen a revolution in our understanding of the structure and function of the human genome, largely thanks to the advent of the polymerase chain reaction (PCR; Rose, 1991). Both disease analysis and genome mapping have expanded enormously in this time (Williamson *et al.*, 1991; Bowcock *et al.*, 1993; Cooper and Schmidtke 1993; Schinzel *et al.*, 1993). DNA-based strategies of disease analysis, gene cloning and mapping the human genome are clearly interdependent processes although they necessarily evolve in parallel (reviewed by Smith and Hood, 1987). One of our strongest motives in attempting to understand the structure and function of the human genome is the expectation that new advances, both practical and conceptual, will find applications in the field of diagnostic medicine. Conversely, the study of the molecular basis of inherited disease and the molecular biology of disease genes is expected to provide fundamental insights into the 'normal' function of the genome and the interaction of its component elements.

The first phase of the Human Genome Project (HGP), the huge international undertaking to map and sequence the $3 \times 10^9$ base pairs (bp) of the human genome, is progressing rapidly (Cantor, 1990; Stepherns *et al.*, 1990; Watson, 1990; Watson and Cook-Deegan, 1991). Before the end of this century, a high resolution physical map of the genome will exist which should then potentiate the next phase, the sequencing of the genome. If this second phase goes according to plan, the sequence of the entire human genome will be available by the year 2005. HGP is coordinated loosely by the Human Genome Organization (HUGO), whose role is to promote cooperation between contributing agencies [e.g. National Institutes of Health (NIH) in the USA, Medical Research Council in the UK, INSERM in France or medical charities such as Généthon or the Centre d'Étude du Polymorphisme Humain (CEPH)].

The strategies and technical approaches being adopted to realize these aims will be outlined briefly in this chapter. First however, we should consider some of the reasons why the sequence of the human genome would be such a valuable resource to those interested in its structure and function:

(i) the future integration of the cytogenetic, genetic, physical and transcriptional maps of the human genome into a single highly informative, multi-level map would greatly facilitate the mapping of disease loci;
(ii) identification of novel genes will reveal novel proteins whose structural analysis will allow predictions to be made regarding their cellular function. Many novel proteins will be of pharmacological importance;
(iii) the characterization of disease genes should provide insights into modes of inheritance and/or explain peculiarities associated with their transmission (e.g. triplet repeats). This should greatly expand the potential for pre-symptomatic/antenatal diagnosis and genetic counselling in human genetic disease;
(iv) the understanding of the function of the proteins encoded by disease genes could potentiate the rational development of appropriate and effective therapies;
(v) comparative mapping in other species will provide information on the relationship between the chromosomal location of genes and their function as well as a better understanding of the principles of chromosomal evolution in mammals.

The size of the human genome has been roughly estimated to be of the order of 3200 Mb (Bodmer, 1981). The size of the (autosomal) genetic map has been estimated to be 2809 centimorgans (cM) in males and 4782 cM in females (Morton, 1991). Thus, on average, 1 cM (equivalent to a 1% recombination probability) on the genetic map is roughly equivalent to 1 Mb on the physical map. A variety of different techniques are being applied individually or in combination to the task of generating maps. The immediate aim is to generate a physical map spanning all human chromosomes and comprising a continuum of overlapping clones which may then be individually sequenced. Mapping the human genome necessarily involves the use of many different analytical methods: however, the resolution of cytogenetic techniques, somatic cell genetics and genetic linkage analysis is limited to approximately $10^6$ bp (1 cM) whilst only the physical mapping techniques of cloning in cosmids and yeast artificial chromosomes (YACs), 'contig mapping' and DNA sequencing are capable of resolving chromosome structure down to the level of the single base pair (Figure 2.1). A combination of these techniques is closing the 'resolution gap' by

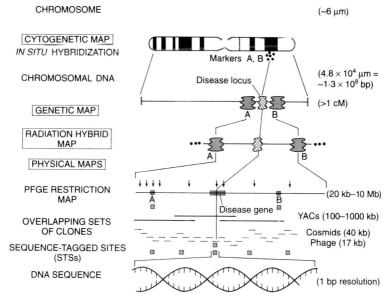

**Figure 2.1.** The hierarchy of approaches to mapping human chromosomes, chromosomal regions and gene/DNA sequences. The cytogenetic map provides the lowest level of resolution: the distance between chromosomal features such as bands and breakpoints can be measured microscopically and the approximate position of markers (including genes) determined by *in situ* hybridization. The genetic map reflects the recombination frequency between linked markers. Radiation hybrid maps are generated by fragmenting chromosomes by X-ray irradiation of somatic cell hybrids. Physical maps may be generated by pulsed-field gel electrophoresis (PFGE) or by the ordering of overlapping sets of clones. The use of sequence-tagged sites (STSs) allows one to link physical and genetic maps and provides a ready starting point for a high resolution map of the DNA sequence.

amalgamating cytogenetic, genetic and physical data into one unified definitive map of the human genome that can be consulted at many different levels.

## 2.2 Markers

Every map requires its own system of markers. Current human gene mapping markers fall into three main categories; genes, anonymous DNA fragments (D-segments) and sequence-tagged sites (STSs). Other markers include fragile sites and chromosome breakpoints. Together, these markers provide 'single-copy landmarks' which can be chromosomally localized and incorporated into both genetic and physical maps. The main categories of marker will now be described.

### 2.2.1 Gene sequences

Several thousand different protein coding genes have been cloned from the human genome (Williamson *et al.*, 1991; Schmidtke and Cooper, 1992;

Bowcock et al., 1993). Some 4670 genes have now been characterized at some level; 3291 of these have been chromosomally mapped (Genome Data Base, June 1994; Fasman et al., 1994). If we accept the best available estimate of human gene number as being between 60 000 and 70 000 (Fields et al., 1994), then some 93% of the total number of human genes still remain to be characterized.

Broadly speaking, the isolation of human disease genes has been accomplished by either *functional* or *positional* cloning. Functional cloning involves the use of information pertaining to either the cDNA sequence (e.g. PCR-based protocol employing degenerate oligonucleotides to isolate a related sequence), the protein product (amino acid sequence data or antibodies) or the function of the protein (receptor–ligand interactions or a selectable marker). Positional cloning refers to the isolation of a specific gene by reference to a genetically and/or physically mapped region (Ballabio, 1993). Examples of functional cloning include the isolation of the genes underlying phenylketonuria (*PAH*) and glucose-6-phosphate dehydrogenase deficiency (*G6PD*; Kwok et al., 1985; Persico et al., 1986), whereas the cloning of the *CFTR* and *NF1* genes underlying cystic fibrosis (Riordan et al., 1989) and neurofibromatosis type I (Wallace et al., 1990), respectively, provide examples of the successful use of the positional cloning approach. Increasingly, however, a combination of the two strategies is being adopted in gene isolation [e.g. as in the cloning of the glycerol kinase (*GK*) gene (Sargent et al., 1993), the Waardenburg syndrome (*PAX3*) gene (Tassabehji et al., 1992) and the amyotrophic lateral sclerosis (*SOD1*) gene (Rosen et al., 1993)]. This combined strategy has been termed the *positional candidate* approach to gene cloning (Ballabio, 1993).

The construction of a transcription map of the genome containing data on the structure, location and expression of human genes is discussed below.

### 2.2.2 DNA polymorphisms

The term polymorphism has been defined (Vogel and Motulsky, 1986) as a "Mendelian trait that exists in the population in at least two phenotypes, neither of which occurs at a frequency of less than 1%". The majority of recognized DNA polymorphisms are neutral single base pair changes detected by virtue of the consequent introduction or removal of a restriction enzyme recognition site and are accordingly termed restriction fragment length polymorphisms (RFLPs). They are inherited as simple Mendelian traits since two alleles are generated as a consequence of the presence or absence of each restriction site.

RFLPs are not rare, being distributed throughout the genome approximately every 200–300 bp (Cooper et al., 1985). Not unexpectedly, the vast majority of RFLPs occur in introns or intergenic regions rather than in coding sequence. RFLPs represent an important tool for genetic mapping studies since they provide the potential to saturate the entire genome with evenly spaced markers (Botstein et al., 1980).

### 2.2.3 D-segments

A D-segment is a region of anonymous DNA (i.e. usually not part of a known gene region) which can be detected using a specific DNA probe. Some 45 838 anonymous D-segments have now been isolated (7418 are polymorphic) and 32 292 of these have been chromosomally mapped (Genome Data Base, June 1994).

### 2.2.4 Sequence-tagged sites

Olson *et al.* (1989) first proposed the idea of using "short tracts of single copy sequence that can be easily recovered by PCR as the landmarks that define the physical map". This type of marker has become known as the STS. STSs offer a number of advantages. Firstly, the different types of probe used in the many different mapping strategies now employed can be made compatible by reference to STSs (Palazzolo *et al.*, 1991; Goold *et al.*, 1993). Indeed, STSs represent the critical link between genetic and physical maps since both genetic markers and physical landmarks can be translated into the 'common language' of the STS (Olson *et al.*, 1989; see Figure 2.2). Secondly, PCR primer data are transmitted between laboratories instead of the clones themselves. For these reasons, STSs represent the fastest expanding category of marker; there are currently 12 959 STSs logged in the Genome Data Base (June 1994 data).

A special type of STS is represented by microsatellite repeats (Beckmann, 1988; Williamson *et al.*, 1991; Beckman and Weber, 1992). Whilst 'classical' genetic polymorphisms are usually made up of two (and occasionally more) alleles per locus, microsatellites are multi-allelic, and thus represent highly informative genetic markers. They are characterized by variable numbers of short clustered repeat units with sequence motifs of 1–5 nt and heterozygosities of at least 70%. Chromosome-specific microsatellite marker loci are now being reported (e.g. Reed *et al.*, 1994).

Sources of STSs include flow-sorted chromosomes (Green *et al.*, 1991), microdissected DNA, end fragments of genomic DNA cloned in YAC vectors (Kere *et al.*, 1992), unique sequences isolated from between frequently occurring interspersed repeats (Brooks-Wilson *et al.*, 1990) and the 3' untranslated regions of cDNA clones (Wilcox *et al.*, 1991). *Expressed sequence tags* (EST; Adams *et al.*, 1992) are essentially STSs associated with cDNAs.

An STS map of human chromosome 11 has been generated using cosmids (Smith *et al.*, 1993). STSs have also been used as landmarks in the construction of overlapping arrays of YAC clones ('contigs') on human chromosome 21 (Chumakov *et al.*, 1992a) and the euchromatic region of the Y chromosome (Foote *et al.*, 1992). It is extremely encouraging that the genetic linkage map of chromosome 21 employing STSs as markers (McInnis *et al.*, 1993) is virtually identical to the physical map of this chromosome derived using a different set of STSs (Chumakov *et al.*, 1992a).

# FUNCTIONAL ANALYSIS OF THE HUMAN GENOME

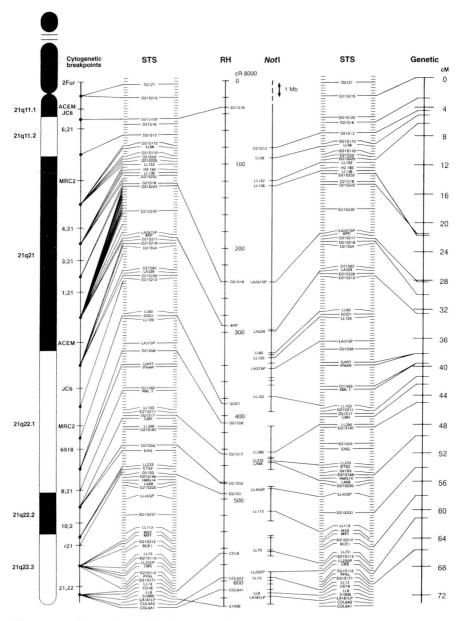

**Figure 2.2.** Comparison of the STS map, the breakpoint panel map, the *Not*I fragment PFGE map and the genetic linkage map of human chromosome 21 (from Chumakov *et al.*, 1992a). The scales shown are centirays (cR) for the radiation hybrid map, megabases (Mb) for the *Not*I PFGE map and centimorgans (cM) for the genetic map.

## 2.2.5 Inter-Alu PCR probes

Inter-*Alu* probes are generated by using oligonucleotide primers, derived from a conserved region (at the 3' end) of the *Alu* consensus sequence, to PCR amplify DNA fragments flanked by these repetititive sequences in inverse orientation (Nelson *et al.*, 1989; Corbo *et al.*, 1990). Since *Alu* sequences are primate-specific, sequences isolated by *Alu*-PCR (whether of cDNA or genomic origin) from somatic cell hybrid DNA are guaranteed to be human in origin. These probes are usually quite complex in that they may contain a variety of other repetitive sequence elements.

### 2.2.6 Allele-specific oligonucleotides

Allele-specific oligonucleotides (ASO) are oligonucleotides which specifically recognize the alternative sequences at a given locus in a hybridization assay. Some 464 ASO markers are logged in the Genome Data Base (June 1994 data).

## 2.3 Cytogenetic mapping

Cytogenetic studies have provided much important information on the chromosomal location of inherited or somatic mutations. In disease mapping, cytogenetic data have been of most use in those cases where initially nothing was known about the structure and function of the protein product of the disease gene and where the chromosomal or sub-chromosomal location of the gene was not known with any precision. For example, cytologically detectable chromosomal deletions associated with Duchenne muscular dystrophy and retinoblastoma provided critical information which potentiated the eventual 'positional cloning' of the underlying genes (Ballabio, 1993). Similarly in tumorigenesis, the isolation and characterization of translocation breakpoints were instrumental in the rapid identification of the genes (*abl* and *bcr*) responsible for chronic myelogenous leukaemia.

Three cytogenetic techniques are worthy of specific mention on account of their particular importance in gene mapping: somatic cell hybrid analysis, radiation hybrid mapping and fluorescence *in situ* hybridization (FISH). These techniques can be used to localize YACs or cosmids to *mapping intervals* of specific regions of a chromosome prior to their being linked together.

### 2.3.1 Somatic cell hybrid analysis

Somatic cell hybrids for use in human gene mapping are made by fusion of a human cell with a rodent cell (reviewed by Cowell, 1992). The initial hybrid cell contains both human and rodent cells but, in subsequent rounds of cell division, human chromosomes are lost preferentially. Stable hybrid cell lines usually contain one or several human chromosomes. A panel of such hybrid cell lines can be used to allocate human DNA sequences including genes to specific chromosomes (Gardiner and Patterson, 1992). However, sub-chromosomal localization is not normally possible unless deletion hybrids made from cells with interstitial deletions or translocation chromosomes are used.

## 2.3.2 Radiation hybrid mapping

The relatively high-resolution physical ordering of genomic DNA fragments can be achieved by radiation hybrid mapping (Goss and Harris, 1975; Benham *et al.*, 1989; Cox *et al.*, 1990; Barrett, 1992). This technique involves the lethal X-ray irradiation of a somatic cell hybrid containing a single human chromosome, thereby fragmenting that chromosome. Subsequent rescue by fusion of the irradiated cells with non-irradiated rodent cells generates viable cell lines containing fragments of specific human chromosomes. These cell lines can then be used as a mapping panel. Different cell hybrid clones are tested for the presence or absence of human markers. Co-transfer of a pair of markers provides information on both order and physical location. The closer two markers are on a chromosome, the less likely they are to be separated by radiation-induced breakage. Analysis of the presence/absence of different markers in a hybrid panel yields the order of the markers on the chromosome.

This approach has been used successfully to generate maps of portions of chromosomes 5 (Warrington *et al.*, 1991), 11 (Richard *et al.*, 1993), 16 (Ceccherini *et al.*, 1992), 21 (Burmeister *et al.*, 1991; see Figure 2.2), 22 (Frazer *et al.*, 1992) and the X chromosome (Gorski *et al.*, 1992) among others. *Alu*-PCR has been used to obtain chromosome- and region-specific probes from radiation hybrids (Brooks-Wilson *et al.*, 1990; Cole *et al.*, 1991; Benham and Rowe, 1992). Problems in the use of such hybrids however include the instability/ heterogeneity of human chromosome fragments (Cole *et al.*, 1991; Sapru *et al.*, 1994).

## 2.3.3 Fluorescence in situ hybridization

Genomic fragments can also be physically ordered on the chromosome using FISH (Montanaro *et al.*, 1991; Korenberg *et al.*, 1992). In this technique, the DNA probe is labelled with a reporter molecule which is then detected after hybridization to metaphase chromosomes by a fluorescently labelled affinity molecule which binds to the reporter. The use of different fluorophores permits the mapping of several clones simultaneously. FISH aids the integration of cytogenetic and physical mapping data and has therefore been used to order YAC (Marrone *et al.*, 1994) and cosmid (Takahashi *et al.*, 1992, 1993, 1994) clones. *Alu*-PCR and L1 (LINE) probes have also been used to identify and characterize human chromosomes in somatic cell hybrid lines (Lichter *et al.*, 1990).

Using FISH, the resolution of the identified map positions of markers is variable and dependent upon both the resolution of chromosome banding and the amount of hybridization signal detected. The maximum resolution of FISH on metaphase chromosomes is about 1–2 Mb. Although chromosomal localization is not possible using interphase chromosomes, the relative order and spacing of different probes can be determined with a resolution of as low as 50 kb (Trask *et al.*, 1991). There is, however, some evidence from studies employing FISH of discrepancies in probe location, particularly in the vicinity of the telomeres of metaphase chromosomes (Marrone *et al.*, 1994)

## 2.3.4 In situ PCR

*In situ* PCR is a relatively new technique which combines PCR with *in situ* hybridization in order to detect specific single copy genes in chromosome spreads (reviewed by Komminoth and Long, 1993).

## 2.4 High-resolution physical mapping

A number of different approaches are routinely used in the generation of high-resolution physical maps (reviewed by Nagaraja, 1992). These will be briefly described.

### 2.4.1 Yeast artificial chromosome cloning

The most popular physical approach to genome mapping involves the use of YAC cloning (reviewed by Schlessinger, 1990). YACs are plasmid-based vectors into which have been placed short DNA sequences (centromeres, telomeres and an autonomously replicating sequence) which perform essential chromosomal functions in the yeast cell (Burke *et al.*, 1987). Large human restriction fragments (0.1–1 Mb) can be cloned into these vectors (Albertsen *et al.*, 1990). The use of YACs has allowed a 3- to 20-fold increase in the size of cloned material over that which is possible with cosmids. YACs thus serve to bridge the gap between the megabase level resolution of genetic/cytogenetic maps and the 50 kb resolution limit of cosmid-based methodology. PFGE (see Section 2.4.3) is often used to generate sub-fragments of the appropriate size for cosmid or plasmid cloning (Anand *et al.*, 1989) or for use as hybridization probes in chromosome 'walking' (Wada *et al.*, 1990). (Chromosome 'walking' involves the isolation of a DNA fragment from the end of a cloned insert and its use to screen a YAC or cosmid library to isolate adjacent clones, thus allowing contig expansion.) YAC libraries can be screened for the presence of single copy genes by hybridization to immobilized arrays of clones (Brownstein *et al.*, 1989).

The main disadvantage of YAC cloning is the high proportion of chimeric clones containing non-contiguous genomic sequences. Other problems include the presence of homologous widely dispersed sequences in the genome which give rise to incorrect contig assembly, the inability to clone some sequences in yeast and the instability of some human sequences once cloned (Bellanné-Chantelot *et al.*, 1992). As a consequence, YAC contigs often contain deletions and rearrangements (Trask *et al.*, 1992).

Mapping the entire human genome is estimated to require a library of perhaps 50 000 YACs of average size 1000 kb. YACs may also be used to determine the order and physical distance between chromosomally localized single copy landmarks. The simplest application of YAC technology has been where two markers are physically linked on a chromosome by virtue of the fact that they hybridize to the same YAC clone [e.g. *F9* gene and cx38.1 (DXS102); Wada *et al.*, 1990]. Mapping larger distances requires the assembly of a series of overlapping YAC clones.

## 2.4.2 Contig assembly

YAC libraries can be screened either by PCR (Riley *et al.*, 1992) or by hybridization (Ross *et al.*, 1992). The latter strategy involves the hybridization of filters containing large numbers of YAC clones with a series of probes to generate 'fingerprints' which can then be compared in order to assemble contigs (Ross *et al.*, 1992). The strategy adopted in contig mapping depends on the probes available, the size of the cloned inserts and the nature of the library to be screened.

Restriction enzyme fingerprinting has been used to establish contigs from cosmid clones (Carrano *et al.*, 1989; Barillot *et al.*, 1991; Trask *et al.*, 1992). Fingerprinting may also be performed by screening YAC libraries with repetitive sequence probes [*Alu* or LINE-1 (L1); Stallings *et al.*, 1990; Green and Olson, 1990a; Bellanné-Chantelot *et al.*, 1991]. Oligomer hybridization can also be used to generate fingerprint data from cosmids or YAC vectors (reviewed by Hoheisel, 1994). When the length of the target region is relatively short, as in the above case, most oligomers bind only once. This technique can be used to generate high-resolution maps of defined regions (Hoheisel *et al.*, 1993).

The isolation of chromosome-specific clones from a human genomic YAC library has been accomplished by *Alu*-PCR (Chumakov *et al.*, 1992b). Briefly, primers flanking a conserved *Alu* consensus sequence were first used to PCR amplify *Alu* sequences from a somatic cell hybrid line containing human chromosome 21. These PCR products were then used as a probe to identify chromosome 21-derived sequences in the YAC library. Screening of a YAC library made from DNA derived exclusively from the region Xq24–q28 enabled the construction of contigs covering most of this region (Schlessinger *et al.*, 1991; Little *et al.*, 1992). *Alu*-PCR has potentiated the rapid isolation and sequencing of ends of YAC clones from the X chromosome (Nelson *et al.*, 1991).

Bellanné-Chantelot *et al.* (1992) assembled more than 1000 contigs from a human YAC library and estimated that 15–20% of the genome was covered with contigs larger than 3 Mb. Cosmid clones are preferable if large-scale DNA sequencing is intended. Contigs are therefore increasingly being generated by screening cosmid libraries with YAC clones before sequencing is initiated.

## 2.4.3 Pulsed-field gel electrophoresis and CpG island mapping

PFGE is an important technique for the physical mapping of the human genome (reviewed by Smith *et al.*, 1988; Bickmore, 1992). Conventional electrophoretic techniques cannot resolve DNA fragments much above 20 kb. PFGE is capable of resolving fragments up to 10 Mb, the size of an average human metaphase band. Separation of such large fragments is made possible by the use of alternating electric fields which force the DNA molecules to reorientate as they pass through the agarose gel. Since the speed at which a molecule reorientates is a function of its molecular weight, the resolution limit is increased.

Although restriction fragment patterns have been used to establish maps of the chromosomes of more primitive organisms (e.g. *Escherichia coli*, yeast, *Caenorhabditis elegans*), this approach has so far only been applied to relatively short defined regions in the human genome. PFGE has been useful in generating restriction maps of 250 kb to 5 Mb regions of human chromosomes using 'rare cutter' restriction enzymes (enzymes with a 6 or 8 bp recognition sequence containing one or more of the rare CpG dinucleotides, e.g. *Not*I) which cleave preferentially within CpG islands (Brown and Bird, 1986). Using a battery of such enzymes, long range restriction maps spanning multiple CpG islands can be generated (e.g. Hardy *et al.*, 1986; van Ommen *et al.*, 1986; Lawrance *et al.*, 1987; Smith *et al.*, 1987; Burmeister *et al.*, 1988; Fulton *et al.*, 1989; Gardiner *et al.*, 1990; reviewed by Barlow and Lehrach, 1987; see Figure 2.2). Good correspondence was found between a *Not*I restriction map of chromosome 21q and a map of ordered YAC clones (Wang and Smith, 1994).

### *2.4.4 Chromosome jumping/linking libraries*

Physical maps of human chromosomes can be generated without having to clone the bulk of the chromosome region under study. This is achieved by mapping recognition sites for rare cutter restriction enzymes which tend to cluster in CpG islands (see Chapter 1; Lindsay and Bird, 1987). The use of jumping/linking libraries allows one to travel a given distance along a particular chromosome from a given location in the genome (Poustka and Lehrach, 1986; Collins, 1988). Clones in a chromosome *jumping library* contain the two ends of *Not*I fragments which have first been circularized by ligation and then cleaved to remove most of the intervening DNA (Collins *et al.*, 1987; Poustka *et al.*, 1987). Fragments originally located some considerable distance apart in the genome thus become juxtaposed. The length of the 'jump' is defined by the selection of restriction enzymes used to cleave the genomic DNA (i.e. *Not*I generates fragments of average size 1000 kb). A *linking library* contains DNA fragments which contain a rare restriction site (e.g. *Not*I). Linking clones can be used to prove that two large restriction fragments are adjacent on the chromosome and therefore represent an important means of ordering large DNA fragments separated by PFGE (Buting *et al.*, 1988; Pohl *et al.*, 1988). In the case of *Not*I, only 10 000 clones are required in either a jumping or a linking library in order for the entire genome to be represented. Obviously, a smaller library is required if flow-sorted chromosomes are used as the source material for constructing the library.

Zabarovsky *et al.* (1990, 1991, 1994c) have described an improved procedure for long range mapping that involves the random sequencing of *Not*I jumping and linking clones followed by contig assembly; this has potentiated the construction of *Not*I maps of human chromosomes (Zabarovsky *et al.*, 1993, 1994b; Wang *et al.*, 1994). Sequences of *Not*I linking clones can also provide STSs (Olson *et al.*, 1989; Zhu *et al.*, 1993) which may be used as anchor points for subsequent fine mapping, thereby permitting the joining of physical and genetic maps. Most (~90%) *Not*I linking clones contain transcribed sequences (Allikmets *et al.*, 1994) and therefore also contribute to the construction of the transcription map (see Section 2.6). Since approximately 20% of human CpG island-containing genes and some 12% of all human genes contain *Not*I restriction sites (Allikmets *et al.*, 1994), this strategy is likely to prove to be a very important means of locating and identifying novel gene sequences.

*Not*I linking clones from human chromosomes have proven very useful in the construction of long range restriction maps (e.g. chromosome 21; Saito *et al.*, 1991; Ichikawa *et al.*, 1992, 1993; Hattori *et al.*, 1993; Wang and Smith, 1994). *Not*I linking libraries are now available which cover the entire human genome (Zabarovsky *et al.*, 1994a).

### *2.4.5 DNA sequencing*

Resolution at the level of the single base pair clearly requires DNA sequencing (Smith and Hood, 1987). A multitude of new methods are now becoming available (Barrell, 1991; Harding and Keller, 1992; Uhlén *et al.*, 1992; Venter *et al.*, 1992). Based upon fluorescent labelling and multiplexing, these automated techniques will be essential if rapid progress in establishing the full sequence of the human genome is to be made.

### *2.4.6 Progress in physical mapping*

Cohen *et al.* (1993) have initiated the construction of what they term a "first generation physical map of the human genome" by employing several different mapping techniques to screen the CEPH YAC library. Some 33 000 YAC clones, of average length 900 kb, were fingerprinted using THE and L1 (LINE) repeat sequence probes, and screened by inter-*Alu* PCR fragment/STS hybridization and FISH. A total of 2100 polymorphic genetically mapped STSs were then used to screen the same set of YAC clones and the STS content was established in about 20% of them. Hybridization of YAC clones to inter-*Alu* PCR products permitted the assignment of over 4000 YAC probes to human chromosomes. Finally, some 500 YACs containing genetically mapped STSs (one every 7.4 cM) were precisely positioned by FISH. This combined use of genetic, physical and cytogenetic methodology promises to provide an integrated multi-level mapping approach which is likely to be less sensitive to the vagaries of any one individual method.

STSs around several disease genes have been used to isolate YACs and to build up contigs [e.g. the cystic fibrosis (*CFTR*) gene on chromosome 7 (Green and Olson, 1990b) and the dystrophin (*DMD*) gene on the X chromosome (Coffey *et al.*, 1992)]. High-resolution physical maps of individual chromosomes are also being constructed: detailed maps of chromosome 21q, the euchromatic region of the Y chromosome and 40% of the X chromosome are now available which consist of continuous arrays of overlapping genomic clones derived from human YAC libraries (Barillot *et al.*, 1992; Chumakov *et al.*, 1992a; Foote *et al.*, 1992; Mandel *et al.*, 1992; see Figure 2.2).

## 2.5 Genetic mapping

The concept of chromosome mapping by genetic means had its origin with Morgan and his colleague Sturtevant in the early years of this century. Working on *Drosophila*, these early geneticists recognized that the frequency with which non-allelic mutants were combined or separated in successive generations by genetic recombination could be used to derive linear maps of the corresponding loci on the chromosome. In humans, the first gene to be mapped to a specific chromosome was that underlying colour blindness, which was deduced very early on to reside on the X chromosome on account of its characteristic inheritance pattern (Wilson, 1911). Bell and Haldane (1937) demonstrated close linkage on the X chromosome between the loci for colour blindness and haemophilia A. The first autosomal linkage was established by Mohr (1951) between secretor factor and the Lutheran blood group, but it was not until 1968 that McKusick and colleagues succeeded in chromosomally assigning the first autosomal gene (Duffy blood group to chromosome 1) by demonstrating linkage between the Duffy locus and a chromosome heteromorphism (Donahue *et al.*, 1968).

In 1980, Botstein *et al.* suggested the possibility of constructing a linkage map of the human genome using RFLPs as genetic markers. These workers appreciated that the utility of a genetic map was dependent not only upon the number and spacing of the markers but also upon their informativity (heterozygosity). Clearly, this strategy depends on having a sufficient number of genetic markers to track the inheritance of every portion of every chromosome but, given these markers, this task can be performed on a limited number of three-generation families (White *et al.*, 1985). It must however be remembered that, since genetic marker order is always probabilistic, it can never be 100% certain.

The first linkage map of the human genome was based on the pattern of inheritance of 403 polymorphic loci through a panel of 21 three-generation CEPH families (Donnis–Keller *et al.*, 1987). This map was still low-resolution with an average spacing of 10–15 cM between markers. In 1992, the NIH/CEPH Collaborative Mapping Group published details of their genetic linkage map of the human genome which contained 1676 genetic markers at 1416 loci, including 279 genes and expressed sequences and 339 microsatellite repeat markers. This map spanned approximately 92% of the length of the autosomes and approximately 95% of the length of the X chromosome. With the exception of chromosomes 3 and 5, each chromosome map exhibited an average marker spacing of less than 5 cM. Specimen genetic linkage maps from this study are shown in Figure 2.3.

A slightly different genetic linkage map of the genome has been reported by CEPH/Généthon (Weissenbach *et al.*, 1992). This map was based upon the segregation of 814 highly polymorphic $(CA)_n$ repeat markers through eight three-generation pedigrees; 74% of markers possessed a heterozygosity of more than 0.7 and 68% of markers could be ordered with likelihood support of more than 1000:1. These markers are about 5 cM apart and together span about 90% of the human genome. Several sub-telomeric regions exhibited a low density of microsatellite markers, whereas clusters of these markers were observed in other regions. The cytogenetic locations of the markers used had not been determined and so correlation between physical and genetic distances was not possible.

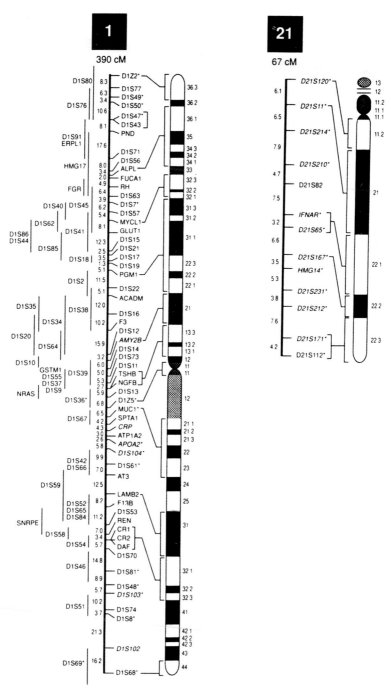

**Figure 2.3.** Linkage maps of the largest and smallest human chromosomes (from the NIH/CEPH Collaborative Mapping Group, 1992).

Despite gaps and some discrepancies evident when comparison was made with the NIH/CEPH map, this microsatellite map promises to be extremely useful for linkage mapping of diseases which remain to be chromosomally allocated. One of the most recent genetic linkage maps of the human genome contains 5840 loci (including 3617 PCR-formatted short tandem repeat polymorphisms and 427 genes) spanning 4000 cM at an average density of 0.7 cM (Murray *et al.*, 1994). Further improvement of individual chromosome maps toward the eventual goal of 1 cM resolution is anticipated as a result of the chromosomal localization of microsatellite markers (Couillin *et al.*, 1994; Reed *et al.*, 1994) and the integration of microsatellite data into existing maps (e.g. Atwood *et al.*, 1994). The latest microsatellite genetic linkage map of the human genome comprises 2066 $(CA)_n$ repeat markers spanning a total distance of 3690 cM (Gyapay *et al.*, 1994).

## 2.6 Transcription map of the human genome

Only a small fraction (<5%) of the human genome is expressed as mRNA but as yet only a small proportion of the estimated 70 000 or so genes have been identified (4670 genes catalogued of which 3291 have been mapped; Genome Data Base, June 1994). The construction of a transcription map of the human genome is important for a number of reasons. Such a map would ideally contain information on the precise chromosomal location, orientation, structure, sequence, splicing pattern and developmental/spatial pattern of expression of human genes. The transcription map would undoubtedly yield new insights into questions of gene number, organization, expression and evolution as well as facilitating the isolation of new genes of clinical interest (reviewed by Hochgeschwender and Brennan, 1991; Hochgeschwender, 1992). The many and varied means by which novel transcribed sequences may be identified, based upon DNA hybridization, DNA sequence or functional properties, are reviewed in detail in subsequent chapters. It is therefore only necessary to summarize them briefly here (Table 2.1).

As far as genome mapping is concerned, the advantage of concentrating on transcribed sequences is obvious; the protein-coding portion of the genome can be rapidly and intensively investigated. However, many of the cDNA-based methods used to identify transcribed sequences do not involve analysis of their site or pattern of expression. This type of approach therefore neglects many transcribed sequences either on account of a low level of expression or because expression is confined to an obscure tissue or even to a small number of cells. Whilst a genomic DNA-based approach will ultimately generate the necessary sequence data, the task is then to use specific algorithms to recognize the characteristics of coding sequences, to pick out the transcribed needle in the genomic haystack (Mural *et al.*, 1992). The correct identification of coding sequences which are very short, and perhaps characterized by unusual codon usage, may however be extremely difficult. Moreover, genes which encode RNAs that are not translated are likely to be missed by this approach. It is likely therefore that the transcription map of the human genome will eventually be completed through a combination of cDNA and genomic DNA mapping/ sequencing strategies.

**Table 2.1.** Approaches to the identification of novel gene transcripts in genomic DNA sequences (adapted from Collins, 1992)

| Methods | References |
| --- | --- |
| *Hybridization-based* | |
| Northern blotting | |
| Inter-species cross-hybridization ('zoo blots') | Monaco *et al.* (1986), Rommens *et al.* (1989) |
| Identification of CpG islands | Bird (1986), Lindsay and Bird (1987), Melmer *et al.* (1990), Bickmore (1992), Melmer and Buchwald (1992), Valdes *et al.* (1994) |
| cDNA library screening/sequencing | Hochgeschwender *et al.* (1989), Wallace *et al.* (1990), Adams *et al.* (1991, 1992, 1993a) |
| Subtractive hybridization | Duguid *et al.* (1988), Hara *et al.* (1991), Swaroop *et al.* (1991) |
| Heteronuclear cDNA library screening | Whitmore *et al.* (1994) |
| Homologous recombination | Vollrath *et al.* (1988), Kurnit and Seed (1990) |
| Direct cDNA selection using immobilized arrays of cosmids or YACs | Lovett *et al.* (1991), Parimoo *et al.* (1991) |
| Reverse transcript PCR | Cheng and Zhu (1994) |
| *Function-based* | |
| Expression of *lac-z* fusion proteins to identify open reading frames | Gray *et al.* (1982) |
| Exon amplification/trapping | Duyk *et al.* (1990), Buckler *et al.* (1991) |
| Enhancer/ poly(A) signal trapping | Weber *et al.* (1984); Rosenthal (1987) |
| Gene transfer and transcript identification | Strathdee *et al.* (1992) |
| *Sequence-based* | |
| Exon prediction | Fearon *et al.* (1990), Brunak *et al.* (1991), Uberbacher and Mural (1991) |
| Inter-specific comparison | Adams *et al.* (1993b), Sargent *et al.* (1993) |
| Differential display | Liang and Pardee (1992) |
| Database similarity searching | Gish and States (1993) |

## 2.7 Comparative gene mapping

Comparative gene mapping data are an important source of information on the structure, function and evolution of the human genome. Indeed, the position or even the existence of a gene may be inferred from such data. Genome maps are currently being developed for nearly 30 different mammalian species (O'Brien and Graves, 1991; Davisson *et al.*, 1991; Dietrich *et al.*, 1992; O'Brien *et al.*, 1993), mostly by means of physical methods such as somatic cell hybrid analysis and *in situ* hybridization. The extent of synteny between different mammalian species reflects both genome conservation and diversification during the adaptive radiation of the mammals.

Comparative gene mapping is also important in studies of human disease since human disease loci may be located by establishing the location of disease loci in other species. By 1993, some 976 homologous loci had been assigned to both human and mouse chromosomes (O'Brien *et al.*, 1993). Some 210 of these are known to be associated with human genetic diseases and 40 with a murine model of a human genetic disease. These data have led to the recognition of 80 syntenic regions on the autosomes (O'Brien *et al.*, 1993).

Comparative mapping allows the study both of the similarities and of the differences between mammalian genomes. *Genomic mismatch scanning* permits the identification of DNA segments that have the same sequence in two genomes (Nelson *et al.*, 1993) whereas *representational difference analysis* can be used to identify DNA fragments that differ between two genomes (Lisitsyn *et al.*, 1993).

## Acknowledgements

I am most grateful to Mike Chipperfield, DNA Data Coordinator of the Genome Data Base, Johns Hopkins University School of Medicine, Baltimore, MD, for provision of statistical information on known loci, probes and polymorphisms and to Drs Paula Hallam and Caroline Formstone for comments on the manuscript.

## References

Adams MD, Kelley JM, Gocayne JD, Dubnick M, Polymeropoulos MH, Xiao H, Merril CR, Wu A, Olde B, Moreno RF, Kerlavage AR, McCombie WR, Venter JC. (1991) Complementary DNA sequencing: expressed sequence tags and human genome project. *Science* **252**: 1651–1656.

Adams MD, Dubnick M, Kerlavage AR, Moreno R, Kelley JM, Utterback TR, Nagle JW, Fields C, Venter JC. (1992) Sequence identification of 2,375 human brain genes. *Nature* **355**: 632–634.

Adams MD, Kerlavage AR, Fields C, Venter JC. (1993a) 3400 expressed sequence tags identify diversity of transcripts in human brain. *Nature Genetics* **4**: 256–267.

Adams MD, Soares MB, Kerlavage AR, Fields C, Venter JC. (1993b) Rapid cDNA sequencing (expressed sequence tags) from a directionally cloned human infant brain cDNA library. *Nature Genetics* **4**: 373–380.

Albertsen HM, Abderrahim H, Cann HM, Dausset J, Paslier DL, Cohen D. (1990) Construction and characterization of a yeast artificial chromosome library containing seven haploid human genome equivalents. *Proc. Natl Acad. Sci. USA* **87**: 4256–4260.

Allikmets RL, Kashuba VI, Petterson B, Gizatullin R, Lebedeva T, Kholodnyuk ID, Bannikov VM, Petrov N, Zakharyev VM, Winberg G, Modi W, Dean M, Uhlen M, Kisselev LL, Klein G, Zabarovsky ER. (1994) *Not*I linking clones as a tool for joining physical and genetic maps of the human genome. *Genomics* **19**: 303–309.

Anand R, Villasente A, Tyler-Smith C. (1989) Construction of yeast artificial libraries with large inserts using fractionation by pulsed-field gel electrophoresis. *Nucleic Acids Res.* **17**: 4325–4333.

Atwood J, Chiano M, Collins A, Donis-Keller H, Dracopoli N, Fountain J, Falk C, Goudie D, Gusella J, Haines J, Armour JAL, Jeffreys AJ, Kwiatkowski D, Lathrop M, Matise T, Northrup H, Pericak-Vance MA, Phillips J, Retief A, Robson E, Shields D, Slaugenhaupt S, Vergnaud G, Weber J, Weissenbach J, White R, Yates J, Povey S. (1994) CEPH Consortium map of chromosome 9. *Genomics* **19**: 203–214.

Ballabio A. (1993) The rise and fall of positional cloning? *Nature Genetics* **3**: 277–279.

Barillot E, Dausset J, Cohen D. (1991) Theoretical analysis of a physical mapping strategy using random single-copy landmarks. *Proc. Natl Acad. Sci. USA* **88**: 3917–3921.

Barillot E, Gesnouin P, Pook S, Vaysseix G, Frelat G, Schmitz A, Sambucy J-L, Bosch A, Estivill X, Weissenbach J, Vignal A, Riethman H, Cox D, Patterson D,

**Gardiner K, Hattori M, Sakaki Y, Ichikawa H, Ohki M, Le Paslier D, Heilig R, Antonarakis S, Cohen D.** (1992) Continuum of overlapping clones spanning the entire human chromosome 21q. *Nature* **359**: 380–387.
**Barlow D, Lehrach H.** (1987) Genetics by gel electrophoresis: the impact of pulsed field electrophoresis on mammalian genetics. *Trends Genet.* **3**: 167–171.
**Barrell B.** (1991) DNA sequencing: present limitations and prospects for the future. *FASEB J.* **5**: 40–45.
**Barrett JH.** (1992) Genetic mapping based on radiation hybrid data. *Genomics* **13**: 95–103.
**Beckmann JS.** (1988) Oligonucleotide polymorphisms: a new tool for genomic genetics. *Bio/Technology* **6**: 1061–1064.
**Beckmann JS, Weber JL.** (1992) Survey of human and rat microsatellites. *Genomics* **12**: 627–631.
**Bell J, Haldane JBS.** (1937) The linkage between the genes for colour blindness and haemophilia in man. *Proc. R. Soc. Biol. Lond.* **123**: 119–150.
**Bellanné-Chantelot C, Barillot E, Lacroix B, Le Paslier D, Cohen D.** (1991) A test case for physical mapping of human genome by repetitive sequence fingerprints: construction of a physical map of a 420kb YAC subcloned into cosmids. *Nucleic Acids Res.* **19**: 505–510.
**Bellanné-Chantelot C, Lacroix B, Ougen P, Billault A, Beaufils S, Bertrand S, Georges I, Gilbert F, Gros I, Lucotte G, Susini L, Codani J-J, Gesnouin P, Pook S, Vaysseix G, Lu-Kuo J, Ried T, Ward D, Chumakov I, Le Paslier D, Barillot E, Cohen D.** (1992) Mapping the whole human genome by fingerprinting yeast artificial chromosomes. *Cell* **70**: 1059–1068.
**Benham F, Rowe P.** (1992) Use of *Alu*-PCR to characterize hybrids containing multiple fragments and to generate new Xp21.3–p22.2 markers. *Genomics* **12**: 368–376.
**Benham F, Hart K, Crolla J, Bobrow M, Francavilla M, Goodfellow PN.** (1989) A method for generating hybrids containing nonselected fragments of human chromosomes. *Genomics* **4**: 509–517.
**Bickmore W.** (1992) Analysis of genomic DNAs by pulsed-field gel electrophoresis. In: *Techniques for the Analysis of Complex Genomes* (ed. R Anand). Academic Press, London, pp. 19–38.
**Bird AP.** (1986) CpG-rich islands and the function of DNA methylation. *Nature* **321**: 209–213.
**Bodmer W.** (1981) Gene clusters, genome organization and complex phenotypes. When the sequence is known, what will it mean? *Am. J. Hum. Genet.* **33**: 664–682.
**Botstein D, White DL, Skolnick M, Davis RW.** (1980) Construction of a genetic linkage map in man using restriction fragment length polymorphisms. *Am. J. Hum. Genet.* **32**: 314–331.
**Bowcock AM, Bakker B, Ceverha P, Chipperfield MA, Minter-Morrison A, Porter CJ, Pearson PL.** (1993) Report of the DNA committee. In: *Human Gene Mapping; a Compendium* (eds. AJ Cuticchia, PL Pearson). Johns Hopkins University Press, Baltimore, MD, pp. 893–972.
**Brooks-Wilson AR, Goodfellow PN, Povey S, Nevanlinna HA, de Jong PJ, Goodfellow PJ.** (1990) Rapid cloning and characterization of new chromosome 10 DNA markers by Alu element-mediated PCR. *Genomics* **7**: 614–620.
**Brown WRA, Bird AP.** (1986) Long-range restriction site mapping of mammalian genomic DNA. *Nature* **322**: 477–481.
**Brownstein BH, Silverman GA, Little RD, Burke DT, Korsmeyer SJ, Schlessinger D, Olson MV.** (1989) Isolation of single copy human genes from a library of yeast artificial chromosome clones. *Science* **244**: 1348–1351.
**Brunak S, Engelbrecht J, Knudsen S.** (1991) Prediction of human mRNA donor and acceptor sites from the DNA sequence. *J. Mol. Biol.* **220**: 49–65.
**Buckler AJ, Chang DD, Graw SL, Brook JD, Haber DA, Sharp PA, Housman DE.** (1991) Exon amplification: a strategy to isolate mammalian genes based on RNA splicing. *Proc. Natl Acad. Sci. USA* **88**: 4005–4009.
**Burke DT, Carle GF, Olson MV.** (1987) Cloning of large exogenous DNA into yeast by means of artificial chromosome vectors. *Science* **236**: 806–812.
**Burmeister M, Momaco AP, Gillard EF, Van Ommen GB, Affara NA, Ferguson-Smith MA, Kunkel LM, Lehrach H.** (1988) A 10-megabase physical map of human Xp21 including the Duchenne muscular dystrophy gene. *Genomics* **2**: 189–202.

Burmeister M, Kim S, Price ER, de Lange T, Tantravahi U, Myers RM, Cox DR. (1991) A map of the distal region of the long arm of chromosome 21 constructed by radiation hybrid mapping and pulsed-field gel electrophoresis. *Genomics* **9**: 19–30.

Buting K, Passarge E, Horsthemke B. (1988) Construction of a chromosome 15-specific linking library and identification of potential gene sequences. *Genomics* **3**: 143–149.

Cantor CR. (1990) Orchestrating the human genome project. *Science* **248**: 49–51.

Carrano AV, Lamerdin J, Ashworth LK, Watkins B, Branscomb E, Slezak T, Raff M, de Jong PJ, Keith D, McBride L, Meister S, Kornick M. (1989) A high resolution, fluorescence-based semi-automated method for DNA fingerprinting. *Genomics* **4**: 129–136.

Ceccherini I, Romeo G, Lawrence S, Breuning MH, Harris PC, Himmelbauer H, Frischauf AM, Sutherland GR, Germino GG, Reeders ST, Morton NE. (1992) Construction of a map of chromosome 16 by using radiation hybrids. *Proc. Natl Acad. Sci. USA* **89**: 104–108.

Cheng J-F, Zhu Y. (1994) Reverse transcription-polymerase chain reaction detection of transcribed sequences on human chromosome 21. *Genomics* **20**: 184–190.

Chumakov IM, Rigault P, Guillou S, Ougen P, Billaut A, Guasconi G, Gervy P, LeGall I, Soularue P, Grinas L, Bougueleret L, Bellanné-Chantelot C, Lacroix B, Barillot E, Gesnouin P, Pook S, Vaysseix G, Frelat G, Schmitz A, Sambucy J-L, Bosch A, Estivill X, Weissenbach J, Vignal A, Riethman H, Cox D, Patterson D, Gardiner K, Hattori M, Sakaki Y, Ichikawa H, Ohki M, Le Paslier D, Heilig R, Antonarakis S, Cohen D. (1992a) Continuum of overlapping clones spanning the entire human chromosome 21q. *Nature* **359**: 380–387.

Chumakov IM, Le Gall I, Billault A, Ougen P, Soularue P, Guillou S, Rigault P, Bui H, De Tand M-F, Barillot E, Abderrahim H, Cherif D, Berger R, Le Paslier D, Cohen D. (1992b) Isolation of chromosome 21-specific yeast artificial chromosomes from a total human genome library. *Nature Genetics* **1**: 222–225.

Coffey AJ, Roberts RG, Green ED, Cole CG, Butler R, Anand R, Giannelli F, Bentley DR. (1992) Construction of a 2.6 Mb contig in yeast artificial chromosomes spanning the human dystrophin gene using an STS-based approach. *Genomics* **12**: 474–484.

Cohen D, Chumakov I, Weissenbach J. (1993) A first-generation physical map of the human genome. *Nature* **366**: 698–701.

Cole CG, Goodfellow PN, Bobrow M, Bentley DR. (1991) Generation of novel sequence tagged sites (STSs) from discrete chromosomal regions using *Alu*-PCR. *Genomics* **10**: 816–826.

Collins FS. (1988) Chromosome jumping. In: *Genome Analysis; a Practical Approach* (ed. KE Davies). IRL Press, Oxford, pp. 73–94.

Collins FS. (1992) Positional cloning: let's not call it reverse anymore. *Nature Genetics* **1**: 3–6.

Collins FS, Drumm ML, Cole JL, Lockwood WK, Vande Woude GF, Iannuzzi MC. (1987) Construction of a general human jumping library, with application to cystic fibrosis. *Science* **235**: 1046–1049.

Cooper DN, Smith BA, Cooke HJ, Niemann S, Schmidtke J. (1985) An estimate of unique DNA sequence heterozygosity in the human genome. *Hum. Genet.* **69**: 201–205.

Cooper DN, Schmidtke J. (1993) Diagnosis of genetic disease using recombinant DNA. fourth edition. *Hum. Genet.* **92**: 211–236.

Corbo L, Maley JA, Nelson DL, Caskey CT. (1990) Direct cloning of human transcripts with hnRNA from hybrid cell lines. *Science* **249**: 652–655.

Couillin P, Le Guern E, Vignal A, Fizames C, Ravisé N, Delportes D, Reguigne I, Rosier MF, Junien C, Van Heyningen V, Weissenbach J. (1994) Assignment of 112 microsatellite markers to 23 chromosome subregions delineated by somatic hybrids: comparison with the genetic map. *Genomics* **21**: 379–387.

Cowell JK. (1992) Somatic cell hybrids in the analysis of the human genome. In: *Human Cytogenetics; a Practical Approach*, Vol. II. *Malignancy and Acquired Abnormalities* (eds DE Rooney, BH Czepulkowski). IRL Press, Oxford, pp. 235–252.

Cox DR, Burmeister M, Price ER, Kim S, Myers RM. (1990) Radiation hybrid mapping: a somatic cell genetic method for constructing high-resolution maps of mammalian chromosomes. *Science* **250**: 245–250.

Davisson MT, Lalley PA, Peters J, Doolittle DP, Hillyard AL, Searle AG. (1991) Report of the comparative committee for human, mouse and other rodents. *Cytogenet. Cell Genet.* **8:** 1152–1189.

Dietrich W, Katz H, Lincoln SE, Shin H-S, Friedman J, Dracopoli NC, Lander ES. (1992) A genetic map of the mouse suitable for typing intraspecific crosses. *Genetics* **131:** 423–447.

Donahue RP, Bias WB, Renwick JH, McKusick VA. (1968) Probable assignment of the Duffy blood group locus to chromosome 1 in man. *Proc. Natl Acad. Sci. USA* **61:** 949–955.

Donis-Keller H, Green P, Helms C, Cartinhour S, Weiffenbach B, Stephens K, Keith TP, Bowden DW, Smith DR, Lander ES, Botstein D, Akots G, Rediker KS, Gravius T, Brown VA, Rising MB, Parker C, Powers JA, Watt DE, Kauffman ER, Bricker A, Phipps P, Muller-Kahle H, Fulton TR, Ng S, Schumm JW, Braman JC, Knowlton RG, Barker DF, Crooks SM, Lincoln SE, Daly MJ, Abrahamson J. (1987) A genetic linkage map of the human genome. *Cell* **51:** 319–337.

Duguid JR, Rohwer RG, Seed B. (1988) Isolation of cDNAs of scrapie-modulated RNAs by subtractive hybridization of a cDNA library. *Proc. Natl Acad. Sci. USA* **85:** 5738–5742.

Duyk GM, Kim S, Myers RM, Cox DR. (1990) Exon trapping: a genetic screen to identify candidate transcribed sequences in cloned mammalian genomic DNA. *Proc. Natl Acad. Sci. USA* **87:** 8995–8999.

Fasman KH, Cuticchia AJ, Kingsbury DT. (1994) The GDB Human Genome Data Base anno 1994. *Nucleic Acids Res.* **22:** 3462–3469.

Fasman ER, Cho KR, Nigro JM, Kern SE, Simons JW, Ruppert JM, Hamilton SR, Preisinger AC, Thomas G, Kinzler KW, Vogelstein B. (1990) Identification of a chromosome 18q gene that is altered in colorectal cancers. *Science* **247:** 49–56.

Fearon ER, Cho KR, Nigro JM, Kern SE, Simons JW, Ruppert JM, Hamilton SR, Preisinger AC, Thomas G, Kinzler KW, Vogelstein B. (1990) Identification of a chromosome 18q gene that is altered in colorectal cancers. *Science* **247:** 49–56.

Fields C, Adams MD, White O, Venter JC. (1994) How many genes in the human genome? *Nature Genetics* **7:** 345–346.

Foote S, Vollrath D, Hilton A, Page DC. (1992) The human Y chromosome: overlapping DNA clones spanning the euchromatic region. *Science* **258:** 60–66.

Frazer KA, Boehnke M, Budarf ML, Wolff RK, Emanuel BS, Myers RM, Cox DR. (1992) A radiation hybrid map of the region on human chromosome 22 containing the neurofibromatosis type 2 locus. *Genomics* **14:** 574–584.

Fulton TR, Bowcock AM, Smith DR, Deneshvar L, Green P, Cavalli-Sforza LL, Donis-Keller H. (1989) A 12 megabase restriction map at the cystic fibrosis locus. *Nucleic Acids Res.* **17:** 271–284.

Gardiner K, Patterson D. (1992) The role of somatic cell hybrids in physical mapping. *Cytogenet. Cell Genetics* **59:** 82–85.

Gardiner K, Horisberger M, Kraus J, Tantravahi U, Korenberg J, Rao V, Reddy S, Patterson D. (1990) Analysis of human chromosome 21: correlation of physical and cytogenetic maps; gene and CpG island distributions. *EMBO J.* **9:** 25–34.

Gish W, States DJ. (1993) Identification of protein coding regions by database similarity search. *Nature Genetics* **3:** 266–269.

Goold RD, diSibio GL, Xu H, Lang DB, Dadgar J, Magrane GG, Dugaiczyk A, Smith KA, Cox DR, Masters SB, Myers RM. (1993) The development of sequence-tagged sites for human chromosome 4. *Hum. Mol. Genet.* **2:** 1271–1288.

Gorski JL, Boehnke M, Reyner EL, Burright EN. (1992) A radiation hybrid map of the proximal short arm of the human X chromosome spanning incontinentia pigmenti 1 (IP1) translocation breakpoints. *Genomics* **14:** 657–665.

Goss SJ, Harris H. (1975) New method for mapping genes in human chromosomes. *Nature* **255:** 680–683.

Gray MR, Colot HV, Guarente L, Rosbash M. (1982) Open reading frame cloning: identification, cloning and expression of open reading frame DNA. *Proc. Natl Acad. Sci. USA* **79:** 6598–6602.

Green ED, Olson MV. (1990a) Systematic screen of yeast artificial chromosome libraries by use of the polymerase chain reaction. *Proc. Natl. Acad. Sci. USA* **87**: 1213–1217.

Green ED, Olson MV. (1990b) Chromosomal region of the cystic fibrosis gene in yeast artificial chromosomes: a model for human genome mapping. *Science* **250**: 94–98.

Green ED, Mohr RM, Idol JR, Jones M, Buckingham JM, Deaven LL, Moyzis RK, Olson MV. (1991) Systematic generation of sequence-tagged sites for physical mapping of human chromosomes: application to the mapping of human chromosome 7 using yeast artificial chromosomes. *Genomics* **11**: 548–564.

Gyapay G, Morissette J, Vignal A, Dib C, Fizames C, Millasseau P, Marc S, Bernardi G, Lathrop M, Weissbach J. (1994) The 1993–94 Généthon human genetic linkage map. *Nature Genetics* **7**: 246–339.

Hara E, Kato T, Nakada S, Sekiya S, Oda K. (1991) Subtractive cDNA cloning using oligo(dT)$_{30}$-latex and PCR: isolation of cDNA clones specific to undifferentiated human embryonal carcinoma cells. *Nucleic Acids Res.* **19**: 7097–7104.

Harding JD, Keller RA. (1992) Single-molecule detection as an approach to rapid DNA sequencing. *Trends Biotechnol.* **10**: 55–66.

Hardy DA, Bell JI, Long EO, Lindsten T, McDevitt HO. (1986) Mapping of the class II region of the human major histocompatibility complex by pulsed-field gel electrophoresis. *Nature* **323**: 453–455.

Hattori M, Toyoda A, Ichikawa H, Ito T, Ohgusu H, Oishi N, Kano T, Kuhara S, Ohki M, Sakaki Y. (1993) Sequence-tagged *Not*I sites of human chromosome 21: sequence analysis and mapping. *Genomics* **17**: 39–44.

Hochgeschwender U. (1992) Toward a transcriptional map of the human genome. *Trends Genet.* **8**: 41–44.

Hochgeschwender U, Brennan MB. (1991) Identifying genes within the genome: new ways for finding the needle in a haystack. *BioEssays* **13**: 139–144.

Hochgeschwender U, Sutcliffe JG, Brennan MB. (1989) Construction and screening of a genomic library specific for mouse chromosome 16. *Proc. Natl Acad. Sci. USA* **86**: 8482–8486.

Hoheisel JD. (1994) Application of hybridization techniques to genome mapping and sequencing. *Trends Genet.* **10**: 79–83.

Hoheisel JD, Maier E, Mott R, McCarthy L, Grigoriev AV, Schalkwyk LC, Nizetic D, Francis F, Lehrach H. (1993) High resolution cosmid and P1 maps spanning the 14 Mb genome of the fission yeast *S. pombe*. *Cell* **73**: 109–120.

Ichikawa H, Shimizu K, Saito A, Wang D, Oliva R, Kobayashi H, Kaneko Y, Miyoshi H, Smith CL, Cantor CR, Ohki M. (1992) Long-distance restriction mapping of the proximal long arm of human chromosome 21 with *Not*I linking clones. *Proc. Natl Acad. Sci. USA* **89**: 23–27.

Ichikawa H, Hosoda F, Arai Y, Shimizu K, Ohira M, Ohki M. (1993) A *Not*I restriction map of the entire long arm of human chromosome 21. *Nature Genetics* **4**: 361–365.

Kere J, Nagaraja R, Mumm S, Cicciodicola A, D'Urso M, Sclessinger D. (1992) Mapping human chromosomes by walking with sequence-tagged sites from end fragments of yeast artificial chromosome inserts. *Genomics* **14**: 241–248.

Komminoth P, Long AA. (1993) *In-situ* polymerase chain reaction; an overview of methods, applications and limitations of a new molecular technique. *Virchows Arch. B Cell. Pathol.* **64**: 67–73.

Korenberg JR, Yang-Feng T, Schreck R, Chen XN. (1992) Using fluorescence *in situ* hybridization (FISH) in genome mapping. *Trends Biotechnol* **10**: 27–32.

Kurnit DM, Seed B. (1990) Improved genetic selection for screening bacteriophage libraries by homologous recombination *in vivo*. *Proc. Natl Acad. Sci. USA* **87**: 3166–3169.

Kwok K, Ledley FD, DiLella AG, Robson KJH, Woo SLC. (1985) Nucleotide sequence of a full-length complementary DNA clone and amino acid sequence of human phenylalanine hydroxylase. *Biochemistry* **24**: 556–561.

Lawrance SK, Smith CL, Srivastava R, Cantor CR, Weissman SM. (1987) Megabase-scale mapping of the HLA gene complex by pulsed field gel electrophoresis. *Science* **235**: 1387–1390.

Liang P, Pardee AB. (1992) Differential display of eukaryotic messenger RNA by means of the polymerase chain reaction. *Science* **257**: 967–971.

Lichter P, Ledbetter SA, Ledbetter DH, Ward DC. (1990) Fluorescence *in situ* hybridization with *Alu* and L1 polymerase chain reaction probes for rapid characterization of human chromosomes in hybrid cell lines. *Proc. Natl Acad. Sci. USA* **87**: 6634–6638.

Lindsay S, Bird AP. (1987) Use of restriction enzymes to select potential gene sequences in mammalian DNA. *Nature* **327**: 336–338.

Lisitsyn N, Lisitsyn N, Wigler M. (1993) Cloning the differences between two complex genomes. *Science* **259**: 946–951.

Little RD, Pilia G, Johnson S, Zucchi I, D'Urso M, Schlessinger D. (1992) Yeast artificial chromosomes spanning 8 Mb and 15 cM of human cytogenetic band. *Proc. Natl Acad. Sci. USA* **89**: 177–181.

Lovett M, Kere J, Hinton LM. (1991) Direct selection: a method for the isolation of cDNAs encoded by large genomic regions. *Proc. Natl Acad. Sci USA* **88**: 9628–9632.

Mandel J-L, Monaco AP, Nelson DL, Schlessinger D, Willard H. (1992) Genome analysis and the human X chromosome. *Science* **258**: 103–109.

Marrone BL, Campbell EW, Anzick SL, Shera K, Campbell M, Yoshida TM, McCormick MK, Deaven L. (1994) Mapping of low-frequency chimeric yeast artificial chromosome libraries from human chromosomes 16 and 21 by fluorescence *in situ* hybridization and quantitative image analysis. *Genomics* **21**: 202–207.

McInnis MG, Chakravarti A, Blaschak J, Petersen MB, Sharma V, Avramopoulos D, Blouin J-L, König U, Brahe C, Matise TC, Warren AC, Talbot CC, Van Broeckhoven C, Litt M, Antonarakis SE. (1993) A linkage map of human chromosome 21: 43 PCR markers at average intervals of 2.5 cM. *Genomics* **16**: 562–571.

Melmer G, Buchwald M. (1992) Identification of genes using oligonucleotides corresponding to splice site consensus sequences. *Hum. Mol. Genet.* **1**: 433–438.

Melmer G, Sood Rommens J, Rego D, Tsui L-C, Buchwald M. (1990) Isolation of clones on chromosome 7 that contain recognition sites for rare-cutting enzymes by oligonucleotide hybridization. *Genomics* **7**: 173–181.

Mohr J. (1951) Search for linkage between Lutheran blood group and other hereditary characters. *Acta Pathol. Microbiol. Scand.* **28**: 207–210.

Monaco AP, Neve RL, Colletti-Feener C, Bertelson CJ, Kurnit DM, Kunkel LM. (1986) Isolation of candidate cDNAs for portions of the Duchenne muscular dystrophy gene. *Nature* **323**: 646–650.

Montanaro V, Casamassimi A, D'Urso M, Yoon JY, Freije W, Schlessinger D, Muenke M, Nussbaum RL, Saccone S, Maugeri S, Santoro AM, Motta S, Valle GD. (1991) *In situ* hybridization to cytogenetic bands of yeast artificial chromosomes covering 50% of human Xq24–Xq28 DNA. *Am. J. Hum. Genet.* **48**: 183–194.

Morton NE. (1991) Parameters of the human genome. *Proc. Natl Acad. Sci. USA* **88**: 7474–7476.

Mural RJ, Einstein JR, Guan X, Mann RC, Uberbacher EC. (1992) An artificial intelligence approach to DNA sequence feature recognition. *Trends Biotechnol.* **10**: 66–69.

Murray JC, Beutow KH, Weber JL, Ludwigsen S, Scherpier-Heddema T, Manion F, Quillen J, Sheffield VC, Sunden S, Duyk GM, Weissenbach J, Gyapay G, Dib C, Morrissette J, Lathrop GM, Vignal A, White R, Matsunami N, Gerken S, Melis R, Albertsen H, Plaetke R, Odelberg S, Ward D, Dausset J, Cohen D, Cann H. (1994) A comprehensive human linkage map with centimorgan density. *Science* **265**: 2049–2054.

Nagaraja R. (1992) Current approaches to long-range physical mapping of the human genome. In: *Techniques for the Analysis of Complex Genomes* (ed. R Anand). Academic Press, London, pp. 1–18.

Nelson DL, Ledbetter SA, Corbo L, Victoria MF, Ramirez-Solis R, Webster TD, Ledbetter DH, Caskey CT. (1989) *Alu* polymerase chain reaction: a method for rapid isolation of human-specific sequences from complex DNA sources. *Proc. Natl Acad. Sci. USA* **86**: 6686–6690.

Nelson DL, Ballabio A, Victoria MF, Pieretti M, Bies D, Gibbs RA, Maley JA, Chinault AC, Webster TD, Caskey CT. (1991) *Alu*-primed PCR for regional assignment of 110

YAC clones from the human X chromosome: identification of clones associated with a disease locus. *Proc. Natl Acad. Sci. USA* **88**: 6157–6161.

Nelson SF, McCusker JH, Sander MA, Kee Y, Modrich P, Brown PO. (1993) Genomic mismatch scanning: a new approach to genetic linkage mapping. *Nature Genetics* **4**: 11–18.

NIH/CEPH Collaborative Mapping Group (1992) A comprehensive genetic linkage map of the human genome. *Science* **258**: 67–86.

O'Brien SJ, Graves JAM. (1991) Report of the Committee on Comparative Gene Mapping. *Cytogenet. Cell Genet.* **58**: 1124–1151.

O'Brien SJ, Peters J, Searle AG, Womack JE, Johnson PA, Graves JAM. (1993) Report of the Committee on Comparative Gene Mapping. In: *Human Gene Mapping; a Compendium* (eds AJ Cuticchia, PL Pearson). Johns Hopkins University Press, Baltimore, MD, pp. 846–892.

Olson M, Hood L, Cantor C, Botstein D. (1989) A common language for physical mapping of the human genome. *Science* **245**: 1434–1435.

van Ommen GJB, Verkerk JMH, Hofker MH, Monaco AP, Kunkel LM, Ray P, Worton R, Wieringa B, Bakker E, Pearson PL. (1986) A physical map of 4 million bp around the Duchenne muscular dystrophy gene on the human X-chromosome. *Cell* **47**: 499–504.

Palazzolo MJ, Sawyer SA, Martin CH, Smoller DA, Hartl DL. (1991) Optimized strategies for sequence-tagged-site selection in genome mapping. *Proc. Natl Acad. Sci. USA* **88**: 8034–8038.

Parimoo S, Patanjali SR, Shukla H, Chaplin DD, Weissman SM. (1991) cDNA selection: efficient PCR approach for the selection of cDNAs encoded in large chromosomal DNA fragments. *Proc. Natl Acad. Sci. USA* **88**: 9623–9627.

Pearson ML, Söll D. (1991) The human genome project: a paradigm for information management in the life sciences. *FASEB J.* **5**: 35–39.

Persico MG, Viglietto G, Martini G, Toniolo D, Paonessa G, Moscatelli C, Dono R, Vulliamy T, Luzzatto L, D'Urso M. (1986) Isolation of human glucose-6-phosphate dehydrogenase (G6PD) cDNA clones: primary structure of the protein and unusual 5' non-coding region. *Nucleic Acids Res.* **14**: 2511–2522.

Pohl TM, Zimmer M, MacDonald ME, Smith B, Bucan M, Poustka A, Volina S, Searle S, Zebetner G, Wasmuth JJ, Gusella J, Lehrach H, Frischauf AM. (1988) Construction of a *Not*I linking library and isolation of new markers close to the Huntington's disease gene. *Nucleic Acids Res.* **19**: 9185–9198.

Poustka A, Lehrach H. (1986) Jumping libraries and linking libraries: the next generation of molecular tools in mammalian genetics. *Trends Genet.* **2**: 174–179.

Poustka A, Pohl T, Barlow DP, Frischauf AM, Lehrach H. (1987) Construction and use of chromosome jumping libraries from *Not*I-digested DNA. *Nature* **352**: 353–355.

Reed PW, Davies JL, Copeman JB, Bennett ST, Palmer SM, Pritchard LE, Gough SCL, Kawaguchi Y, Cordell HJ, Balfour KM, Jenkins SC, Powell EE, Vignal A, Todd JA. (1994) Chromosome-specific microsatellite sets for fluorescence-based, semi-automated genome mapping. *Nature Genetics* **7**: 390–394.

Richard CW, Cox DR, Kapp L, Murnane J, Cornelis F, Julier C, Lathrop GM, James MR. (1993) A radiation hybrid map of human chromosome 11q22–q23 containing the ataxia-telangiectasia disease locus. *Genomics* **17**: 1–5.

Riley JH, Ogilvie D, Anand R. (1992) Construction, characterization and screening of YAC libraries. In: *Techniques for the Analysis of Complex Genomes* (ed. R Anand). Academic Press, London, pp. 59–79.

Riordan JR, Rommens JM, Kerem BS, Alon N, Rozmahel R, Grzelczak Z, Zielenski J, Lok S, Plavsic N, Chou JL, Drumm ML, Iannuzzi ML, Collins FS, Tsui LC. (1989) Identification of the cystic fibrosis gene: cloning and characterization of complementary DNA. *Science* **245**: 1066–1073.

Rommens JM, Iannuzzi MC, Kerem B, Drumm ML, Melmer G, Dean M, Rozmahel R, Cole JL, Kennedy D, Hidaka N, Zsiga M, Buchwald M, Riordan JR, Tsui L-C, Collins FS. (1989) Identification of the cystic fibrosis gene: chromosome walking and jumping. *Science* **245**: 1059–1065.

Rose EA. (1991) Applications of the polymerase chain reaction to genome analysis. *FASEB J.* **5**: 46–54.

Rosen DR, Siddique T, Patterson D, Figlewicz DA, Sapp P, Hentati A, Donaldson D, Goto J, O'Regan JP, Deng H-X, Rahmani Z, Krizus A, McKenna-Yasek D, Cayabyab A, Gaston SM, Berger R, Tanzi RE, Halperin JJ, Herzfeldt B, Van den Bergh R, Hung W-Y, Bird T, Deng G, Mulder DW, Smyth C, Laing NG, Soriano E, Pericak-Vance MA, Haines J, Rouleau GA, Gusella JS, Horvitz HR, Brown RH. (1993) Mutations in Cu/Zn superoxide dismutase gene are associated with familial amyotrophic lateral sclerosis. *Nature* **362**: 59–62.

Rosenthal N. (1987) Identification of regulatory elements of cloned genes with functional assays. *Methods Enzymol.* **152**: 704–720.

Ross MT, Hoheisel JD, Monaco AP, Larin Z, Zehetner G, Lehrach H. (1992) High-density gridded YAC filters: their potential as genome mapping tools. In: *Techniques for the Analysis of Complex Genomes* (ed. R Anand). Academic Press, London, pp. 137–153.

Saito A, Abad JP, Wang D, Ohki M, Cantor CR, Smith CL. (1991) Construction and characterization of a *Not*I linking library of human chromosome 21. *Genomics* **10**: 618–630.

Sapru M, Gu J, Gu X, Smith D, Yu C-E, Wells D, Wagner M. (1994) A panel of radiation hybrids for human chromosome 8. *Genomics* **21**: 208–216.

Sargent CA, Affara NA, Bentley E, Pelmear A, Bailey DMD, Davey P, Dow D, Leversha M, Aplin H, Besley GTN, Ferguson-Smith MA. (1993) Cloning of the X-linked glycerol kinase deficiency gene and its identification by sequence comparison to the *Bacillus subtilis* homologue. *Hum. Mol. Genet.* **2**: 97–106.

Schinzel A, McKusick VA, Francomano C, Pearson PL. (1993) Report of the committee for clinical disorders and chromosome aberrations. In: *Human Gene Mapping; a Compendium* (eds AJ Cuticchia, PL Pearson). Johns Hopkins University Press, Baltimore, MD, pp. 735–772.

Schlessinger D. (1990) Yeast artificial chromosomes: tools for mapping and analysis of complex genomes. *Trends Genet.* **6**: 248–258.

Schlessinger D, Little RD, Freije D, Abide F, Zucchi I, Porta G, Pilia G, Nagaja R, Johnson SK, Yoon JY, Srivastava A, Kere J, Palmieri G, Ciccodicola A, Montanaro V, Romano G, Casamassimi A, D'Urso M. (1991) Yeast artificial chromosome-based genome mapping: some lessons from Xq24–28. *Genomics* **11**: 783–793.

Schmidtke J, Cooper DN. (1992) A comprehensive list of cloned human DNA sequences—1991 update. *Nucleic Acids Res.* **20** (Suppl.): 2181–2198.

Smith L, Hood L. (1987) Mapping and sequencing the human genome: how to proceed. *Bio/Technology* **5**: 933–939.

Smith CL, Econome JG, Schutt A, Kloc S, Cantor CR. (1987) A physical map of *Escherichia coli* K12 genome. *Science* **236**: 1448–1453.

Smith CL, Kloc SR, Cantor CR. (1988) Pulsed-field gel electrophoresis and the technology of large DNA molecules. In: *Genome Analysis; a Practical Approach* (ed. KE Davies). IRL Press, Oxford, pp. 41–72.

Smith MW, Clark SP, Hutchinson JS, Wei YH, Churukian AC, Daniels LB, Diggle KL, Gen MW, Romo AJ, Lin Y, Selleri L, McElligott DL, Evans GA. (1993) A sequence-tagged site map of human chromosome 11. *Genomics* **17**: 699–725.

Stallings RL, Torney DC, Hildebrand CE, Longmire JL, Deaven LL, Jett JH, Dogget NA, Moysis RK. (1990) Physical mapping of human chromosomes by repetitive sequence fingerprinting. *Proc. Natl Acad. Sci. USA* **87**: 6218–6222.

Stephens JC, Cavanaugh ML, Gradie MI, Mador ML, Kidd KK. (1990) Mapping the human genome: current status. *Science* **250**: 237–244.

Strathdee CA, Gavish H, Shannon WR, Buchwald M. (1992) Cloning of cDNAs for Fanconi's anaemia by functional complementation. *Nature* **356**: 763–767.

Swaroop A, Xu J, Agarwal N, Weissman SM. (1991) A simple and efficient cDNA library subtraction procedure: isolation of human retina-specific cDNA clones. *Nucleic Acids Res.* **19**: 1954.

Takahashi E, Yamakawa K, Nakamura Y, Hori T. (1992) A high-resolution cytogenetic map of human chromosome 3: localization of 291 new cosmid markers by direct R-banding fluorescence *in situ* hybridization. *Genomics* **13**: 1047–1055.

Takahashi E, Hitomi A, Nakamura Y. (1993) A high-resolution cytogenetic map of human chromosome 5: localization of 206 new cosmid markers by direct R-banding fluorescence *in situ* hybridization. *Genomics* **17**: 234–236.

Takahashi E, Koyama K, Hitomi A, Itoh H, Nakamura Y. (1994) A high-resolution cytogenetic map of human chromosome 9: localization of 203 new cosmid markers by direct R-banding fluorescence *in situ* hybridization. *Genomics* **19**: 373–375.

Tassabehji M, Read AP, Newton VE, Harris R, Balling R, Gruss P, Strachan T. (1992) Waardenburg's syndrome patients have mutations in the human homologue of the *Pax-3* paired box gene. *Nature* **355**: 635–636.

Trask BJ, Massa H, Kenwrick S, Gitschier J. (1991) Mapping of human chromosome Xq28 by two-color fluorescence *in situ* hybridization of DNA sequences to interphase cell nuclei. *Am. J. Hum. Genet.* **48**: 1–15.

Trask B, Christensen M, Fertitta A, Bergmann A, Ashworth L, Branscomb E, Carrano A, van den Engh G. (1992) Fluorescence *in situ* hybridization mapping of human chromosome 19: mapping and verification of cosmid contigs formed by random restriction enzyme fingerprinting. *Genomics* **14**: 162–167.

Uberbacher EE, Mural RJ. (1991) Locating protein-coding regions in human DNA sequences by a multiple sensor-neural network approach. *Proc. Natl. Acad. Sci. USA* **88**: 11261–11265.

Uhlén M, Hultman T, Wahlberg J, Lundeberg J, Bergh S, Pettersen B, Holmberg A, Ståhl S, Moks T. (1992) Semi-automated solid phase DNA sequencing. *Trends Biotechnol.* **10**: 52–55.

Valdes JM, Tagle DA, Collins FS. (1994) Island rescue PCR: a rapid and efficient method for isolating transcribed sequences from yeast artificial chromosomes and cosmids. *Proc. Natl. Acad. Sci. USA* **91**: 5377–5381.

Venter JC, Adams MD, Martin-Gallardo A, McCombie WR, Fields C. (1992) Genome sequence analysis: scientific objectives and practical strategies. *Trends Biotechnol.* **10**: 8–11.

Vogel F, Motulsky AG. (1986) *Human Genetics – Problems and Approaches*, 2nd edn. Springer, Berlin.

Vollrath D, Davis RW, Connelly C, Hieter P. (1988) Physical mapping of large DNA by chromosome fragmentation. *Proc. Natl Acad. Sci. USA* **85**: 6027–6031.

Wada M, Little RD, Abid F, Porta G, Labella T, Cooper T, Della Valle G, D'Urso M, Schlessinger D. (1990) Human Xq24–28: approaches to mapping with yeast artificial chromosomes. *Am. J. Hum. Genet.* **46**: 95–106.

Wallace MR, Marchuk DA, Andersen LB, Letcher R, Odeh HM, Saulino AM, Fountain JW, Brereton A, Nicholson J, Mitchel AL, Brownstein BH, Collins FS. (1990) Type 1 neurofibromatosis gene: identification of a large transcript disrupted in three NF1 patients. *Science* **249**: 181–186.

Wang D, Smith CL. (1994) Large-scale structure conservation along the entire long arm of human chromosome 21. *Genomics* **20**: 441–451.

Wang J-Y, Zabarovsky ER, Talmadge C, Berglund P, Chan KWK, Pokrovskaya ES, Kshuba VI, Zhen D-K, Boldog F, Zabarovskaya VI, Kisselev LL, Stanbridge EJ, Klein G, Sumegi J. (1994) Somatic cell hybrid panel and *Not*I linking clones for physical mapping of human chromosome 3. *Genomics* **20**: 105–113.

Warrington JA, Hall L, Hinton LM, Miller JN, Wasmuth JJ, Lovett M. (1991) Radiation hybrid map of 13 loci on the long arm of chromosome 5. *Genomics* **11**: 701–708.

Watson JD. (1990) The Human Genome Project: past, present and future. *Science* **248**: 44–49.

Watson JD, Cook-Deegan RM. (1991) Origins of the Human Genome Project. *FASEB J.* **5**: 8–11.

Weber F, de Villiers J, Schaffner W. (1984) An SV40 "enhancer trap" incorporates exogenous enhancers or generates enhancers from its own sequences. *Cell* **36**: 983–992.

Weissenbach J, Gyapay G, Dib C, Vignal A, Morissette J, Millasseau P, Vaysseix G, Lathrop M. (1992) A second-generation linkage map of the human genome. *Nature* **359**: 794–801.

White R, Leppert M, Bishop DT, Barker D, Berkowitz J, Brown C, Callahan P, Holm T, Jerominski L. (1985) Construction of linkage maps with DNA markers for human chromosomes. *Nature* **313**: 101–105.

Whitmore SA, Apostolou S, Lane S, Nancarrow JK, Phillips HA, Richards RI, Sutherland GR, Callen DF. (1994) Isolation and characterization of transcribed sequences from a chromosome 16 hn-cDNA library and the physical mapping of genes and transcribed sequences using a high-resolution somatic cell panel of human chromosome 16. *Genomics* **20**: 169–175.

Wilcox AS, Khan AS, Hopkins JA, Sikela JM. (1991) Use of 3' untranslated sequences of human cDNAs for rapid chromosome assignment and conversion to STSs: implications for an expression map of the genome. *Nucleic Acids Res.* **19**: 1837–1843.

Williamson R, Bowcock A, Kidd K, Pearson P, Schmidtke J, Ceverha P, Chipperfield M, Cooper DN, Coutelle C, Hewitt J, Klinger K, Langley K, Weber J, Beckman J, Tolley M, Maidak B. (1991) Report of the DNA Committee and catalogues of cloned and mapped genes, markers formatted for PCR and DNA polymorphisms. *Cytogenet. Cell Genet.* **58**: 1190–1832.

Wilson EB. (1911) The sex chromosomes. *Arch. Mikrosk. Anat. Entwicklungsmech.* **77**: 249–271.

Yu J, Hartz J, Xu Y, Gemmill RM, Korenberg JR, Patterson D, Kao FT. (1992) Isolation, characterization, and regional mapping of microclones from a human chromosome 21 microdissection library. *Am. J. Hum. Genet.* **51**: 263–272.

Zabarovsky ER, Boldog F, Thompson T, Scanlon D, Winberg G, Marcsek Z, Erlandsson R, Stanbridge EJ, Klein G, Sumegi J. (1990) Construction of a human chromosome 3 specific *Not*I linking library using a novel cloning procedure. *Nucleic Acids Res.* **18**: 6319–6324.

Zabarovsky ER, Boldog F, Erlandsson R, Kashuba V, Allikmets RL, Marcsek Z, Kisselev LL, Stanbridge E, Klein G, Sumegi J, Winberg G. (1991) New strategy for mapping the human genome based on a novel procedure for construction of jumping libraries. *Genomics* **11**: 1030–1039.

Zabarovsky ER, Kashuba VI, Pokrovskaya ES, Zabaroska V, Wang J-Y, Berglund P, Boldog F, Stanbridge E, Sumegi J, Klein G, Winberg G. (1993) *Alu*-PCR approach to isolate *Not*I linking clones from the 3p14–3p21 region frequently deleted in renal cell carcinoma. *Genomics* **16**: 713–719.

Zabarovsky ER, Allikmets R, Kholodnyuk I, Zabarovska VI, Paulsson N, Bannikov VM, Kashuba VI, Dean M, Kisselev LL, Klein G. (1994a) Construction of representative *Not*I linking libraries specific for the total human genome and for human chromosome 3. *Genomics* **20**: 312–316.

Zabarovsky ER, Kashuba VI, Kholodnyuk ID, Zabarovska VI, Stanbridge EJ, Klein G. (1994b) Rapid mapping of *Not*I linking clones with differential hybridization and *Alu*-PCR. *Genomics* **21**: 486–489.

Zabarovsky ER, Kashuba VI, Pettersson B, Petrov N, Zakharyev V, Gizatullin Lebedeva T, Bannikov V, Pokrovskaya ES, Zabarovska VI, Allikmets R, Erlandsson R, Domninsky D, Sumegi J, Stanbridge EJ, Winberg G, Uhlén M, Kisselev LL, Klein G. (1994c) Shot-gun sequencing strategy for long-range genome mapping: a pilot study. *Genomics* **21**: 495–500.

Zhu Y, Cantor CR, Smith CL. (1993) DNA sequence analysis of human chromosome 21 *Not*I linking clones. *Genomics* **18**: 199–205.

# 3

# Cloning the transcribed portion of the genome

Paul Towner

## 3.1 Introduction

The genomic DNA of chromosomes within cell nuclei is a repository of encoded information of which a substantial proportion appears to serve no useful purpose. However, some sequences, known as promoters, are able to bind to transcription factors in the upstream portion of transcription units of the DNA and give rise to RNA via transcription. The most important of these is RNA polymerase which reads the antisense strand of the gene and produces a sense-encoding RNA copy using nucleotide triphosphates of the bases adenine, uracil, cytosine and guanine. Total cellular RNA is composed mainly of small nuclear uridine-rich RNA (snRNA), transfer RNA (tRNA) and ribosomal RNA (rRNA) which are transcribed from high copy number genes and are never translated. These types of RNA will not be discussed here; rather, attention will be paid to the very small proportion of RNA which is transcribed in the nucleus of the cell from single or low copy number genes and is destined to be translated in the cytoplasm to form proteins. This type of RNA is transcribed by RNA polymerase II, is known as messenger RNA (mRNA) and, in its nascent form, can be several hundred thousand nucleotides long. This RNA matures quickly, a process which involves capping and methylation of its 5' end, addition of a 3' polyadenosine tail and a decrease in size to typically 1000 – 10 000 nucleotides, before being translocated to the cytoplasm for subsequent expression. The size decrease is due to the removal of intervening sequences or introns, allowing the coding portions, or exons, to become spliced together to form a contiguous open reading frame with untranslated regions (UTRs) at the 5' and 3' ends, as depicted in Figure 3.1.

Capping is ultimately necessary for recognition of the mRNA by the small ribosomal subunit, to enable translation, whereas polyadenylation stabilizes the transcript in the cytoplasm. This latter feature potentiates the efficient isolation of mRNA from a pool of mixed RNA species which can be reverse transcribed to produce a complementary DNA (cDNA) which is more amenable to manipulation and can be used either to prepare a cDNA library or to isolate

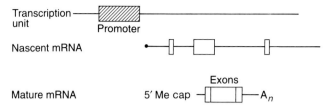

**Figure 3.1.** Generation of nascent and mature mRNA species

the gene of interest by polymerase chain reaction (PCR)-based techniques. Many of the ideas on which the techniques described in this chapter are based upon years of research work. However, there are now many commercial suppliers who offer self-explanatory kits so that the implementation of the technology is very simple.

## 3.2 Gene detection

The cloning of specific genes from the transcribed portion of the genome is generally straightforward if attention is paid to two key elements. Foremost is the requirement for fresh tissue in which it is known or anticipated that the gene of interest will be expressed. This will ideally provide a plentiful supply of mRNA in which no degradation of the mRNA encoding the gene will have occurred. Secondly, a suitable assay must be available which can identify the cDNA encoding the gene of interest. The most unambiguous method is to use a functional assay to detect the gene product. This relies upon a suitable expression system in which the entire correctly folded protein can be generated in a fully active form which can then be readily detected. If the expressed protein is incomplete or inactive (which is frequently the case), then it can only be detected by screening a cDNA library using an antibody. Any clones which are identified must be very carefully selected since, because of the huge number of epitopes available from other cloned inserts, it is very easy to choose clones which do not encode the gene. By far the most commonly used route for the detection of specific genes is by hybridization using a labelled complementary nucleic acid sequence under stringent conditions. The clones which have been identified and isolated can then be sequenced and the amino acid sequence of the encoded protein deduced. This DNA can be further manipulated and constructs prepared from which functional proteins can be isolated and characterized to show that the gene of interest has indeed been isolated. The identification of the true clone which encodes the gene of interest is frequently the major hurdle because it is very easy to be misled by strong hybridization signals during screening. Consequently, any false positives are recognized only at subsequent steps, either during plaque purification, Southern blotting of inserts or, occasionally, only after sequencing. Such a waste of effort can be avoided if more than one nucleic acid probe is used under stringent hybridization conditions. A major recent advance which circumvents this problem in library screening is to use PCR. This allows the gene of interest to be targeted from the outset and will at least provide a unique DNA sequence with which to probe a library and can even be utilized to isolate the entire gene without ever requiring a cDNA library.

## 3.3 Preparation of target material

### 3.3.1 Isolation of total RNA

Unlike DNA, RNA is relatively unstable by virtue of the presence of a 2' hydroxyl (OH) group juxtaposed to the 3'–5' phosphodiester bond which links the nucleotides of the RNA. At alkaline pH, the 2' hydroxyl group acts as a nucleophile, leading to transesterification to form a free 5' hydroxyl and a 2'-3' phosphodiester bond, as shown in Figure 3.2. At neutral pH, the RNA can also be degraded by virtue of contaminating RNases which may be present from within the starting tissue or introduced via skin contact with the operator. Generally, RNA should be prepared and stored immediately in a form which cannot degrade: either in a guanidinium solution or after precipitation using ethanol, storage being at –70°C. It is often more useful to copy the mRNA into cDNA immediately after its isolation, thus avoiding any problems associated with its degradation. From the outset, it is worthwhile being fastidious with the reagents and equipment which are going to be in contact with the RNA. With experience, it will become clear that only a few essential rules need to be followed. As a guideline, have available a separate set of Gilson pipettes for use only with RNA; their barrels and plungers should be kept clean and occasionally soaked in 2% diethylpyrocarbonate (DEPC). Heat treat non-plastics at 180°C for 2 h and ensure that a flat-bed agarose gel electrophoresis tank is dedicated for RNA work. Finally, prepare all buffers which come into contact with RNA using DEPC-treated water.

The most effective method for the isolation of RNA from a range of tissues uses guanidinium thiocyanate and acidic phenol (Chomczynski and Sacchi, 1987). This procedure has been adopted by many workers and forms the basis of many commercial RNA extraction kits. Briefly, tissue is collected and stored at –70°C, then pulverized in the presence of liquid $N_2$. The powdered tissue is

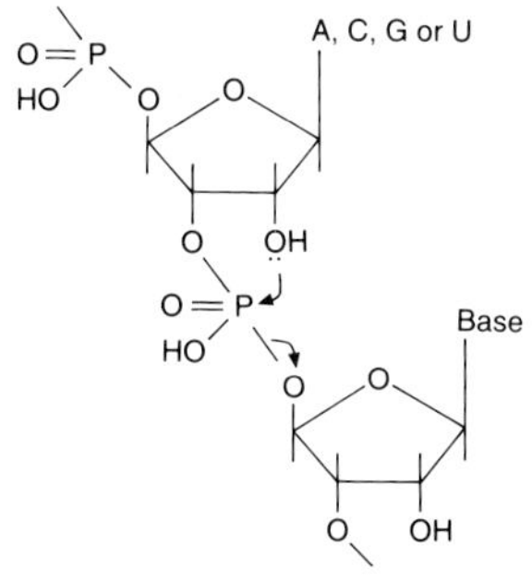

**Figure 3.2.** The susceptible phosphodiester linkage in RNA.

mixed with the guanidinium/phenol solution, extracted with chloroform and the RNA in the aqueous phase and precipitated. The quality of the preparation can easily be determined by electrophoresis on a denaturing agarose gel. A good sample of RNA will have all RNA species intact and can be distinguished by the presence of two clearly defined bands of large and small rRNAs of 4700 and 2000 nucleotides respectively, with a low-molecular-weight broad band of sn- and tRNAs of 75–120 nucleotides. The mRNA will not be visible since it accounts for only 1–5% of the total RNA, but it can be assumed to be intact if the rRNA bands are sharp with no smearing, indicating that they have not degraded. This material is very suitable for use in Northern blots and RNase protection analyses and can be employed directly for the preparation of cDNA if it is to be used solely for PCR. If the cDNA is to be used to prepare a library, it will be first necessary to isolate the poly($A^+$) RNA and use this as starting material.

### 3.3.2 Isolation of mRNA

The majority of mRNAs have a polyadenosine tract at their 3' ends. This feature allows straightforward purification by hybridization to a complementary polythymidine tract of 15–25 nucleotides which is attached to an inert support (Aviv and Leder, 1972). The most common supports are based on cellulose or the inert surface of coated paramagnetic spheres. Purification using cellulose-based supports utilizes normal column chromatographic techniques for the purification steps and work best with copious amounts (>2 mg) of RNA. Magnetic beads can be used with very small quantities of RNA (50 µg) and allow for the extensive washing of the bound mRNA before release from the support (Jakobsen *et al.*, 1990). For efficient separation, the RNA is dissolved in LiCl and heated to 65°C to disrupt secondary structures, mixed with the oligo(dT) paramagnetic beads and allowed to hybridize. The beads are immobilized using a magnet and the supernatant changed to facilitate washing and elution of the mRNA. The sample can then be used immediately in cDNA synthesis, frozen at –70°C for storage, or precipitated using ethanol, and then stored.

### 3.3.3 Preparation of cDNA

The enzyme reverse transcriptase is able to use the 3' end of short double-stranded regions on mRNA to initiate the synthesis of a first-strand cDNA copy of the mRNA (Baltimore, 1970; Temin and Mizutani, 1970). The double-stranded region is provided by adding oligodeoxyribonucleotides which are allowed to anneal to the mRNA. These can either be oligo(dT) which is complementary to the 3' poly(A) terminus, or 8mer oligos of random sequence which can base-pair at many sites along the mRNA sequence. If the mRNA starting material is undegraded, then it is possible to prepare cDNA of identical length if a high quality reverse transcriptase is utilized [e.g. Superscript (Gibco–BRL)] which is devoid of any RNase activity (Retsel *et al.*, 1980). The product of the reaction is a heteroduplex of cDNA and mRNA; this material is highly suitable for use in PCR but is more often used immediately for the synthesis of a second strand of cDNA. If a cDNA library is desired, then the synthesis of a second DNA strand is obligatory and is prepared by means of a combination of DNA polymerase I and RNase H; whilst RNase H nicks the remaining mRNA strand, DNA polymerase I uses the RNA fragments to initiate second strand synthesis (Gubler and Hoffman, 1983).

### 3.3.4 Selection of specific genes

Genes are generally selected and isolated from gene libraries or by PCR-based techniques. The most appropriate route ultimately depends upon how much is known about the gene itself or its protein product. In some very special circumstances, a clone may be selected by complementation using a strain of *E. coli* which is deficient in a specific enzyme, making bacterial growth in selective media possible only by functional expression of the missing gene (Chang *et al.*, 1978). However, this is seldom applicable. Alternatively, batches of size-selected mRNA can be injected into *Xenopus* oocytes, where the protein will be expressed, allowing the selection of small aliquots of mRNA based on functional activity (Hediger, 1987a,b; Masu *et al.*, 1987).

## 3.4 Library-based cDNA cloning strategies

The traditional procedure of isolating a clone encoding a specific gene is to prepare double-stranded cDNA and use this to prepare a cDNA library which is then screened with a biomolecule which can identify the gene of interest. Libraries of cDNA can be based on plasmid vectors (Okayama and Berg, 1983). However, these should be avoided if possible because the efficiency of transformation of *E. coli* is low in comparison to transfection using a λ-based phage. This can result in rare cDNA sequences not being represented in the library. Many λ-based vector kits are commercially available, from which large numbers of recombinant clones can easily be obtained, so that low copy number genes have a significant chance of being present. Some vectors are highly versatile and allow the easy manipulation of the identified clone such that it can be auto-excised as a plasmid and used in sequencing and the preparation of mRNA for expression studies.

The mRNA used in this primary step of preparing a library involves a substantial investment of time and resources and should fulfil certain criteria. If a nucleic acid probe is available for use as a hybridization probe in the screening process, it is worth using it at this stage on a Northern blot to detect the presence and determine the size of the transcript (Alwine *et al.*, 1977; Dyson, 1990). If the gene of interest is of low abundance, then it is expedient to perform a nuclease protection assay if a reasonable size probe derived from PCR experiments is available (Smith, 1990). The results of these analyses will give an indication of the number of plaques from the library which will need to be screened to find a positive signal and may also suggest that a different tissue or stage of development would be a more profitable choice. Occasionally the whole prospect of preparing a library might be delayed until an enriched source of starting material, prepared by subtractive hybridization, is available (see Section 3.4.2).

### 3.4.1 Construction of a cDNA library

The most efficient route to obtaining a clone of a specific gene encoded by cDNA is to ligate linkers on to the ends of double-stranded cDNA and use this material in the preparation of a cDNA library. The cDNA can then be ligated into the arms of a λ-based vector and packaged into the λ phage using high quality packaging extracts. A high titre cDNA library can be used to transfect *E. coli* where the ensuing plaques will offer a very good chance of identifying the required clone. One of the most successful strategies is to prepare cDNA following a directional cloning protocol which relies upon the use of a modified cDNA synthesis primer. The primer consists of a polythymidine tract ($T_{17}$) at its 3' end, flanked with a sequence of between 28 and 38 nucleotides which is about 50% GC and includes several restriction sites. The primer is used in place of the poly(T) primer in first strand synthesis and results in cDNA which has a known sequence incorporated into its 5' end (i.e. the 3' end of the original mRNA). The second cDNA strand is synthesized and linkers ligated to the double-stranded cDNA, which is then digested with two restriction enzymes, one of which cuts in the linker and the other at a site within the cDNA primer. If methyl-dCTP is used in the first strand synthesis reaction, then the methylation-sensitive enzymes used in the digestion will not recognize any internal restriction sites. The product is unidirectional, with a high proportion of full-length transcripts, and can be cloned in a specific orientation within the λ phage arms so that the 5' end of the cDNA is under the control of an inducible promoter such as *LacZ* which can be activated by the gratuitous inducer isopropyl thiogalactosidase (IPTG). A portion of the library can be used to infect the host bacteria, mixed with molten low gelling temperature nutrient agarose and poured on to a 254 x 254 mm nutrient agar plate, which can accommodate up to $10^5$ phage. The bacterial lawn which forms will be interspersed with tiny holes, or plaques, indicating phage growth. Nylon membrane, to which both bacteria and phage adhere, is then layered over the lawn. The membrane can then be processed and screened by hybridization using nucleic acid probes, or antibodies which would cross-react with the gene product, if IPTG had been present in the growth media.

### 3.4.2 Screening cDNA libraries

*Identification of gene sequences in cDNA libraries using antibodies.* If a protein product has been isolated, it can be used to raise antibodies even from a few micrograms of protein within a polyacrylamide gel. A polyclonal antibody will recognize several epitopes on the protein from which it was prepared, which makes it susceptible to cross-reactivity with many types of protein present on the filter which is being screened. Moreover, even though *E. coli* tolerates the presence of foreign nucleic acid, the expression of some protein sequences affects the growth of the bacteria which harbour them, possibly causing under-representation of the gene. Overall, the potential for selection of false positives is very high, although it must be stated that this technique was once a mainstay in screening protocols (Kemp and Cowman, 1981a,b). The basis of the technique has been further developed to improve the selection criteria of clones by placing cDNAs into a phage-based vector such that the corresponding gene product is incorporated into the phage coat protein (Hogrefe et al., 1993). This method of *phage display* was originally devised to obtain functional antibody fragments (Winter and Milstein, 1991) but has been further manipulated to allow the isolation of a phage which encodes virtually any cDNA. Consequently, antibodies are still a very useful basis from which to detect genes in libraries, but it is frequently easier and cheaper to use nucleic acid probes which are, in general, more reliable. Bear in mind that an antibody has to be prepared from either a purified protein or peptide fragments, either of which could easily be microsequenced, and the amino acid sequence used as a basis for oligonucleotide synthesis.

*Identification of gene sequences in cDNA libraries using a nucleic acid probe.* Three types of nucleic acid probe can be used in library screening, depending on the quality and quantity of information available from the gene of interest. The best probe will be a gene-encoding fragment and will probably have been isolated using PCR. Alternatively, an entire gene or fragment encoding a supposedly closely related sequence can be used to screen for homologous genes (Grunstein and Hogness, 1975; Hanahan and Meselson, 1980). If an amino acid sequence from the protein is available, either from the N terminus or internal peptide fragments, oligonucleotides can be prepared and used in screening. It is wise to restrict the number of codons chosen to represent amino acid positions otherwise they would be so redundant that they would become non-specific. In general, longer oligomers provide better results; the choice of 15- and 17mers a decade ago was based on cost rather than because they were superior to 30-, 40- and 50mers. The use of inosine in place of other nucleotide mixtures is unnecessary. It is important to keep the redundancy level low by selecting favoured codons and it is crucial to use two oligonucleotide probes when possible. The probes can be detected via enzyme-linked and chemical-based methods although the original procedure of using [$^{32}$P]phosphate radiolabelling has not yet been surpassed in sensitivity.

*Identification of gene sequences in cDNA libraries using subtracted material.* Under some circumstances, no information about the gene of interest is available. Phenotypic changes in a cell during development, or in response to a hormone, may be the only clue that the activity of specific genes has changed. The subtle differences in mRNA populations between two cell types may be the only handle from which to glean information which will ultimately lead to the identification of the genes responsible for the change of phenotype (Wang and Brown, 1991). Experimentally, this can be done by using the technique of differential display (see Section 3.5.6) or by preparing either a subtracted library or a subtracted probe. In each case, mRNA is prepared from two different tissues, or cell or tissue which has undergone a phenotypic change, one of which is converted to cDNA, and its parent mRNA strand destroyed using 0.5 M NaOH. The sense mRNA and antisense single-stranded cDNA are hybridized and removed from solution. Abundant mRNAs are expressed in most cells most of the time and so are removed by this process. However, a few unique populations of mRNA, characteristic of phenotypic change or tissue, remain and can be converted to cDNA and analysed. If the sample has been prepared using cDNA attached to Dynabeads (Hara *et al.*, 1991), then a 10-fold enrichment of specific genes can be anticipated (Rodriguez and Chader, 1992). Samples prepared by this route would be of no use as nucleic acid probes because far too many non-specific genes would be present. However, the sample would be enriched and thus offer an advantage in cDNA library construction if an additional probe for use in screening was available. If the subtracted nucleic acid sample is expressly required for use as a nucleic acid probe in screening a cDNA library, then the subtraction process must be highly specific. At present, this is only viable by using chemical cross-linking reagents such as DZQ (2,5 diaziridinyl-1,4-benzoquinone) which are able to covalently modify and link hybrids of RNA and DNA, allowing a very high enrichment of rare transcripts (Hampson *et al.*, 1992).

### 3.4.3 Manipulation of identified cDNA sequences

An unambiguously identified plaque in a cDNA library screen must be isolated from the parent plate and diluted, then re-plated and re-screened in order to check that it is pure enough to provide a stock of phage containing the gene of interest. If a vector such as ZAP II (Stratagene) is used, then the insert can be auto-excised to provide a so-called *phagemid* (i.e. a plasmid which possesses the F' single-stranded replication origin and can be used to recover single-stranded DNA packaged as an M13 phage). Either the double-stranded plasmid or the single-stranded phage can be used in sequencing. Additionally, the insert is flanked by T3 and T7 promoters and can be used for the preparation of sense and antisense RNA for translation and use as a probe, respectively.

## 3.5 PCR-based isolation of genes from cDNA

Owing to the inherent problems in identifying and isolating specific cDNA species from libraries in the absence of a suitable probe, it is now common to isolate gene fragments using PCR-based techniques (Innis *et al.*, 1990). At the very least, this will provide a highly specific probe with which to identify the gene in a library, or it could provide the gene in its entirety suitable for cloning and expression. The resources required for successful PCR are minimal and, in the primary stages, only require first-strand synthesized cDNA prepared from small amounts of tissue from which the mRNA has been neither purified nor checked for the presence of the gene by Northern analysis. For PCR to be successful, oligonucleotide primers need to be synthesized based on amino acid sequences from the protein whose gene is required. If a number of amino acid sequences of peptides derived from the protein are known, then a wide selection of potential sites on which to base the primers should be available. In the absence of such amino acid sequences, it is still often possible to design primers to regions of amino acid sequence similarity which are common to a closely related group of proteins.

It is impossible in all cases to choose a nucleotide sequence which perfectly matches the gene because of the degeneracy of the genetic code. Consequently, a pool of oligonucleotide primers is synthesized in which most, if not all, sequence possibilities are present. These primers would be totally useless as nucleic acid probes in screening a library but their successful utilization in PCR will give a DNA fragment which, upon sequencing, will be shown to encode a portion of the gene of interest. This sequence information is invaluable; it can be used as a probe in library screening and, additionally, as a basis for the synthesis of gene-specific primers which can be used in additional PCR reactions to obtain the 3′ and 5′ region of the gene, from which the sequence of the entire gene can be deduced without ever having to prepare or screen cDNA libraries.

The DNA fragments which arise in PCR should always be electrophoresed on agarose or acrylamide gels to estimate the size and distribution of products. Care should also be exercised to provide adequate controls in which either of the primers or the target cDNA are absent from the reaction. This should result in the absence of the specifically amplified product compared to the test reactions. Occasionally, PCR reactions yield no products which encode the desired gene. It then becomes worthwhile to use control primers which are known to amplify other abundant genes successfully. The test primers should be checked by end-labelling and electrophoresis to determine their size. Even if they are full length, they are frequently improved if they are extracted with phenol and then precipitated using ethanol. If the PCR products are of the size anticipated, then their authenticity can be identified by Southern transfer (Southern, 1975) and hybridization using an additional end-labelled primer whose sequence should be represented in the PCR product. Unambiguous identification will require the sequencing of the product after cloning, or directly using one of the two primers from which it was derived, if they are non-redundant. A contiguous open reading frame (ORF) should be apparent which is characteristic of the gene required. In the first round of PCR, which will utilize redundant primers, amplification products of just a few hundred nucleotides should be aimed for. The gene-specific primers can be used to amplify larger fragments of around 1.5 kbp. However, even much longer cDNAs should pose no problems with the introduction of PCR protocols which can be used to obtain 20 kbp or more (Cheng *et al.*, 1994).

### *3.5.1 Primer design*

The length of primers (oligomers) which have been useful in PCR range from 18mers, as gene-specific primers, 20–30mers for redundant or mixed-pool primers, and up to 50- or 60mer where a gene-specific sequence of 20–30 bases is flanked by other sequences encoding restriction sites. The preparation of gene-specific primers should follow some simple rules which will enhance the chance of success. Choose primers which are:

(i) directed to regions where the number of A and T bases equals the number of C and G bases;
(ii) unable to anneal to themselves or each other, otherwise the majority of the amplified product will be based on primer extension products;
(iii) have a 3' C or G residue;
(iv) have similar melting temperatures so that they anneal with similar kinetics.

There are several computer software packages available which will remove the guesswork in designing a primer pair. Occasionally, a specific region of cDNA, such as the regions flanking an ORF, will require amplification with combinations of primers which the computer software may suggest are incompatible but which are unavoidable. However, the desired product will probably be present, although additional amplification products may also arise.

### 3.5.2 Mixed-pool or redundant oligonucleotide primers

The advances in microsequencing of peptides in recent years has meant that just a few micrograms of protein or peptide is sufficient to determine accurately a dozen or more amino acid positions. Not all of the sequence may be useful. Of the 20 amino acids, three (Leu, Ser, Arg) are encoded by six codons; five (Pro, Thr, Val, Ala, Gly) are encoded by four codons; nine (Phe, Tyr, Cys, His, Gln, Asn, Lys, Asp, Glu) are encoded by two codons; Ile by three codons and Trp and Met by unique codons. The advantage of an amino acid sequence encoded by single or twin codons restricts the total possible codon selections, whereas an oligonucleotide based on a peptide sequence rich in six-codon amino acids will be redundant. However, even in these cases, the primers will still function if their 3' end can be chosen in regions which give a non-degenerate G–C rich sequence. In the absence of an amino acid sequence, the gene of interest may belong to a group of proteins in which certain motifs of sequence similarity (not identity) occur at specific regions. The amino acid sequence on which the primer is based is then itself redundant, but if unique amino acids are present in the aligned sequence, and can be placed so that they correspond to the 3' terminus of the primer, then a suitable primer can still be designed and used successfully in PCR.

### 3.5.3 Primary PCR reaction

The original technique to isolate genes using PCR-based methods relied on RACE (rapid amplification of cDNA ends; Frohman *et al.*, 1988; Frohman and Martin, 1989; Frohman, 1993). This technique was based on the preparation of cDNA with a modified oligo(dT) primer, similar to the primer used to prepare cDNA for directional cloning, except that this flanking sequence can be 30–40 nucleotides long. Thus, all cDNAs have this unique known sequence at their 5' ends (i.e. 3' on the original mRNA). This sequence is far more useful than relying on a cDNA poly(T) tract because primers with higher annealing temperatures can be designed to complement the outer region (Ro) or the inner region (Ri) of the sequence. In 3' RACE, an upstream sense-encoding gene-specific primer is used with Ro and a portion of the reaction re-amplified with a downstream sense-encoding gene-specific primer and Ri which, upon electrophoresis, should reveal a fragment encoding the desired gene. This protocol is not always reliable because long PCR products arising from long 3' UTRs can be difficult to amplify using redundant oligonucleotide primers. An alternative, and frequently more successful, PCR reaction is to use two redundant gene-specific primers initially, one of which is sense-encoding and the other antisense. The PCR product may be anticipated to be smaller and will be based entirely on an ORF and offer increased reliability in recognition of the sequence encoding the true gene (Laughton *et al.*, 1994; Towner and Gärtner 1994). The steps involved in isolating PCR fragments encoding genes are depicted in Figure 3.3.

### 3.5.4 Isolation of the 3' end of a cDNA

The sequence data from the central portion of the gene can be used for the synthesis of additional primers. If the cDNA was prepared using a modified poly(T) primer, then a sense-encoding gene-specific primer can be used in conjunction with the primer which is complementary to the flanking sequence of the cDNA synthesis primer. The fragment which arises will have sequence overlap with that which is already known and will extend to the terminus of the ORF and then through the 3' UTR, which should have a recognizable polyadenylation signal prior to the poly(A) sequence. Occasionally more than one product arises in this reaction, none of which encodes the genuine 3' terminus. This is due to the cDNA synthesis primer annealing to short spans of adenosine within the 3' UTR and initiating synthesis of truncated products. The sequences obtained can be used as a basis for the synthesis of additional primers, and used in PCR to isolate additional products which have clearly identifiable sequence overlap so that it is possible to *walk* toward the genuine 3' end of the gene.

### 3.5.5 Isolation of the 5' end of a cDNA

To isolate the 5' end of a cDNA, it is first necessary to provide a sequence motif which can be exploited by PCR. Generally this is done by incubating first-strand cDNA with terminal deoxynucleotidyl transferase (TdT) and dATP which will add a string of dA residues to its 3' end, which represents the 5' end of the gene. This material is used in primer extension with the $T_{17}$ RiRo cDNA synthesis primer and then in a PCR reaction with an antisense gene-specific primer and the RiRo primer which flanks the poly(T) sequence. The sequence of the product which arises will overlap with the known sequence and should have a clearly recognizable initiation codon which specifies the start of the gene. The TdT enzyme is inhibited by oligonucleotides and consequently

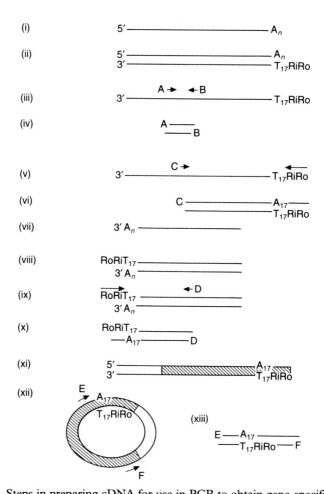

**Figure 3.3.** Steps in preparing cDNA for use in PCR to obtain gene-specific products. (i) mRNA is used for cDNA synthesis. (ii) The cDNA synthesis primer RoRiT$_{17}$ is used initially to obtain single-stranded cDNA. The mRNA strand is still present. (iii) An upstream sense-encoding redundant oligodeoxynucleotide primer (A) and a downstream antisense redundant primer (B) are used in PCR. (iv) A gene-specific PCR product arises whose sequence is determined. (v) A gene-specific primer (C) is used in PCR with a primer complementary to the RoRi sequence of the cDNA synthesis primer. (vi) A gene-specific product encoding the carboxyl end of the ORF of the gene and the 3' UTR arises. (vii) First-strand cDNA, preferably synthesized from a gene-specific primer, is homopolymer tailed using dATP and terminal deoxynucleotidyl transferase (TdT). (viii) RoRiT$_{17}$ is used in primer extension. (ix) An antisense-encoding gene-specific primer (D) and the primer complementary to the sequence of RoRi is used in PCR. (x) A gene-specific product encoding the amino terminus of the ORF of the gene and the 5' UTR arises. (xi) The 5' end of the gene can also be obtained by using double-stranded cDNA from the material in step (ii) after second strand synthesis; the central portion and 3' end of the gene are obtained as described above. (xii) Double-stranded cDNA is blunt-ended and self-ligated, then used in PCR with two gene-specific primers, a sense oligonucleotide from near the 3' terminus of the gene (E) and an antisense primer based on sequence from the known 5' portion of the sequence (F). (xiii) A PCR product arises which encodes the 3' terminus of the gene, RoRiT$_{17}$, the 5' UTR, followed by the amino terminus end of the ORF. The known sequence of the double-stranded material in (xi) and (xii) is hatched for clarity.

the first-strand cDNA must be purified by passing through a size exclusion column to remove traces of the cDNA synthesis primer (taking care to avoid losing the small amount of cDNA). An alternative strategy, which obviates the need to remove the cDNA synthesis primer, is to capture the mRNA on the oligo(dT$_{25}$) tracts present on Dynabeads and to use this material for cDNA synthesis and dA tailing. Primer extension of dA-tailed cDNA immobilized on Dynabeads using the T$_{17}$ RiRo cDNA synthesis primer can be followed by extensive washing, and the newly synthesized strand released by incubation in 0.2 M NaOH, then neutralized and used in PCR.

An alternative method of using PCR amplification to obtain the 5' end of genes relies on using two gene-specific primers with double-stranded cDNA which has been self-ligated to form circular molecules (Towner and Gärtner, 1992). The sense-encoding primer is based on a sequence slightly upstream of the polyadenylation signal and the antisense-encoding primer is based on penultimate sequences available in the 5' region of the gene. However, the technique also works with gene-specific primed cDNA with appropriate primers for the PCR reaction (Zeiner and Gehring, 1994). The ensuing PCR product will have two regions of overlap with the previously known sequence and will also encode the remainder of the N-terminal region and 5' UTR. It could also prove effective to perform PCR on a gene library using a gene-specific primer in combination with primers complementary to the cloning sites of the vector. Regardless of which method is used to identify the gene, sequence information will be available upon which to base primers to amplify the entire ORF for subsequent cloning and expression.

### 3.5.6 Gene identification by differential display

This technique is very attractive at a practical level because it allows the isolation of differentially expressed genes from a range of tissues and depends solely on the recognition of different sized PCR products which arise from a limited set of primers (Liang and Pardee, 1992; Liang *et al.*, 1993). cDNA is prepared from samples using a short oligo(dT) primer of 11 bases which has a 3'-terminal A, C or G. This same primer is used with its cognate cDNA in a PCR reaction with one of a series of additional primers of arbitrary sequence, but with similar melting temperature. The PCR reaction can be laced with an [$\alpha$-$^{32}$P]dNTP radiolabel to enhance detection of products following agarose gel electrophoresis. The majority of products from a specific primer set will be identical from any starting sample, but other DNA fragments will arise from tissues where an additional gene is being transcribed. This product can be used directly in library screening or cloned and sequenced and used to obtain the remainder of the cDNA whose function can be assessed after expression.

## 3.6 Expression systems

The unambiguous assignment of a gene product to a particular function requires a suitable expression system for the protein to allow its subsequent characterization. The simplest method is to prepare mRNA from the cloned gene using T3, T7 or SP6 polymerase and to use this as the template for protein synthesis in a cell-free lysate. The amount of protein produced is insufficient to characterize directly but with some proteins, such as transcription factors, there is ample material for use in gel-shift assays to analyse its binding properties (Schleif, 1988). If the mRNA had been prepared with a 5' methyl cap and was polyadenylated, it could be injected into *Xenopus* oocytes to produce active polypeptides (Gurdon and Wickens, 1983). This is especially attractive for membrane proteins whose activity can only be measured by active transport or electrophysiological techniques (Barnard et al., 1982). If pure samples of protein were required, then a purpose-designed vector and appropriate host should be employed to maximize protein yield and facilitate its purification. Virtually all of the prokaryotic expression systems are based on *E. coli*, whereas eukaryotic expression systems utilize yeast, insect cells and mammalian cell lines with a range of vectors based on plasmids and viral systems.

### 3.6.1 Expression using E. coli

There has been a resurgence in the use of *E. coli* in expression systems, since the much heralded eukaryotic expression systems have not been as successful as had been hoped. This, in combination with improvements to the range of vectors tailored for *E. coli* and the enormous quantities of protein available, has led to a number of mammalian genes being expressed in *E. coli* very effectively. It must be borne in mind that *E. coli* is unable to modify proteins post-translationally and that some native structures and proteins in excess of 80 kDa are unlikely to be obtained. This system of expression is considerably cheaper than any other and so it is worthwhile to utilize a few different vectors if the first ones tried do not work. This is because it is impossible to predict which is the best vector for a particular gene. For effective expression, the initiation codon of the gene should be juxtaposed to the ribosome-binding site, which is present in many expression vectors, so that a good initiation complex is formed. The transformed bacteria should be allowed to grow to mid-log phase before being induced and then grown for a further period before being harvested. Samples of the bacteria can be removed at any stage in the process and dissolved in buffer suitable for SDS–gel electrophoresis, enabling optimum times and temperatures of growth to be correlated with protein expression.

The protein can be released from the cells by sonication, or treatment with lysozyme, then mild detergent; the soluble proteins can then be purified by chromatographic procedures. If protein is produced but is in the insoluble cellular debris, then it would be appropriate to attempt expression as a fusion protein. This would also be the advice if the protein could not be detected. If the protein is present in the soluble fraction of the bacterial extract but has a smaller molecular weight than anticipated, then the DNA sequence should be checked for pseudo-initiation regions and clusters of rarely used codons, either of which could be responsible for the truncated protein.

If the protein is expressed as a fusion product, its purification is much easier, since the N-terminal portion of fusion vectors has been designed to allow reversible binding to a chromatographic matrix and offers a specific recognition site at the fusion junction which allows precise cleavage and subsequently pure samples. The most widely used commercially available systems are based on maltose-binding protein and glutathione S-transferase which have factor Xa and thrombin proteolytic cleavage sites respectively, although many additional commercially developed systems are continually being introduced. There can be wide variation in the efficiency of cleavage, depending on whether the site is occluded by the protein. Consequently a linker sequence of several glycines is sometimes inserted upstream of the cleavage site to aid its exposure. A much smaller moiety can be fused to the gene product of either prokaryotic or eukaryotic vectors using an encoded hexa-histidine motif. This hexapeptide is small enough not to interfere with the activity of the protein but it allows a very easy purification step using an immobilized nickel support from which the protein can be eluted using imidazole (Janknecht *et al.*, 1991). If the moiety is fused to the C terminus of the protein, then the advantage of isolating full-length translation products is also offered.

### 3.6.2 Eukaryotic expression systems

The most suitable expression system for human genes is in a mammalian cell host, where faithful post-translational modifications will be assured. If non-processed proteins are required and an *E. coli*-based system has been unsuitable, there are a range of other systems worth investigating. There have been many improvements to yeast expression systems. These include novel and precise inducible expression by glucocorticoids in *Saccharomyces cerevisiae* (Schena *et al.*, 1991) and the exploitation of the methylotrophic organism *Pichia pastoris* (Ratner, 1989; Scorer *et al.*, 1994). Also, insect cells infected with recombinant baculovirus can be usefully exploited but will not necessarily process the protein as precisely as a mammalian cell.

Mammalian cells can be infected transiently or, if a selectable marker is included, a stably maintained cell line can be prepared. The DNA can be introduced directly into cells as a plasmid, or a number of very sophisticated viral transfer protocols followed. These range from vaccinia- to retroviral-mediated transfer. The latter system offers precise control of gene copy number. A selectable marker can be included and the gene can be constitutively expressed from the promoter activity of the viral flanking sequence. Further improvements to these type of vectors, such as the introduction of multiple ribosome entry sites allowing the translation of polycistronic mRNA, make this a very attractive method of gene delivery (Zitvogel *et al.*, 1994).

## References

**Alwine JC, Kemp DJ, Stark GR.** (1977) Method for detection of specific RNAs in agarose gels by transfer to diazobenzyloxy-methyl paper. *Proc. Natl Acad. Sci. USA* **74:** 5350–5354.

**Aviv H, Leder P.** (1972) Purification of biologically active globin mRNA by chromatography on oligothymidylic acid cellulose. *Proc. Natl Acad. Sci. USA* **69**: 1409–1412.

**Baltimore D.** (1970) RNA-dependent DNA polymerases in virions of RNA tumour viruses. *Nature* **226**: 1209-1210.

**Barnard EA, Miledi R, Sumikawa K.** (1982) Translation of exogenous messenger RNA coding for nicotinic acetylcholine receptors produces functional receptors in *Xenopus* oocytes. *Proc. R. Soc. Lond. (Ser. B)* **215**: 241–246.

**Chang ACY, Nunberg JH, Kaufman RJ, Erlich HA, Schimke RT, Cohen SN.** (1978) Phenotypic expression in *E. coli* of a DNA sequence coding for mouse dihydrofolate reductase. *Nature* **275**: 617–624.

**Cheng S, Fockler C, Barnes WM, Higuchi R.** (1994) Effective amplification of long targets from cloned inserts and human genomic DNA. *Proc. Natl Acad. Sci. USA* **91**: 5695–5699.

**Chomczynski P, Sacchi N.** (1987) A single-step method of RNA isolation by acid guanidinium thiocyanate-phenol-chloroform extraction. *Anal. Biochem.* **162**: 156–159.

**Dyson NJ.** (1990) Immobilisation of nucleic acids and hybridisation analysis. In: *Essential Molecular Biology* (ed. TA Brown). Oxford University Press, Oxford, pp. 111–156.

**Frohman MA.** (1993) Rapid amplification of complementary DNA ends for generation of full-length complementary DNAs: thermal RACE. *Methods Enzymol.* **218**: 340–356.

**Frohman MA, Martin GR.** (1989) Rapid amplification of cDNA ends using nested PCR. *PCR Tech.* **1**: 165–170.

**Frohman MA, Dush MK, Martin GR.** (1988) Rapid production of full-length cDNAs from rare transcripts: amplification using a single gene-specific oligonucleotide primer. *Proc. Natl Acad. Sci. USA* **85**: 8998–9002.

**Grunstein M, Hogness DS.** (1975) Colony hybridisation: a method for the isolation of cloned DNAs that contain a specific gene. *Proc. Natl Acad. Sci. USA* **72**: 3961-3965.

**Gubler U, Hoffman BJ.** (1983) A simple and very efficient method for generating cDNA libraries. *Gene* **25**: 263–269.

**Gurdon JB, Wickens MP.** (1983) The use of *Xenopus* oocytes for the expression of cloned genes. *Methods Enzymol.* **101**: 370–386.

**Hampson IN, Pope L, Cowling GJ, Dexter TM.** (1992) Chemical cross-linking subtraction (CCLS) a new method for the generation of subtractive hybridisation probes. *Nucleic Acids Res.* **20**: 2899.

**Hanahan D, Meselson M.** (1980) Plasmid screening at high colony density. *Gene* **19**: 147–151.

**Hara E, Kato T, Nakada S, Sekiya S, Oda K.** (1991) Subtractive cDNA cloning using oligo(dT)30 latex and PCR: isolation of cDNA clones specific to undifferentiated human embryonal carcinoma cells. *Nucleic Acids Res.* **19**: 7097–7104.

**Hediger MA, Coady MJ, Ikeda TS, Wright EM.** (1987a) Expression cloning and cDNA sequencing of the $Na^+$/glucose co-transporter. *Nature* **330**: 379–381.

**Hediger MA, Ikeda TS, Coady MJ, Gundersen CB, Wright EM.** (1987b) Expression of size selected mRNA encoding the intestinal $Na^+$/glucose co-transporter in *Xenopus laevis* oocytes. *Proc. Natl Acad. Sci. USA* **84**: 2634–2637.

**Hogrefe HH, Mullinax RL, Lovejoy AE, Hay BN, Sorge JA.** (1993) A bacteriophage λ vector for the cloning and expression of immunoglobulin Fab fragments on the surface of filamentous phage. *Gene* **128**: 119–126.

**Innis MA, Gelfand DH, Sninsky JJ, White TJ.** (1990) *PCR Protocols: a Guide to Methods and Applications.* Academic Press, New York.

**Jakobsen KS, Breivold E, Hornes E.** (1990) Purification of mRNA directly from crude plant tissues in 15 minutes using magnetic oligo dT microspheres. *Nucleic Acids Res.* **18**: 3669.

**Janknecht R, de Martynhof G, Lou J, Hipskind RA, Nordheim A, Stunnenberg HG.** (1991) Rapid and efficient purification of native histidine-tagged protein expressed by recombinant vaccinia virus. *Proc. Natl Acad. Sci. USA* **88**: 8972–8976.

**Kemp DJ, Cowman AF.** (1981a) Detection of expressed polypeptides by direct immunoassay of colonies. *Methods Enzymol.* **79**: 622–630.

**Kemp DJ, Cowman AF.** (1981b) Direct immunoassay for detecting *Escherichia coli* colonies that contain polypeptides encoded by cloned DNA segments. *Proc. Natl Acad. Sci. USA* **78**: 4520–4524.

**Laughton DL, Amar M, Thomas P, Towner P, Harris P, Lunt GG, Wolstenholme AJ.** (1994) Cloning of a putative inhibitory amino acid receptor subunit from the parasitic nematode *Haemonchus contortus*. *Recep. Channels* **2**: 155–163.

**Liang P, Pardee AB.** (1992) Differential display of eukaryotic messenger RNA by means of the polymerase chain reaction. *Science* **257**: 967–971.

**Liang P, Auerboukh L, Pardee AB.** (1993) Distribution and cloning of eukaryotic mRNAs by means of differential displays: refinements and optimisation. *Nucleic Acids Res.* **21**: 3269-3275.

**Masu Y, Nakayama K, Tamaki H, Harada Y, Kuno M, Nakanishi S.** (1987) cDNA cloning of bovine substance-K receptor through oocyte expression system. *Nature* **329**: 836–838.

**Okayama H, Berg P.** (1983) A cDNA cloning vector that permits expression of cDNA inserts in mammalian cells. *Mol. Cell. Biol.* **2**: 151–170.

**Ratner M.** (1989) Protein expression in yeast. *Bio/Technology* **7**: 1129–1133.

**Retsel EF, Collett MS, Faras AJ.** (1980) Enzymatic synthesis of deoxribonucleic acid by avian retrovirus reverse transcriptase *in vitro*: optimal conditions required for transcription of large ribonucleic acid templates. *Biochemistry* **19**: 513–518.

**Rodriguez IR, Chader GJ.** (1992) A novel method for the isolation of tissue-specific genes. *Nucleic Acids Res.* **20**: 3528.

**Schena M, Picard D, Yamamoto KR.** (1991) Vectors for constitutive and inducible gene expression in yeast. *Methods Enzymol.* **194**: 389–398.

**Schleif R.** (1988) DNA binding by proteins. *Science* **241**: 1182–1187.

**Scorer CA, Clare JJ, McCombie WR, Romanos MA, Sreekrishna K.** (1994) Rapid selection using G418 of high copy number transformants of *Pichia pastoris* for high-level foreign gene expression. *Bio/Technology* **12**: 181–184.

**Smith CP.** (1990) Methods for mapping transcribed DNA sequences. In: *Essential Molecular Biology* (ed TA Brown). Oxford University Press, Oxford, pp. 237–252.

**Southern E.** (1975) Detection of specific sequences among DNA fragments separated by gel electrophoresis. *J. Mol. Biol.* **98**: 503–517.

**Temin HM, Mizutani S.** (1970) RNA-dependent DNA polymerase in virions of Rous sarcoma virus. *Nature* **226**: 1211–1213.

**Towner P, Gärtner W.** (1992) cDNA cloning of 5′ terminal regions. *Nucleic Acids Res.* **20**: 4669–4670.

**Towner P, Gärtner W.** (1994) The primary structure of mantid opsin. *Gene* **143**: 227–231.

**Wang Z, Brown DD.** (1991) A gene expression screen. *Proc. Natl Acad. Sci. USA* **88**: 11505–11509.

**Winter G, Milstein C.** (1991) Man-made antibodies. *Nature* **349**: 293–299.

**Zeiner M, Gehring U.** (1994) Cloning of 5′ terminal regions by inverse PCR. *BioTechniques* **17**: 151–153.

**Zitvogel L, Tahara H, Cai Q, Storkus G, Muller G, Wolf SF, Gately M, Robbins PD, Lotze MT.** (1994) Construction and characterization of retroviral vectors expressing biologically active human interleukin-2. *Human Gene Therapy* **5**: 1493–1506.

# 4

# Retroviral insertional mutagenesis

Farzin Farzaneh, Joop Gäken and Shu-Uin Gan

## 4.1 Introduction

The first retrovirus was discovered by Paton Rous as a filterable agent which could transmit cancer in chickens (Rous, 1911). The oncogenic property of Rous sarcoma virus (RSV) and many other retroviruses has since been shown to be encoded by activated viral homologues (v-*onc*) of normal cellular proto-oncogenes, c-*onc* (see Weiss *et al.*, 1982, 1985; Bishop, 1983, 1987; Varmus, 1988; Bouton and Parsons, 1993; Sugden, 1993). The study of these rapidly transforming retroviruses has resulted in the identification of a large number of such oncogenes, the normal cellular homologues of genes which encode important regulatory products such as growth factor receptors, growth and differentiation factors, components of signal transduction and transcription factors (see Bishop, 1983, 1987; Nusse, 1986, 1991; Varmus, 1988; Bouton and Parsons, 1993; Sugden, 1993).

Other, slow transforming retroviruses, which lack their own transforming genes, are still able to induce malignant transformation of the host cells, as a result of their chance integration in the vicinity of regulatory genes important for the control of cellular growth, differentiation and programmed cell death (Hayward *et al.*, 1981; Selten *et al.*, 1984, 1985; Nusse *et al.*, 1984; Peters *et al.*, 1986; Moreau-Gachelin, 1988; Morishita *et al.*, 1988; Bear *et al.*, 1989; Buchberg *et al.*, 1990; Ben-David *et al.*, 1990, 1992; Askew *et al.*, 1991, 1994; Haupt *et al.*, 1991; Levy *et al.*, 1992; Tremblay *et al.*, 1992; Kreider *et al.*, 1993; Habets *et al.*, 1994). The disrupted regulation of these genes is caused by the loss of normal control of transcription by their cognate regulatory sequences or the disruption of their coding sequences resulting in the synthesis of mutated (usually truncated) gene products.

A large number of mammalian genes important in malignant cell transformation have thus been identified by the study of the natural history of these viruses and of tumours produced as a by-product of the normal viral life cycle (see Varmus, 1988; Nusse, 1991). Retroviruses, and especially recombinant retroviral vectors, may also be employed for experimental insertional mutagenesis in order to study the genetic basis of any phenotype which can be adequately selected. Such studies (Varmus et al., 1981; Goff, 1987; Kung et al., 1991) have resulted in the identification of a number of novel genes including *Int*-1 (Nusse et al., 1984), *Int*-2 (Dickson et al., 1984), *Pim*-1 (Cuypers et al., 1984), *Evi*-1 (Morishita et al., 1988; Kreider et al., 1993), *Spi*-1 (Moreau-Gachelin et al., 1988), *Evi*-2 (Buchberg et al., 1990), *His*-1 and *His*-2 (Askew et al., 1991, 1994), *flvi*-2 (Levy and Lobelle-Rich, 1992), *bcar*-1 (Dorssers et al., 1993) and *Tiam*-1 (Habets et al., 1994). The identification of these genes was made possible by the characteristic integration of the double-stranded DNA copy of the viral RNA genome into the chromosomal DNA of the host cells (Varmus, 1982, 1988; Askew et al., 1994; Finnegan, 1994; Boris-Lawrie and Temin, 1994). Although the provirus DNA sequences at the site of chromosomal integration are precisely determined and characteristic of each virus, the vast majority of integrations are the product of non-homologous recombinations resulting in the pseudo-random disruption of the genome (Brown et al., 1987; Fujiwara and Mizuuchi, 1988; Bowerman et al., 1989; Brown et al., 1989; Fujiwara and Craigie, 1989; Roth et al., 1989; Mooslehner et al., 1990; Coffin, 1992a,b; Goff, 1992; Whitcomb and Hughes, 1992; Kulkosky and Skalka, 1994). Retroviral vectors can therefore act as agents of insertional mutagenesis. The major advantage of retroviral insertional mutagenesis over other mutagenesis strategies (e.g. chemical- or radiation-induced) is the presence of the proviral DNA sequences at or near the affected locus. These viral sequences can provide a very useful tag for the subsequent identification of the genetic target responsible for the generation of the selected phenotype.

## 4.2 The retroviral life cycle

The RNA genome of retroviruses encodes three open reading frames, *gag*, *pol* and *env*, the products of which are required for viral replication (see Miller, 1992). *gag* encodes the structural proteins of the viral capsid, *pol* the enzymes required for genomic processing, including reverse transcription of the viral RNA genome into a double-stranded DNA copy (provirus) and the integration of the provirus into the host cell genome (integrase, IN). The *env* gene encodes viral envelope proteins which determine its host range. These coding sequences are flanked by regions which contain signals that are essential for the expression of the viral genes. In the RNA genome, the coding sequences are preceded by regions called R and U5 at the 5' end, and U3 and R at the 3' end (Figure 4.1). The U3 region contains several eukaryotic transcription enhancer motifs and binding sites for the cellular RNA polymerase II.

Viral replication requires a double-stranded provirus DNA copy as its template, which is integrated in the host cell genome, and is initiated by RNA polymerase II-mediated synthesis of a transcript which starts at the R region of the 5' long terminal repeat (LTR) and extends to the polyadenylation site within the U3 region in the 3' LTR (see Boris-Lawrie and Temin, 1994;

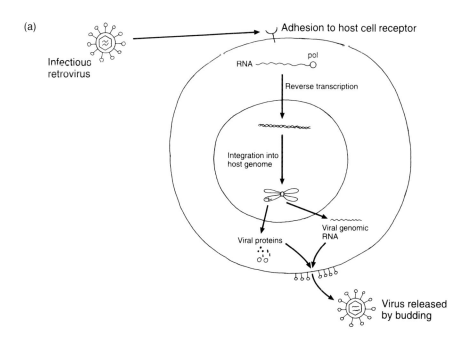
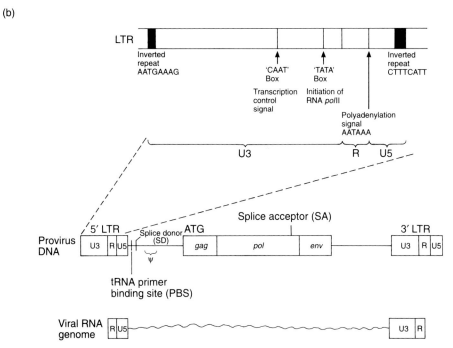

**Figure 4.1.** (a) Retroviral life cycle. (b) The structure of Moloney murine leukaemia virus.

Miller, 1992). The presence of the ψ sequence downstream of the 5' LTR (nucleotides 215–565 in Moloney murine leukaemia virus) permits the packaging of the full-length viral RNA genome into the viral capsid which is composed of two RNA strands complexed with *pol* gene products, host cell tRNA molecules and the core structure which is encoded by gag proteins (Mann *et al.*, 1983; Mann and Baltimore, 1985; Adam and Miller, 1988). The position of the ψ sequence, downstream of the splice donor site, ensures that only the full-length viral RNA is packaged (Rein, 1994). The spliced viral RNAs, or for that matter host cell-derived RNA molecules other than tRNA, which lack the ψ sequence are thus excluded. The capsid is surrounded by a lipid bilayer, derived from the host cell membrane, and glycoproteins, encoded by the viral *env* gene, which are assembled together as the virus is released from the host cell membrane. Virus attachment to its host cell is then determined by interactions between the viral coat proteins and host cell surface receptors. This is followed by virus entry by receptor-mediated endocytosis.

In the host nucleus, the two identical viral RNA strands and reverse transcriptase are used in an intriguingly complex set of manoeuvres to generate a double-stranded DNA copy of the viral genome (see Figure 4.2). The integrase function (IN) of the viral *pol* gene product then facilitates the non-homologous recombination and integration of the provirus into the host cell chromosomal DNA (Schwartzberg *et al.*, 1984; Panganiban and Temin, 1984; Donehower and Varmus, 1984).

During reverse transcription, the single copies of U3 and U5 in the viral RNA genome are duplicated to generate the two LTR sequences (Grandgenett and Mumm, 1990), each of which is composed of U3–R–U5. As depicted in Figure 4.2, the U3 region of the 3' LTR, which does not participate in the promotion of viral gene transcription, serves as the template for the synthesis of the U3 region in the 5' LTR of the next generation of provirus. Therefore mutations introduced into the enhancer and/or promoter sequences contained in the U3 region of the 3' LTR will not affect the synthesis of the viral gene products (including the RNA genome of the virus itself) from that template. Such mutations will, however, affect the LTR-controlled transcription in the provirus which is generated by the reverse transcription of the viral RNA genome made from this template. This intriguing feature of the retroviral life cycle has enabled the development of so-called self-inactivating retroviral vectors (Yu *et al.*, 1986; Marty *et al.*, 1990; Soriano *et al.*, 1991). These are potentially useful gene delivery vehicles which do not carry the risk of insertional activation of cellular genes in the vicinity of their site of integration since their provirus copies lack promoter/enhancer activity. However, although these vectors remain useful experimental tools, their relatively low titre (about $10^4$/ml) has prevented them from gaining popularity as delivery vehicles. The fact that the enhancer/promoter function of the U3 region in the 3' LTR is not required for viral replication, but prevents expression of the virus-encoded genes in the cells infected by such a self-inactivating virus, has aided the development of vectors which have proved extremely useful in a specific application of retroviral insertional mutagenesis, namely the promoter/enhancer trap (discussed in detail in Chapter 5).

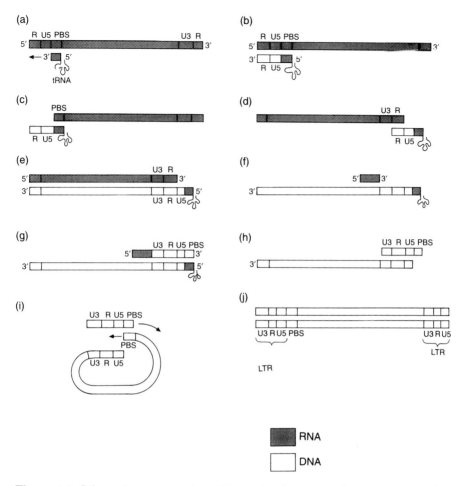

**Figure 4.2.** Schematic representation of the molecular events of reverse transcription. (a) The retrovirus provides +ve strand RNA. (b) tRNA primes DNA synthesis of the –ve strand. (c) RNase H activity of the viral *pol* gene product removes the reverse-transcribed RNA. (d) Intermolecular 'jump'. DNA hybridizes with the 3' R region of the other viral RNA genome. (e) The –ve DNA strand is extended. (f) RNase H removes most of the viral RNA. (g) 3' end of the –ve DNA strand is synthesized. (h) RNA and tRNA are removed. (i) Intramolecular 'jump'. (j) Synthesis of the double-stranded DNA copy of the virus is completed.

## 4.3 Host range

The host range of retroviral vectors is determined by the viral envelope glycoproteins and the target cell surface receptors to which they bind. The cellular receptors for a number of retroviral vectors have been identified. For example, the susceptibility of rodent cells to infection by ecotropic murine leukaemia viruses is determined by binding of the virus envelope glycoprotein gp70 to the ecotropic receptor (Albritton *et al.*, 1989; Wang *et al.*, 1991).

This protein encodes a cationic amino acid channel which has multiple membrane-spanning domains (Kavanaugh et al., 1994a). Expression of gp70 and its binding to this ecotropic envelope receptor in infected murine cells prevents infection by additional vectors expressing the same envelope protein. Other species are refractory to infection by ecotropic vectors because of the absence of this protein on their cell surface. Cells from a broad spectrum of species can, however, be infected by viruses expressing an amphotropic envelope protein which binds to another membrane protein, a sodium-dependent phosphate symporter (Kavanaugh et al., 1994b). Expression of the ecotropic receptor in cells naturally lacking this protein results in these cells becoming capable of being infected by ecotropic virus. Thus the host range of retroviral vectors can be altered either by changing the viral envelope proteins or by expressing the appropriate receptor on the surface of the target cells (Albritton et al., 1989); additional cell-encoded accessory factors are, however, required (Wang et al., 1991).

## 4.4 Replication-defective retroviral vectors

Many of the earlier studies of retroviral insertional mutagenesis, particularly their involvement in oncogenesis, employed wild-type retroviruses. However, replication-competent vectors suffer from a number of limitations, including the uncontrolled spread of infection from the original targets to new cells with which they come into contact, and a high frequency of recombination resulting in the generation of mixed virus stocks containing vectors with unknown properties (see Boris-Lawrie and Temin, 1994; Finnegan, 1994). In addition, since the number of infections per cell is controlled not by the relative virus titre, but by the saturation of the host cell receptors, the number of integrations per cell cannot be adequately controlled. This makes it very difficult to obtain cells with single provirus integrations. These factors make the interpretation of data difficult and pose potentially serious safety problems. These problems are avoided in replication-defective retroviral vectors (Mann et al., 1983; Miller and Buttimore, 1986; Markowitz et al., 1988a,b; Danos and Mulligan, 1988; Morgenstern and Land, 1990) which, although able to infect their host cells and integrate into their genome, are unable to replicate in these cells, resulting in a 'one-way' infection, with no virus being produced by the host cells.

Retrovirus replication is dependent on a number of *cis*- and *trans*-acting factors. The *cis*-acting elements include the promoter/enhancer elements of the LTR, splice donor and splice acceptor sites, the polyadenylation signal in the U3 region of the LTR, the $\psi$ packaging sequence and the tRNA primer binding site. The *trans*-acting factors are provided by *gag*, *pol* and *env* gene products. Recombinant retroviral vectors containing *cis*-acting elements, but lacking these essential viral genes, can therefore be propagated in the presence of other replication-competent vectors which provide these genes (helper virus). However, propagation in the presence of helper virus suffers from the same limitations inherent in the use of replication-competent vectors. The alternative is to propagate the vector in cell lines which are engineered to express only the *trans*-acting genes, *gag*, *pol* and *env*.

## 4.5 Packaging cell lines

The first generation of packaging cells, such as ψ2 cells, were produced by infection of rodent cells (NIH/3T3 mouse fibroblasts) with mutant viruses with a deleted packaging signal (Mann *et al.*, 1983; Cepko *et al.*, 1984). The mutant virus could therefore not be packaged, but these cells could provide the required *trans*-acting factors necessary for the packaging of a replication-defective vector which contained its own packaging sequence. However, relatively rare events, such as homologous recombinations between the ψ-containing vector and the mutant provirus sequences could rescue the defect and result in the production of helper-independent replication-competent virus. The next generation of packaging cells were produced by transfection of NIH/3T3 cells with plasmid expression vectors encoding only *gag*, *pol* and *env* genes (Miller and Buttimore, 1986). However, even in these cells, prolonged passaging of the producer cells could lead to generation of replication-competent virus at a low but detectable frequency (Temin, 1990; Cornetta *et al.*, 1991; Donahue *et al.*, 1992; Nienhuis *et al.*, 1993). The exact cause of this is not known but could be due to recombination between sequences used to make the packaging cells and the endogenous retrovirus-like elements found in most eukaryotic cells (see Chapter 1), or recombination between homologous regions of the vector and the defective virus used to make the packaging cells (Otto *et al.*, 1994).

The most reliable packaging cells now available have been generated by transfection of the three viral genes in two separate DNA fragments (Figure 4.3), thus making the recombination-mediated generation of replication-competent vector dependent on multiple recombination events within the same cell (Danos and Mulligan, 1988; Markowitz *et al.*, 1988a,b; Morgenstern and Land, 1990). Extensive safety studies have failed to detect the generation of helper-independent virus in the ecotropic packaging cell lines GP+E-86 (Markowitz *et al.*, 1988a), ψ-CRE (Danos and Mulligan, 1988) and WE (Morgenstern and Land, 1990) or in the amphotropic GP+*env*AM12 (Markowitz *et al.*, 1988b) and ψ-CRIP (Danos and Mulligan, 1988) packaging cells. In addition, reduction of regions of homology in recombinant vectors destined for human gene therapy has further decreased the chance of recombinational events resulting in the generation of helper virus (Miller and Buttimore, 1986; Danos and Mulligan, 1988; Markowitz *et al.*, 1988a,b; Morgenstern and Land, 1990).

The titre of the vector produced by packaging cells, as determined by infection of NIH/3T3 cells, can be routinely as high as $10^6$/ml of culture supernatant; the actual titre is dependent on a number of factors including the choice of the disabled retroviral vector, sequences cloned into the vector, etc. Higher titres can be obtained in a number of ways such as culturing the producer cells at 32°C, centrifugation of culture supernatant to concentrate

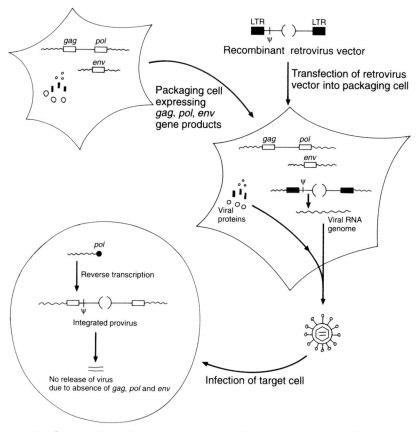

**Figure 4.3.** Construction of the virus producer cell line and infection of target cells.

the vector which can then be resuspended at higher densities, or freeze drying (Kotani *et al.*, 1994). Another means of increasing the virus titre is to co-culture amphotropic and ecotropic producer cells together. Virus from either cell type can infect the other without interference, resulting in theoretically limitless back-and-forth ('ping-pong') vector replication (Kozak and Kabat, 1990). Whilst substantially increasing the levels of vector-encoded proteins, this does, however, increase the risk of generating replication-competent virus (Lynch and Miller, 1991; Donahue *et al.*, 1992), and is therefore not recommended for most applications.

An important point to consider is the presence of a number of resident retroviruses within the genome of murine (Stoye *et al.*, 1988) as well as human (Krieg *et al.*, 1992; see Chapter 1) cell lines. Introduction into these cells of factors which are required, in *trans* (*gag, pol* and *env*), for the production of the desired vector could also result in the packaging and release of resident viruses (Scadden *et al.*, 1990). The packaging of the resident virus genomes does of course require the presence of the packaging ($\psi$) sequence. However, because of the absence of a selectable marker in the resident virus genomes, these viruses are harder to detect. This raises a potentially serious problem for retroviral insertional mutagenesis. The mutation caused by the integration of the known recombinant provirus may not be the cause of the selected phenotype, rather the insertional mutation may have been caused by contaminating virus produced by endogenous provirus present in the genome of the packaging cell line. However, strategies which use the LTR sequences for recovery of the cellular DNA flanking the site of provirus integration may overcome this problem to some extent because of sequence similarity between the LTR and many mouse endogenous proviruses. Therefore, approaches such as inverse PCR (Silver and Keerikatte, 1989) which use the LTR sequences to clone flanking genomic DNA may allow the amplification of target cell DNA sequences which flank not only the known retroviral vector, but also those which may have been the target of insertional mutagenesis by other, unknown, contaminating vectors which share LTR sequence homology (see below). This also means that insertional mutagenesis studies which use retroviral vectors produced by mouse packaging cell lines to infect and isolate mutant cell lines from another species (e.g. human cell lines) are more likely to enable the identification of the genes mutated by provirus integration.

## 4.6 Conditions required for efficient mutagenesis

Infection of target cells and the integration of provirus DNA into the host cell genome can only take place in the S phase of the cell cycle (Varmus *et al.*, 1977; Hsu and Taylor, 1982; see also Weiss *et al.*, 1985; Varmus, 1988). Therefore, the rapid proliferation of target cells is an important prerequisite (Roe *et al.*, 1993). In addition, there is clear evidence for the preferential integration of provirus in actively transcribed genomic loci (Scherdin *et al.*, 1990). Whilst this may be an advantage in some circumstances, it is important to note that the selection of the desired mutant phenotype must take place simultaneously with the infection procedure. This ensures that the appropriate genes are in an active transcriptional configuration and therefore accessible for provirus integration.

Another important requirement is a relatively low rate of spontaneous mutation and a substantial increase in the mutation frequency after retroviral infection. In the absence of a clear difference between the spontaneous and induced rates of mutation, one cannot be sure that the selected phenotype is in fact the product of a retrovirus insertion, rather than spontaneous mutations in other genes. Obviously in such circumstances, the identification of the site of provirus integration will not lead to the identification of mutated genes responsible for the generated phenotype. Finally, the simultaneous selection of multiple phenotypes, for instance vector-encoded drug resistance and the desired mutant characteristic (e.g. resistance to induction of differentiation), will reduce the risk of isolating spontaneous mutants. This is particularly useful when the multiplicity of infection is less than 1 (i.e. less than one provirus per cell in the target population).

## 4.7 Mechanisms involved in retroviral insertional mutagenesis

Provirus integration in the host cell genome can result in both gene activation and gene inactivation (Figure 4.4). Activation is mediated by the insertion of powerful promoter/enhancer elements, contained within the viral LTR structures. This could be due to the effect of enhancer elements contained within the LTR sequences which stimulate the transcription of genes placed either upstream or downstream of the site of provirus integration. Such genes may be located up to several kilobases away. Transcriptional activation could also be the result of insertion of the viral promoter sequences contained within the 5' and 3' LTR. Transcripts of the viral genome initiated at the R region of the 5' LTR could read through to adjacent downstream cellular genes. In such transcripts, the viral splice donor site may facilitate splicing of the viral RNA to the transcript of the adjacent genomic sequence. Alternatively, the 3' LTR could promote the transcription of adjacent downstream genes with transcripts which are initiated in the R region of the 3' LTR. Insertional activation may also result from disruption of the 5' regulatory sequences responsible for the inhibition of transcription under some physiological conditions. Provirus integration resulting in the disruption of 3' untranslated A–U rich RNA instability motifs, which are present in short-lived mRNA sequences, would increase the steady-state level of RNA, resulting in increased expression of the gene product.

Provirus integration could also decrease the expression of a gene by disruption of either its cognate regulatory sequences or its coding sequences, thereby resulting in transcription of mRNAs which encode proteins truncated in either the N- or C-terminal regions. Such inactivating mutations have generated hypoxanthine–guanine phosphoribosyltransferase (*HPRT*; King *et al.*, 1985) and $\beta_2$-microglobulin deletion mutants (Frankel *et al.*, 1985). Insertional inactivation may also result from provirus integration within intronic sequences. Although this may have no phenotypic effect due to the splicing out of the viral transcript, interference with RNA splicing could inhibit the expression of a normal gene product. It is important to note that the disruption of the coding sequences may result in expression of a truncated gene product (e.g. a growth factor receptor depleted of its ligand-binding domain and therefore in a constitutively active state independently of the presence of the ligand, or other regulatory factors) associated with a phenotype with characteristics similar to or even more severe than increased expression of a normal gene product.

In addition to these direct *cis*-acting effects, there can be indirect *trans* regulatory effects resulting from the presence of the viral genome within the cell, but independent of its position of integration. This could be the product of genes or other regulatory elements encoded by the virus. The position-independent effects of provirus integration, which are not induced by mutation of cellular genes, would be easy to identify since they would be present either in all cells or at least in a vastly larger number of cells than would be compatible with low-frequency integrations into specific genomic domains.

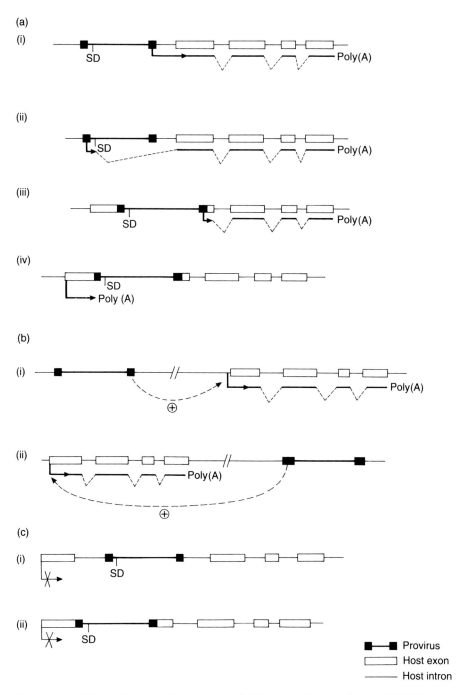

**Figure 4.4.** Effect of retroviral integration. (a) Insertional activation. (i) 3' LTR promoter activation; (ii) 5' LTR promoter activation: provirus integration at the 5' end of the gene, resulting in a readthrough transcript; (iii) provirus integration in the coding sequence of the host gene, resulting in the expression of a 5'-truncated transcript; 3, expression of the 3'-truncated transcript. (b) Enhancer activation. (i) Upstream enhancer effect; (ii) downstream enhancer effect. (c) Insertional inactivation. (i) Provirus integration into an intron; (ii) provirus integration into the coding sequence.

## 4.8 Mutation frequency

The frequency of insertional mutations is of course related to the number of integration events and the size of the target sequence. The target size is determined by the total size of the genome and the average size of the domain within which provirus integration would generate the selected phenotype.

Mammalian cells contain $2-3 \times 10^6$ kb of DNA per haploid genome, and most genes span between 2 and 50 kb of DNA. Based on the assumption that the integration of the provirus anywhere within this domain would result in the insertional inactivation of that gene, a single copy gene would represent a target approximately $10^{-5}-10^{-6}$ of the total DNA. Therefore, the random integration of each provirus could increase the inactivating mutation frequency in a gene of interest by $10^{-5}-10^{-6}$. However, the actual frequency of insertional mutations is often lower than this predicted rate. For example, in a population of super-infected F9 embryonal carcinoma cells containing an average of 25 integrations per cell, inactivating mutations in the *HPRT* gene are increased from a spontaneous frequency of approximately $10^{-7}$ in the uninfected cells to $10^{-6}$ in the infected population; about 50-fold lower than expected. This may be because the sites of provirus integration are not truly random (Wilson and Cohen, 1988; Shih *et al.*, 1988; Craigie, 1992). The *HPRT* locus therefore represents a 'cold-spot' for retrovirus integration. On the other hand, because the majority of the provirus integrations are into the actively transcribed genomic sequences (Scherdin *et al.*, 1990), the frequency for other genomic sites, particularly the actively transcribed domains, may in practice be higher than the predicted rate.

As in other forms of mutagenesis, it is difficult to isolate diploid cells with inactivating mutations in co-dominant genes, unless there is some form of actual or *functional haploidy* in that gene. This is because with an average mutation frequency of approximately $10^{-6}$ per haploid copy, the frequency of mutations in both alleles would be about $10^{-12}$. However, if only one functional copy of the gene is available in the cell (functional haploidy; e.g. the *HPRT* gene in female mammalian cells caused by X chromosome inactivation), if there is a significant gene dosage effect (in the absence of dosage compensation) or if there is partial genotypic haploidy (as in a number of established cell lines; e.g. the human myeloid cell line, HL-60), it would be possible to isolate such insertionally inactivated mutants. There are, however, examples of co-dominant genes in diploid cells in which the spontaneous frequency of mutation is very much higher than the mathematically predicted rate. An example of this is the autosomal gene adenosine phosphoribosyltransferase (*APRT*). Interestingly, the frequency of both spontaneous and retroviral insertional inactivation of the *APRT* locus in mouse embryonal carcinoma cells is only 100-fold lower than the frequency for the *HPRT* locus (King *et al.*, 1985). This suggests that either the *APRT* locus is a 'hot spot' for mutagenesis, or that there is some form of functional haploidy in this autosomal gene. The frequency of insertional activations, resulting in the expression of a previously silent gene, would also be affected by similar factors. However, because insertional activations are dominant, the co-integration of the provirus in both copies of a diploid gene is not necessary.

## 4.9 Multiplicity of infection

In principle, there are two types of retroviral vectors which could be used for insertional mutagenesis: replication defective and replication competent. With replication-competent vectors, only a few integrations per cell can be obtained (Kozak, 1985). This is because the expression of the viral envelope gene in the infected cell results in the saturation of the cellular receptors for the viral envelope protein, thereby eventually preventing the re-infection of the cell with additional copies of the virus. An exception to this general rule is presented by embryonal carcinoma cell lines which, in general, are refractory to the expression of viral genes (Sarma *et al.*, 1967), thus permitting super-infection to very high copy numbers (Stewart *et al.*, 1982). By contrast to the replication-competent retroviruses, the replication-defective vectors lack the envelope gene and can therefore be used for super-infection of target cells, thus resulting in multiple provirus integrations in a single cell.

The super-infection of target cells increases the mutation frequency and thus facilitates the isolation of mutants. However, this poses another problem, namely the subsequent difficulty in identifying, amongst all the integrated copies present, the integration which has resulted in the activation, or inactivation, of the gene/s responsible for the generation of the selected phenotype. This problem will be discussed further in Section 4.12.

## 4.10 Mutant selection procedures

The availability of a reliable and efficient means of mutant selection is very important. This usually means either the selective killing of the non-mutagenized cells or the growth advantage of the mutated cells. Examples include the isolation of HPRT-deficient cells by their ability to grow in the presence of the nucleotide analogues (e.g. 8-azaguanine; Hooper, 1985), or the selection of growth factor-independent cells from a population of cells dependent on the exogenous supply of these factors (Stocking *et al.*, 1988). The selection of insertional mutants in genes whose products are expressed on the cell surface can also be achieved by immune lysis of the wild-type cells, as for the isolation of mutants in the $\beta_2$-microglobulin locus using allele-specific antibodies (Frankel, 1985). Although, in theory, the isolation of mutants by screening procedures may also be possible, in practice this will usually be laborious and often impractical. A notable exception to this would be the isolation of mutants which are products of insertional activation, or inactivation of genes encoding cell surface markers. Such mutants could be isolated by fluorescence-activated cell sorting, panning with the aid of antibody-coated plates, or antibody-coated cell separation columns.

## 4.11 Cloning of the sites of provirus integration

There are two commonly used strategies for the cloning of cellular sequences which have been disrupted by the insertion of provirus sequences; these are described overleaf.

### 4.11.1 Construction of genomic libraries

Provided that the vector used for the mutagenesis carries within its genome a functional tRNA suppressor gene (e.g. the *supF* tyrosine suppressor tRNA gene), it will be possible to make sub-genomic libraries which are highly enriched for the cellular sequences flanking the sites of provirus integration (Reik *et al.*, 1984; Loebel *et al.*, 1985). For this purpose, total genomic DNA is isolated from clonal populations of the mutagenized cells and digested with restriction enzymes in such a way as to generate intact DNA fragments containing the *supF* gene and flanking cellular sequences. This is achieved by the digestion of the genomic DNA with either a vector non-cutter, or with single cutters which cut between the *supF* gene and the end of the 5' or 3' LTR. The mutant cell DNA is then cloned in phage vectors which carry amber mutations in genes which are essential for the phage proliferation (e.g. λgtwes or EMBL 3A). In such a vector, in a suppressor-free host, only those clones which contain a functional suppressor gene (*supF*) in the insert will proliferate. Thus each of the phage colonies generated will contain cellular sequences flanking one of the sites of provirus integration.

### 4.11.2 PCR-mediated amplification

The polymerase chain reaction (PCR) is a powerful method for the primer-directed amplification of specific DNA sequences *in vitro* (Saiki *et al.*, 1985). Using a procedure first described by Silver and Keerikatte (1989), and referred to as *inverse PCR*, it is possible to 'capture' one of the two cellular flanking sequences into the adjacent LTR (see Figure 4.5); PCR can then be used specifically to amplify cellular sequences flanking the site of provirus integration. For this purpose, mutant cell DNA is first digested with a frequently cutting restriction endonuclease and self-ligated, at low DNA concentrations to favour self-ligation. This is followed by redigestion of the sample with another endonuclease with a recognition site within the LTR, at a position between the site of the frequent cutter and the cellular site of integration. The cellular DNA originally bordering this LTR will now be contained within the two LTR fragments generated by the infrequent cutter; this will not be true of the cellular sequences flanking the other LTR. The use of primers directed against provirus sequences 5' and 3' to the second restriction enzyme then allows the amplification of the genomic DNA which originally flanked one of the two LTR sequences, depending on the choice of enzymes used. Therefore, for each integrated copy of the provirus, one genomic DNA fragment, representing one of the two cellular flanking sequences, will be amplified. The presence of appropriate restriction enzyme recognition sequences within the PCR primers would allow the simple cloning and sequencing of the amplified fragment.

## 4.12 Identification of the gene of interest

The precise nature of the strategy for identifying genes mutated by retroviral insertion would of course depend, to a very large extent, on the predicted

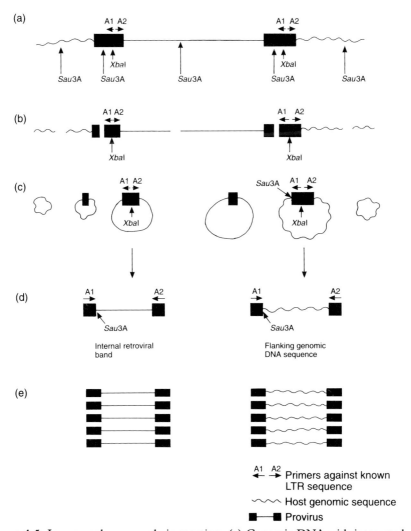

**Figure 4.5.** Inverse polymerase chain reaction. (a) Genomic DNA with integrated provirus. (b) Digestion with frequent cutter, e.g. *Sau*3A. (c) Self-ligation. (d) Digestion with appropriate 6–8 cutter, e.g. *Xba*I. (e) PCR using an appropriate primers (A1 and A2).

nature of the gene and the characteristics of the function which is being studied. Thus far, this has largely depended on the identification of common sites of provirus integration in independent clones of cells sharing a selected phenotype, and/or previous knowledge of the gene being mutated, as was the case in the isolation of mutants both in the *HPRT* (King *et al.*, 1985; Kuehn *et al.*, 1987) and $\beta_2$-microglobulin loci (Frankel *et al.*, 1985). However, in a mutant cell with multiple copies of provirus integration, the identification of the single provirus insertion resulting in the generation of the isolated phenotype is a particularly difficult task. Nuclear run-on screening of the genomic sequences flanking the different sites of provirus integration could prove a useful tool for distinguishing between these sites and highlighting the integration event most likely to be responsible for the generation of the selected phenotype (Gäken and Farzaneh, 1991). This is a particularly useful approach for the identification of genes involved in inducible cellular functions such as differentiation, damage-induced DNA repair, etc.

### 4.12.1 Identification of common sites of provirus integration

The identification of the integration event which has resulted in the appearance of the selected phenotype could be aided by the detection of a common site of provirus integration in independently arising clones of cells with the selected phenotype (King et al., 1985; Dickson et al., 1984). The identification of such a site may be possible by screening genomic DNA prepared from independently isolated clones of cells sharing a common mutant phenotype. However, integration in regions several kilobases apart may affect the same target gene product. Pulsed-field gel electrophoresis, which allows the resolution of very large DNA fragments, may prove a more effective means of screening the mutant cell genomic DNA sequences for common sites of provirus integration.

### 4.12.2 Library screening by nuclear run-on probes

In appropriately isolated nuclei, it is possible to elongate and radioactively label to a high specific activity, RNA transcripts which have already been initiated at the time of nuclei isolation; no new transcripts are initiated. This procedure enables the synthesis of *run-on* probes to all actively transcribed genes and offers a powerful means for the screening and identification of genes that are differentially expressed during inducible cellular processes (e.g. the induction of differentiation). A combination of two such probes made from the wild-type cells before and after treatment with the inducing agent could be used for the screening of the corresponding sub-genomic library containing the flanking sequences. The differential pattern of hybridization to the two probes would then allow the rapid identification of the integration events which are likely to have resulted in the appearance of the selected phenotype (Gäken and Farzaneh, 1991). The detected pattern could reveal valuable information about the nature of the cellular sequences disrupted by provirus integration. For instance, those clones in the sub-genomic library which do not hybridize to either of the two probes are likely to represent provirus integrations into non-transcribed, silent genomic sequences. Similarly, a clone which hybridizes to both probes would represent a constitutively active gene which would be unlikely to be involved in the inducible cellular function being studied. However, clones hybridizing to only one of the two probes would represent likely candidates for differentially regulated genes which may have been disrupted by provirus integration. This approach provides a powerful tool for the identification of gene(s) of interest amongst the many sites of provirus integration in mutant cell lines generated by super-infection. A similar approach has been used for the demonstration of preferential integration of provirus into actively transcribed genomic domains (Scherdin et al., 1990).

## References

**Adam MA, Miller AD.** (1988) Identification of a signal in a murine retrovirus that is sufficient for packaging of nonretroviral RNA into virions. *J. Virol.* **62:** 3802–3806.
**Albritton LM, Tseng L, Scadden D, Cunningham JM.** (1989) A putative murine ecotropic retrovirus receptor gene encodes multiple membrane-spanning protein and confers susceptibility to virus infection. *Cell* **57:** 659–666.
**Askew DS, Bartholomew C, Buchberg AM, Valentine MB, Jenkins NA, Copeland NG, Ihle JN.** (1991) *His*-1 and *His*-2: identification and chromosomal mapping of two commonly rearranged sites of viral integration in a myeloid leukemia. *Oncogene* **6:** 2041–2047.
**Askew DS, Li J, Ihle JN.** (1994) Retroviral insertions in the murine *His*-1 locus activate the expression of a novel RNA that lacks an extensive open reading frame. *Mol. Cell. Biol.* **14:** 1743–1751.
**Bear SE, Bellacosa A, Lazo PA, Jenkins NA, Copeland NG, Hanson C, Levan G, Tsichlis PN.** (1989) Provirus insertion in *Tpl*-1, an *Ets*-1 related oncogene, is associated with tumour progression in Moloney murine leukemia virus induced rat thymic lymphomas. *Proc. Natl Acad. Sci. USA* **86:** 7495–7499.
**Ben-David Y, Giddens EB, Bernstein A.** (1990) Identification and mapping of a common proviral integration site *Fli*-1 in erythroleukemia cells induced by Friend murine leukemia virus. *Proc. Natl Acad. Sci. USA* **87:** 1332–1336.
**Ben-David Y, Bani MR, Chabot B, De Koven A, Bernstein A.** (1992) Retroviral insertions downstream of the heterogeneous nuclear ribonucleoprotein A1 gene in erythroleukemia cells: evidence that A1 is not essential for cell growth. *Mol. Cell. Biol.* **12:** 4449–4455.
**Bishop JM.** (1983) Cellular oncogenes and retroviruses. *Annu. Rev. Biochem.* **52:** 301–354.
**Bishop JM.** (1987) The molecular genetics of cancer. *Science* **235:** 305–311.
**Boris-Lawrie K, Temin HM.** (1994) The retroviral vector. Replication cycle and safety considerations for retrovirus-mediated gene therapy. *Ann. NY Acad. Sci.* **716:** 59–70.
**Bouton AH, Parsons JT.** (1993) Retroviruses and cancer: models for cancer in animals and humans. *Cancer Invest.* **11:** 70–79.
**Bowerman B, Brown PO, Bishop JM, Varmus HE.** (1989) A nucleoprotein complex mediates the integration of retroviral DNA. *Genes Dev.* **3:** 469–478.
**Breuer ML, Cuypers HT, Berns A.** (1989) Evidence for the involvement of *pim*-2, a new common proviral insertion site, in progression of lymphomas. *EMBO J.* **8:** 743–747.
**Brown PO, Bowerman B, Varmus HE, Bishop JM.** (1989) Correct integration of retroviral DNA *in vitro*. *Cell* **49:** 347–356.
**Buchberg AN, Bedigian HB, Jenkins NA, Copeland NG.** (1990) *Evi*-2, a common integration site involved in murine myeloid leukemogenesis. *Mol. Cell. Biol.* **10:** 4658–4666.
**Cepko CL, Roberts BE, Mulligan RC.** (1984) Construction and application of a highly transmissible murine retrovirus shuttle vector. *Cell* **37:** 1053–1062.
**Coffin JM.** (1992a) Retroviral DNA integration. *Dev. Biol. Stand.* **76:** 141–151.
**Coffin JM.** (1992b) Genetic diversity and evolution of retroviruses. *Curr. Top. Microbiol. Immunol.* **176:** 143–164.
**Cornetta K, Morgan RA, French Anderson W.** (1991) Safety issues related to retrovirus-mediated gene transfer in humans. *Hum. Gene Ther.* **2:** 5–14.
**Craigie R.** (1992) Hotspots and warm spots: integration specificity of retroelements. *Trends Genet.* **8:** 187–190.
**Cuypers HT, Selten G, Quint W, Zijlstra M, Robanus Maandag E, Boelens W, Van Wezenbeek P, Melief C, Berns A.** (1984) Murine leukaemia virus-induced T-cell

lymphomagenesis: integration of proviruses in a district chromosomal region. *Cell* **37**: 141–150.

**Danos O, Mulligan RC.** (1988) Safe and efficient generation of recombinant retroviruses with amphotropic and ecotropic host ranges. *Proc. Natl Acad. Sci. USA* **85**: 6460–6464.

**Dickson C, Smith R, Brookes S, Peters G.** (1984) Tumorigenesis by mouse mammary tumour virus: provirus activation of a cellular gene in the common integration region *Int-2*. *Cell* **37**: 529–536.

**Donahue RE, Kessler SW, Bodin D, McDonagh K, Dunbar C, Goodman S, Agricola B, Byrne E, Raffeld M, Moen R, Bacher J, Zsebo KM, Neinhuis AW.** (1992) Helper virus induced T cell lymphoma in nonhuman primates after retroviral-mediated gene transfer. *J. Exp. Med.* **176**: 1125–1135.

**Donehower LA, Varmus HE.** (1984) A mutant murine leukaemia virus with a single missense codon in *pol* is defective in a function affecting integration. *Proc. Natl Acad. Sci. USA* **81**: 6461–6465.

**Dorssers LCJ, van Agthoven T, Dekker A, van Agthoven A, Kok EM.** (1993) Induction of antiestrogen resistance in human breast cancer cells by random insertional mutagenesis using defective retroviruses: identification of *bcar*-1, a common integration site. *Mol. Endocrinol.* **7**: 870–878.

**Finnegan DJ.** (1994) Retroviruses and transposons. Wandering retroviruses. *Curr. Biol.* **4**: 641–643.

**Frankel W, Potter TA, Rosenberg N, Lenz J, Rajan TV.** (1985) Retroviral insertional mutagenesis of a target allele in a heterozygous murine cell line. *Proc. Natl Acad. Sci. USA* **82**: 6600–6604.

**Fujiwara T, Craigie R.** (1989) Integration of mini-retroviral DNA: a cell-free reaction for biochemical analysis of retrovirus integration. *Proc. Natl Acad. Sci. USA* **86**: 3065–3069.

**Fujiwara T, Mizuuchi K.** (1988) Retroviral integration: structure of an integration intermediate. *Cell* **54**: 497–504.

**Gäken J, Farzaneh F.** (1991) Retroviral vectors as insertional mutagens. In: *Practical Molecular Virology*, Vol. 8, *Methods in Molecular Biology* (ed. MKL Collins). Humana Press, Clifton, NJ, pp. 111–130.

**Goff SP.** (1987) Gene isolation by retroviral tagging. *Methods Enzymol.* **152**: 469–481.

**Goff SP.** (1992) Genetics of retroviral integration. *Annu. Rev. Genet.* **26**: 527–544.

**Grandgenett DP, Mumm SR.** (1990) Unravelling retrovirus integration. *Cell* **60**: 3–4.

**Habets GGM, Scholtes EHM, Zuydgeest D, van der Kammen RA, Stam JC, Berns A, Collard JG.** (1994) Identification of an invasion-inducing gene, *Tiam*-1, that encodes a protein with homology to GDP–GTP exchangers for Rho-like proteins. *Cell* **77**: 537–549.

**Haupt Y, Alexander WS, Barri G, Klinken SP, Adams JM.** (1991) Novel zinc finger gene implicated as *myc* collaborator by retrovirally accelerated lymphomagenesis in Eµ–*myc* transgenic mice. *Cell* **65**: 753–763.

**Hayward WS, Neel BG, Astrin SM.** (1981) Activation of a cellular *onc* gene by promoter insertion in ALV-induced lymphoid leukosis. *Nature* **290**: 475–480.

**Hooper ML.** (1985) *Mammalian Cell Genetics*. John Wiley, New York.

**Hsu TW, Taylor JM.** (1982) Effect of aphidicolin on avian sarcoma virus replication. *J. Virol.* **44**: 493–498.

**Kavanaugh MP, Wang H, Boyd CA, North RA, Kabat D.** (1994a) Cell surface receptor for ecotropic host-range mouse retroviruses: a cationic amino acid transporter. *Arch. Virol.* **9** (Suppl.): 485–494.

**Kavanaugh MP, Miller DG, Zhang W, Law W, Kozak SL, Kabat D, Miller AD.** (1994b) Cell-surface receptors for gibbon ape leukemia virus and amphotropic murine retrovirus are inducible sodium-dependent phosphate symporters. *Proc. Natl Acad. Sci. USA* **91**: 7071–7075.

**King W, Patel MD, Lobel LI, Goff SP, Nguyen-Huu MC.** (1985) Insertion mutagenesis of embryonal carcinoma cells by retroviruses. *Science* **228**: 554–558.

**Kotani H, Newton PB, Zhang S, Chiang YL, Otto E, Weaver L, Blaese RM, French Anderson W, McGarrity GJ.** (1994). Improved methods of retroviral vector transduction and production for gene therapy. *Hum. Gene Ther.* **5**: 19–28.

**Kozak SL, Kabat D.** (1990) Ping-pong amplification of a retroviral vector achieves high-level gene expression: human growth hormone production. *J. Virol.* **64:** 3500–3508.
**Kreider BL, Orkin SH, Ihle JN.** (1993) Loss of erythropoietin responsiveness in erythroid progenitors due to expression of the *Evi*-1 myeloid-transforming gene. *Proc Natl. Acad. Sci. USA* **90:** 6454–6458.
**Krieg AM, Gourley MF, Perl A.** (1992) Endogenous retroviruses: potential etiologic agents in autoimmunity. *FASEB J.* **6:** 2537–2544.
**Kuehn MR, Bradley A, Robertson EJ, Evans MJ.** (1987) A potential animal model for Lesch–Nyhan syndrome through introduction of HPRT mutations into mice. *Nature* **326:** 295–298.
**Kulkosky J, Skalka AM.** (1994) Molecular mechanism of retroviral DNA integration. *Pharmacol. Ther.* **61:** 185–203.
**Kung HJ, Vogt PK.** (1991) (eds) *Retroviral Insertion and Oncogene Activation*. Springer, Berlin.
**Kung HJ, Boerkoel C, Carter TH.** (1991) Retroviral mutagenesis of cellular oncogenes: a review with insight into the mechanism of insertional activation. *Curr. Top. Microbiol. Immunol.* **171:** 1–25.
**Levy LS, Lobelle-Rich PA.** (1992) Insertional mutagenesis of *flvi*-2 in tumour induced by infection with LC-FeLV, a *myc*-containing strain of feline leukemia virus. *J. Virol.* **66:** 2885–2892.
**Loebel LI, Patel M, King W, Nguyen-Huu MC, Goff SP.** (1985) Construction and recovery of viable retroviral genomes carrying a bacterial suppressor transfer RNA gene. *Science* **228:** 329–332.
**Lynch CM, Miller AD.** (1991) Production of high-titre helper virus-free retroviral vectors by cocultivation of packaging cells with different host ranges. *J. Virol.* **65:** 3887–3890.
**Mann R, Baltimore D.** (1985) Varying the position of a retrovirus packaging sequence results in the encapsidation of both unspliced and spliced RNAs. *J. Virol.* **54:** 401–407.
**Mann R, Mulligan RC, Baltimore D.** (1983) Construction of a retrovirus packaging mutant and its use to produce helper free defective retrovirus. *Cell* **33:** 149–153.
**Markowitz D, Goff S, Bank A.** (1988a) A safe packaging line for gene transfer: separating viral genes on two different plasmids. *J. Virol.* **62:** 1120–1142.
**Markowitz D, Goff S, Bank A.** (1988b) Construction and use of a safe and efficient amphotropic packaging cell line. *Virology* **167:** 400–406.
**Marty L, Roux P, Royer M, Piechaczyk M.** (1990) MoMuLV-derived self-inactivating retroviral vectors possessing multiple cloning sites and expressing the resistance to either G418 or hygromycin B. *Biochimie* **72:** 885–887.
**Miller AD.** (1992) Retroviral vectors. *Curr. Top. Microbiol. Immunol.* **158:** 1–24.
**Miller AD, Buttimore C.** (1986) Redesign of retrovirus packaging cell lines to avoid recombination leading to helper virus production. *Mol. Cell. Biol.* **6:** 2895–2902.
**Mooslehner K, Karls U, Harbers K.** (1990) Retroviral integration sites in transgenic Mov mice frequently map in the vicinity of transcribed DNA regions. *J. Virol.* **64:** 3056–3058.
**Moreau-Gachelin F, Tavitian A, Tambourin P.** (1988) *Spi*-1 is a putative oncogene in virally induced murine erythroleukemias. *Nature* **331:** 277–280.
**Morgenstern JP, Land H.** (1990) Advanced mammalian gene transfer: high titre retroviral vectors with multiple drug selection markers and a complementary helper-free packaging cell line. *Nucleic Acids Res.* **18:** 3587–3596.
**Morishita K, Parker DS, Mucenski ML, Jenkins NA, Copeland NG, Ihle JN.** (1988) Retroviral activation of a novel gene encoding a zinc finger protein in IL-3-dependent myeloid leukaemia cell lines. *Cell* **54:** 831–840.
**Nienhuis AW, Walsh CE, Liu J.** (1993) Viruses as therapeutic gene transfer vectors. In: *Viruses and Bone Marrow* (ed. NS Young). Marcel Dekker, New York, pp. 353–414.
**Nusse R.** (1986) The activation of cellular oncogenes by retroviral insertion. *Trends Genet.* **2:** 244–247.
**Nusse R.** (1991) Insertional mutagenesis in mouse mammary tumorigenesis. *Curr. Top. Microbiol. Immunol.* **171:** 44–65.
**Nusse R, van Ooyen A, Cox D, Fung YKT, Varmus H.** (1984) Mode of provirus activation of a putative mammary oncogene (*Int*-1) on mouse chromosome 15. *Nature* **307:** 131–136.

Otto E, Jones-Trower A, Vanin EF, Stambaugh K, Mueller SN, French Anderson W, McGarrity GJ. (1994) Characterisation of a replication-competent retrovirus resulting from recombination of packaging and vector sequences. *Hum. Gene Ther.* **5**: 567–575.

Panganiban AT, Temin HM. (1984) The retrovirus *pol* gene encodes a product required for DNA integration: identification of a retroviral *int* locus. *Proc. Natl Acad. Sci. USA* **81**: 7885–7889.

Peters G, Lee AE, Dickson C. (1986) Concerted activation of two potential proto-oncogenes in carcinomas induced by mouse mammary tumour virus. *Nature* **320**: 628–631.

Reik W, Weiher H, Jaenisch R. (1984) Replication-competent Moloney murine leukemia virus carrying a bacterial suppressor tRNA gene: selective cloning of proviral and flanking host sequences. *Proc. Natl Acad. Sci. USA* **82**: 1141–1145.

Rein A. (1994) Retroviral RNA packaging: a review. *Arch. Virol.* **9** (Suppl.): 513–522.

Roe T, Renolds TC, Yu G, Brown PO. (1993) Integration of murine leukemia virus DNA depends on mitosis. *EMBO J.* **12**: 2099–2108.

Roth MJ, Schwartzberg PL, Goff SP. (1989) Structure of the termini of DNA intermediates in the integration of retroviral DNA: dependence on IN function and terminal DNA sequence. *Cell* **58**: 47–54.

Rous P. (1911) A sarcoma of the fowl transmissible by an agent separable from the tumour cell. *J. Exp. Med.* **13**: 397–411.

Saiki RK, Scharf S, Faloona F, Mullis KB, Horn GT, Erlich HA, Arnheim N. (1985) Enzymatic amplification of β-globin genomic sequences and restriction site analysis for diagnosis of sickle cell anemia. *Science* **230**: 1350–1354.

Sarma PS, Cheong MP, Hartley JW. (1967) A viral influence test for mouse leukemia viruses. *Virology* **33**: 180–184.

Scadden DT, Fuller B, Cunningham JM. (1990) Human cells infected with retrovirus vectors acquire an endogenous murine provirus. *J. Virol.* **64**: 424–427.

Scherdin U, Rhodes K, Breindl M. (1990) Transcriptionally active genome regions are preferred targets for retrovirus integration. *J. Virol.* **64**: 907–912.

Schwartzberg P, Colicelli J, Goff SP. (1984) Construction and analysis of deletion mutations in the *pol* gene of Moloney murine leukemia virus: a new viral function required for productive infection. *Cell* **37**: 1043–1052.

Selten G, Cuypers HT, Zijlstra M, Melief C, Berns A. (1984) Involvement of c-*myc* in MuLV induced T cell lymphomas in mice: frequency and mechanisms of activation. *EMBO J.* **4**: 1793–1798.

Selten G, Cuypers HT, Berns A. (1985) A proviral activation of the putative oncogene *pim*-1 in MuLV induced T-cell lymphomas. *EMBO J.* **4**: 1793–1798.

Shih C-C, Stoye JP, Coffin JM. (1988) Highly preferred targets for retrovirus integration. *Cell* **53**: 531–537.

Silver J, Keerikatte V. (1989) Novel use of polymerase chain reaction to amplify cellular DNA adjacent to an integrated provirus. *J. Virol.* **63**: 1925–1928.

Soriano P, Friedrich G, Lawinger P. (1991) Promoter interactions in retrovirus vectors introduced into fibroblasts and embryonic stem cells. *J. Virol.* **65**: 2314–2319.

Stewart CL, Stuhlmann H, Jahner D, Jaenisch R. (1982) *De novo* methylation, expression, and infectivity of retroviral genomes introduced into embryonal carcinoma cells. *Proc. Natl Acad. Sci. USA* **79**: 4098–4102.

Stocking C, Löliger C, Kawai M, Suciu S, Gough N, Ostertag W. (1988) Identification of genes involved in growth autonomy of haematopoietic cells by analysis of factor-independent mutants. *Cell* **53**: 869–879.

Stoye JP, Coffin JM. (1988) Polymorphism of murine endogenous proviruses revealed by using virus class-specific oligonucleotide probes. *J. Virol.* **62**: 168–175.

Sugden B. (1993) How some retroviruses got their oncogenes. *Trends Biochem. Sci.* **18**: 233–235.

Temin HM. (1990) Safety considerations in somatic gene therapy of human disease with retrovirus vectors. *Hum. Gene Ther.* **1**: 111–123.

Tremblay PJ, Kozak CA, Jolicoeur P. (1992) Identification of a novel gene, *Vin*-1 in murine leukaemia virus-induced T-cell leukemias by provirus insertional mutagenesis. *J. Virol.* **66**: 1344–1353.

**Varmus HE.** (1982) Form and function of retroviral proviruses. *Science* **216:** 812–820.
**Varmus HE.** (1988) Retroviruses. *Science* **240:** 1427–1435.
**Varmus HE, Padgett T, Heasley S, Simon G, Bishop JM.** (1977) Cellular functions are required for the synthesis and integration of avian sarcoma virus-specific DNA. *Cell* **11:** 307–319.
**Varmus HE, Quintrell N, Oritz S.** (1981) Retroviruses as mutagens: insertion and excision of a nontransforming provirus alters expression of a resident transforming provirus. *Cell* **25:** 23–36.
**Wang H, Paul R, Burgeson RE, Keene DR, Kabat D.** (1991) Plasma membrane receptors for ecotropic murine retroviruses require a limiting accessory factor. *J. Virol.* **65:** 6468–6477.
**Weiss RA, Teich N, Varmus J, Coffin J** (1982) (eds) *Molecular Biology of Tumor Viruses,* Vol. 1, *RNA Tumor Viruses.* Cold Spring Harbor Laboratory Press, Cold Spring Harbor, NY.
**Weiss RA, Teich N, Varmus J, Coffin J** (1985) (eds) *Molecular Biology of Tumor Viruses,* Vol. 2, *RNA Tumor Viruses.* Cold Spring Harbor Laboratory Press, Cold Spring Harbor, NY.
**Whitcomb JM, Hughes SH.** (1992) Retroviral reverse transcription and integration: progress and problems. *Annu. Rev. Cell Biol.* **8:** 275–306.
**Wilson AF, Cohen JC.** (1988) Hypothesis for testing deviations from random integration: evidence for nonrandom retroviral integration. *Genomics* **3:** 137–142.
**Yu SF, von Ruden T, Kantoff PW, Garber C, Seiberg M, Ruther U, French Anderson W, Wagner EF, Gilboa E.** (1986) Self-inactivating retroviral vectors designed for transfer of whole genes into mammalian cells. *Proc. Natl Acad. Sci. USA* **83:** 3194–3198.

# 5

# Gene entrapment

Harald von Melchner and H. Earl Ruley

## 5.1 Introduction

The molecular analysis of mammalian genomes aims to provide insights into gene organization and function that will assist efforts to isolate genes important in human disease. Whereas in the genomic analysis of lower organisms genetic approaches have been extremely successful, these methods are not well suited for mammals owing to: (i) large genome size, (ii) long reproduction cycles, and (iii) small numbers of offspring.

Consequently, methods such as the development of detailed physical maps (Chapman and Nadeau, 1992; Group, 1992), use of molecular strategies to identify candidate genes (Bird, 1986; Auch and Reth, 1990; Adams *et al.*, 1991; Buckler *et al.*, 1991) and improved coverage of chromosomal regions by cosmid and yeast artificial chromosome (YAC) clones (Foote *et al.*, 1992; Mandel *et al.*, 1992; Antonarakis, 1993), will continue to dominate efforts to understand the structure and function of mammalian genes.

Despite the remarkable progress made with physical methods, the analysis of the mammalian genome is still hindered by two major problems. First, even with improved maps, it is difficult to isolate genes responsible for disease phenotypes, largely owing to problems associated with locating genes within large regions of DNA. Consequently, genes responsible for organismal phenotypes have been identified far more frequently by testing candidate genes than by strict positional cloning (Copeland *et al.*, 1993) and therefore only a small fraction of known genes has been mapped. Second, the number of mutant alleles that can be characterized by physical methods is relatively small. For example, of the 5000 genetic diseases and traits that have been identified in man (McKusick and Amberger, 1993), most are not amenable to mapping. Furthermore, of the 1300 genetic traits that have been mapped in mice (Green, 1989), only a small number have been maintained. Finally, most characterized mutations in both mice and men are dominant and involve phenotypes that affect postnatal life; relatively few are recessive or interfere with embryonic development (Reith and Berstein, 1991).

These considerations make it likely that the functional analysis of most mammalian genes will rely on improved methods: (i) to identify and map candidate genes, and (ii) to mutagenize gene functions *in vivo*.

To this end, several types of vectors have been developed for mammalian cell insertional mutagenesis, allowing the selection of mutagenized cells in which the vector has integrated into expressed genes. This chapter will discuss a specific application of retroviral insertional mutagenesis for the isolation of transcriptionally active mammalian genes. One major difference between the gene trap strategy described in this chapter and the more conventional retroviral insertional mutagenesis (see Chapter 4) is that gene traps allow the identification of genes independently of their function. Retroviral insertional mutagenesis, by contrast, has to be linked to the selection of phenotypes which are the product of the mutation.

## 5.2 Gene trap vectors

Fusions of promoterless and/or enhancerless genes that encode an easily assayable gene product with the controlling elements of other genes, have proved valuable for studying gene regulation in both prokaryotic and eukaryotic cells (Fried *et al.*, 1983; Shapiro, 1983; Weber *et al.*, 1984; Santangelo *et al.*, 1986, 1988; Hiller *et al.*, 1988; Berg *et al.*, 1989). In mammalian cells, gene fusions were first generated by ligation of DNA fragments *in vitro* and then introduced into test cells. However, this strategy has several limitations. First, introduced fusion genes are frequently amplified in cells surviving selection. This increases background and necessitates the screening of multiple clones or the performance of secondary transfections in order to identify clones containing single copy genes. Second, potential promoter/enhancer elements identified after DNA transfer are not expressed in their normal chromosomal location. To address these problems, vectors have been developed that insert a reporter gene into a large collection of mostly random chromosomal sites, including transcriptionally active regions. By selecting for gene expression, clones are obtained in which the reporter gene is fused to the regulatory elements of an endogenous gene. Table 5.1 lists some of the reporter genes that have been used in these studies. In principle, reporter genes should be innocuous, should allow selection of mutagenic insertions, and should accurately reflect the activity of an endogenous gene.

Several investigators have used enhancerless reporter genes to select for integrations in the vicinity of cellular enhancers (Hamada, 1986; Bhat *et al.*, 1988; Gossler *et al.*, 1989; Okamoto *et al.*, 1990). Since their activation requires the acquisition of a cellular enhancer, the vectors have been termed 'enhancer traps'.

Enhancer traps have been used to identify developmentally regulated genes in both *Drosophila* (Bellen *et al.*, 1989; Wilson *et al.*, 1989) and mice (Allen *et al.*, 1988; Kothary *et al.*, 1988; Gossler *et al.*, 1989). Results from several studies indicate that up to 65% of all integrations allow expression of the transduced vectors (Bellen *et al.*, 1989; Wilson *et al.*, 1989; Sablitzky *et al.*, 1993). This suggests that the activating integration target is fairly large for

**Table 5.1.** Reporter and selectable marker genes used in gene trap vectors

| Gene[a] | Threshold of detection[b] | Positive selection | Negative selection | References |
|---|---|---|---|---|
| neo | Low | G418 | None | von Melchner et al. (1992) |
| lacZ | Medium/low | FACS, X-gal | FACS | Reddy et al. (1991), Cohen et al. (1989), Gossler et al. (1989) |
| tk | Medium | HAT | Gancyclovir | Chang et al. (1993) |
| lacZneo | Medium/low | G418, X-gal | FACS | Friedrich and Soriano (1991) |
| tkneo | Low | G418 | Gancyclovir | Friedel, Hicks, Ruley and von Melchner, unpubl. obs. |
| hisD | High | L-histidinol | None | von Melchner and Ruley (1989) |
| hygro | Medium/high | Hygromycin | None | Chang et al. (1993) |

[a] Reporter genes and the products they encode are as follows: *neo*, neomycin phosphotransferase; *lacZ*, *E. coli* β-galactosidase; *tk*, HSV-2 thymidine kinase; *lacZneo*, fusion protein between *E. coli* β-galactosidase and neomycin phosphotransferase; *tkneo*, fusion protein between HSV-2 thymidine kinase and neomycin phosphotransferase; *hisD*, *Salmonella* histidinol dehydrogenase; *hygro*, hygromycin phosphotransferase.
[b] Estimates were obtained in NIH3T3, CHO and ES cells. Detection sensitivities may be different for other cells.

enhancer traps. However, since enhancers can activate transcription over distances of several kilobases and irrespective of their position relative to the gene, expressed inserts may lie some distance from transcribed genes. This may be advantageous in studies where limited numbers of recombinants are available (i.e. chimeric and transgenic mice). On the other hand, the large activating target reduces the mutagenic potential of enhancer traps and greatly hinders later efforts to characterize disrupted genes and control elements from adjacent cellular DNA.

Gene trap vectors contain a reporter gene lacking both promoter and enhancer. Activation requires integration into expressed genes in the same transcriptional orientation relative to the gene. This interrupts the coding sequence of an endogenous gene and generates a fusion transcript between 5' cellular sequences and the reporter gene (Figure 5.1).

Gene trap vectors were originally referred to as promoter traps by some groups, since it was conceivable that these vectors might also be activated by cryptic promoters not associated with genes. However, recent evidence indicates that in those cases that we (von Melchner *et al.*, 1992; DeGregori *et al.*, 1994) and others (Skarnes *et al.*, 1992) have examined, integration interrupted *bona fide* genes, and we now refer to these vectors as gene traps.

The activating mechanism of gene trap vectors is dependent on the availability of splice acceptor sequences inserted upstream of the reporter gene (Figure 5.1). Vectors without splice acceptor sequences are self-inactivating retroviruses with a reporter gene inserted in the *U3* region of a Moloney murine leukaemia virus (MMLV) long terminal repeat (LTR). Virus replication, involving duplication of the reporter gene along with terminal

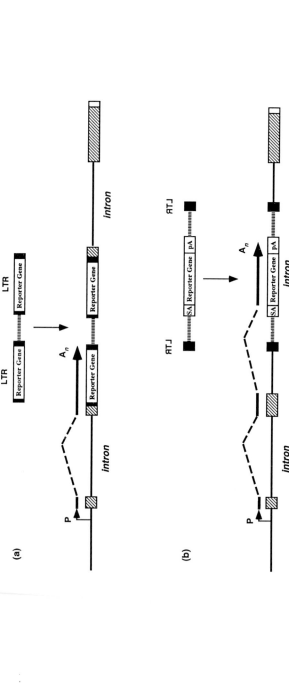

**Figure 5.1.** Retroviral gene trap vectors. (a) Gene trap with a reporter gene in the U3 region of an enhancerless Moloney murine leukaemia (MMLV) virus long terminal repeat (LTR). Retrovirus replication places the reporter gene only 30 nucleotides from the flanking cellular DNA. Integration into 5′ exons (hatched boxes) of expressed cellular genes generates fusion transcripts (heavy horizontal line) initiating at a flanking promoter (P) and terminating in the R region of the 5′ proviral LTR which contains the polyadenylation signal. The mRNA removed by splicing is shown by an interrupted line. In the illustrated case, the gene trap disrupts the second exon of the cellular gene. (b) Gene trap with a splice acceptor sequence upstream of a reporter gene. The reporter gene fused to the splice acceptor sequence and a polyadenylation site downstream (pA) is inserted into the body of an enhancerless MMLV in the reverse transcriptional orientation relative to the provirus. Activation results from integrations into introns of expressed genes and requires splicing from an upstream exon (hatched boxes) to the reporter gene. The sequences removed by splicing include the 3′ end of the provirus and are shown by an interrupted line. Fusion transcripts initiate at cellular promoters (P) and terminate in the polyadenylation site downstream of the reporter gene. In the processed mRNA (heavy horizontal line), exon sequences upstream of the integration site are spliced to the reporter gene.

control sequences, places the reporter gene just 30 nucleotides from the flanking cellular DNA. Integrations into or near expressed 5′ exons generate cell–provirus fusion transcripts from which the reporter gene is translated (Figure 5.1a, Table 5.2; von Melchner and Ruley, 1989). Vectors with splice acceptor sequences are activated from integrations into introns and require splicing from the endogenous gene to the reporter gene (Figure 5.1b; Brenner et al., 1989; Gossler et al., 1989).

**Table 5.2.** Examples of cellular genes disrupted by *U3* gene trap vectors

| Gene trap | Gene | Appended mRNA | | | Virus relative to AUG | References |
|---|---|---|---|---|---|---|
| | | Total | Exon | Intron | | |
| U3His | REX-1 | 271 | 271 | 0 | 5′ (96) | Hosler et al. (1989) |
| U3Neo | Histone H1 | 134 | 134 | 0 | 5′ (132) | Breuer et al. (1989) |
| U3TkNeo | Histone H3.3 | 98 | 98 | 0 | 5′ (24) | Wells et al. (1987) |
| U3Neo | hnRNP A2 | 974 | 813 | 160 | 3′ (657) | Burd et al. (1989) |
| U3TkNeo | P0 protein | 126 | 126 | 0 | 3′ (53) | Krowczynska et al. (1989) |
| U3TkNeo | S12 protein | 189 | 189 | 0 | 3′ (99) | Ayane et al. (1989) |
| U3TkNeo | Ke-3 protein | 177 | 146 | 31 | 3′ (70) | McMurray and Shin (1992) |
| U3Neo | L27 Rib.SU | 306 | 90 | 216 | 3′ (64) | Wool et al. (1990) |
| U3Neo | Fug1 | 226 | 226 | 0 | 5′ (275) | DeGregori et al. (1994) |
| U3Neo | L19 | 169 | 57 | 112 | 3′ (111) | Tsutsui (1993) |
| U3Neo | hnRNPU | 1765 | 321 | 1444 | 3′ (645) | Nakamura and Onno (1990) |
| U3Neo | HumOrf 16 | 287 | 287 | 0 | 5′ (19) | Nomura et al. (1993) |

The *U3* genes are expressed as fusion transcripts extending from the cellular DNA into the provirus. Adjacent intron sequences are appended only when the virus inserts into introns; otherwise, introns are spliced out. The table lists gene targets that have been characterized. The amount (in nucleotides) of cellular sequence appended to each *U3* fusion transcript (total, exon-derived and intron-derived) is listed. The position of each provirus relative (5′ or 3′) to the translation initiation codon of disrupted cellular genes is indicated. The distance between the initiation codon and the proviral integration site is provided in parentheses.

In most cases, gene trap vectors are transduced as recombinant retroviruses, although electroporated DNA has also been used. Retroviruses have the advantage of integrating by a precise process throughout the genome, causing little if any alteration to the surrounding host DNA (Varmus, 1988). Virtually all cells are susceptible to retrovirus infection, owing to the broad host range of available retrovirus vector systems (Danos and Mulligan, 1988; Albritton et al., 1989; Innes et al., 1990; Lynch and Miller, 1991; Landau and Littman, 1992; Burns et al., 1993; Pear et al., 1993). However, genes expressed in non-proliferating cells cannot be isolated, although even this problem may be solved by using retroviruses expressing nuclear localization signals (Bukrinsky et al., 1993).

Several experiments using *U3* gene trap vectors indicate that most expressed genes can be targeted by gene trap selection. The number of target genes in the genome that allow expression of different *U3* genes has been estimated by the frequencies with which gene trap retroviruses disrupted single copy genes (von Melchner and Ruley, 1989; Reddy et al.,

1991; Chang *et al.*, 1993). For example, by using a gene trap with the herpes simplex virus (HSV) thymidine kinase gene in *U3* (U3Tk) (Table 5.1), we found that 1/200 integrated proviruses expressed the *U3* gene. Fusion transcripts had an average of 250 nucleotides (nt) of cellular mRNA appended, which corresponds to the size of the region within a gene that allows *U3* gene expression. Since the maximum target size for viral integration is $6 \times 10^9$ nt in the diploid genome, and only integrations in the same transcriptional orientation (50%) as the disrupted gene are scored, the target that allows U3Tk expression is $1.5 \times 10^7$ nt ($6 \times 10^9 \times 0.5 \times 1/200$). With a target size per gene of 250 nt, the number of genes in the genome that activate U3Tk expression is approximately $6 \times 10^4$. This calculation assumes that retroviruses integrate randomly throughout most of the genome (Coffin *et al.*, 1989; Goff, 1990; Sandmeyer *et al.*, 1990; Withers-Ward *et al.*, 1994) and possible sites of integration (Shih *et al.*, 1988) neither preferentially permit nor exclude *U3* gene expression.

To examine this further, the frequency with which gene trap retroviruses integrate and inactivate hemizygous loci was estimated in an experiment that tested the ability of a U3Hygro gene trap retrovirus to disrupt single copy *tk* genes expressed from different sites of the genome (and introduced by a U3Tk gene trap; Table 5.1). Between $2 \times 10^4$ and $1 \times 10^5$ U3Hygro activating integrations (representing a total of $4 \times 10^6$–$2 \times 10^7$ integrations) were required to disrupt the *tk* gene (Chang *et al.*, 1993). This represents the maximum number of genes in the genome that can activate *U3* gene expression and is comparable to the number of expressed genes estimated from RNA renaturation experiments (Galau *et al.*, 1974; Lewin, 1975; Colber *et al.*, 1977). Therefore, in a total population of approximately $10^7$ infected cells, each with a single copy of the integrated provirus, there will be a U3Hygro activating integration in one allele of each transcriptionally active gene.

These results indicate that even weak promoters can activate *U3* gene expression, as most genes are expressed at relatively low levels (Lewin, 1975). However, as is to be expected, the efficiency with which weak promoters are identified varies greatly with the choice of reporter genes and selection protocols (Table 5.1). In principle, the threshold for detection of a reporter protein correlates inversely with the frequency of expressing gene trap integrations. For example, in 3T3 cells infected with U3His or U3Neo vectors (both containing independently expressed selectable marker genes), the activation frequency of U3His was approximately two orders of magnitude lower than that of U3Neo (von Melchner and Ruley, 1989; Friedel and von Melchner, unpublished observations). Thus, whereas a few molecules of neomycin phosphotransferase appear sufficient to confer neomycin resistance, relatively high levels of histidinol dehydrogenase are required to confer resistance to histidinol, a competitive inhibitor of histidyl-tRNA synthetase (Hartmann and Mulligan, 1988). Consequently, levels of cell–provirus fusion transcripts are abundant in histidinol-resistant clones and relatively sparse in neomycin-resistant clones. Moreover, the average cell–U3His provirus fusion transcript contains two- to fourfold less appended cellular sequence than the average U3Neo transcript, which indicates that strong selection protocols are likely to enrich for integrations close to transcriptional start sites (von Melchner and Ruley, 1989; von Melchner *et al.*, 1990, 1992). In summary, the size of activating integration targets correlates inversely with stringency of selection protocols and increases with promoter strength.

The size of the activating integration target is also dependent on available splice acceptor sites within gene traps. For example, electroporated gene traps are expressed 50 times more frequently if they contain a splice acceptor site (Friedrich and Soriano, 1991). This reflects the fact that vectors with splice acceptor sites can be expressed from anywhere within introns, and vectors lacking splice acceptor sites are activated only from integrations in or near exons (Figure 5.2). Since the average intron is 10–20 times larger than the average exon (Hawkins, 1988), the size of the activating integration target is larger for gene traps activated by splicing.

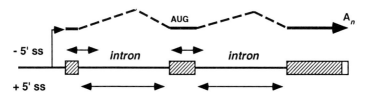

**Figure 5.2.** Genomic integration target for retroviral gene trap vectors with (+ 5′ ss) or without (– 5′ ss) splice acceptor sequences. Activating integrations are in exons (hatched boxes) or introns and on either side of translation initiation codons (AUG). For further explanation, see text.

However, activation targets for retroviral gene traps appear less dependent on available splice sites. Studies involving comparable retroviral gene trap vectors with and without splice acceptor sites indicated that activation frequencies are similar, 11 and 8% respectively. Moreover, in most cases, activating integrations occurred within 500 nt of the endogenous transcript. This may suggest some preference for retrovirus integrations near the 5′ end of genes (Friedrich and Soriano, 1991; Achilles, Zahn and von Melchner, unpublished observations), or selection against fusion genes that append larger amounts of cellular RNA (Kozak, 1991). Interestingly, the frequency of activating integrations obtained with retroviral vectors is twice that obtained following electroporation of similar constructs (Friedrich and Soriano, 1991) and is also greater than the fraction of the genome that is transcribed at any one time. This may reflect some bias in favour of integrations into transcribed chromatin (Rohdewohld *et al.*, 1987; Scherdin *et al.*, 1990) and may reduce the number of gene trap integrations into non-expressed genes. However, the bias is probably small (Withers-Ward *et al.*, 1994) and will not adversely affect the utility of gene trap retroviruses.

## 5.3 Cloning and analysis of flanking sequences

Cellular sequences flanking gene trap proviruses can be cloned by a number of methods. Good results have been obtained with a relatively small number of clones by using inverse polymerase chain reaction (PCR; von Melchner et al., 1990) or PCR amplification of cell–provirus fusion transcripts (5′ RACE – rapid amplification of cDNA ends; see Chapter 3) (Frohman et al., 1988; Skarnes et al., 1992; DeGregori et al., 1994). Large numbers of flanking sequences are best isolated by plasmid rescue following integration of a gene trap shuttle vector (see Section 5.4).

Inverse PCR is particularly suited for *U3* retrovirus gene traps because the selectable marker is positioned within 30 nt of the flanking cellular DNA. Approximately 50% of flanking sequences isolated by inverse PCR hybridize to transcripts of the disrupted genes (von Melchner et al., 1992) and can be used directly to isolate cDNAs of the genes. In addition, with the U3His vector, 50% of upstream sequences contain cellular promoters (von Melchner et al., 1990), thus providing a simple alternative for isolating cellular promoters which, unlike previous methods, does not require prior isolation of the transcribed gene. However, the relatively short inverse PCR products are less likely to contain promoters if derived from U3Neo and U3LacZ vector integrations, since these can be expressed even when positioned further into the gene. This is probably because less protein is required for selection and therefore the fusion genes can tolerate more translational suppression from appended cellular RNA.

5′ RACE offers several advantages over inverse PCR. First, it is not restricted to *U3* gene vectors and works equally well with gene traps that are activated by splicing (Skarnes et al., 1992). Second, unlike genomic sequences recovered by inverse PCR, full-length 5′ RACE products always contain transcribed exon sequences. Third, only expressed fusion genes are amplified from cell clones containing multiple proviruses.

Flanking sequences may be used as a source of primers to clone directly the host gene from cellular mRNA. Using a 3′ RACE protocol (Frohman et al., 1988), we have been able to amplify two endogenous transcripts of 2.5 and 1 kb, respectively (von Melchner and Ruley; Thorey and von Melchner, unpublished observations). This strategy provides full-length cDNA from the average mRNA of 2–3 kb, and circumvents the screening of cDNA libraries. Although up to now direct cloning of mRNAs larger than 3–4 kb has been difficult, recent improvements in PCR technology which allow the amplification of up to 35 kb of genomic sequence are likely to resolve this problem (Barnes, 1994; Cohen, 1994).

## 5.4 Isolation and use of promoter-tagged sites

To simplify the process of isolating cellular genes disrupted by gene trap mutagenesis, we have developed a retrovirus gene trap shuttle vector, U3NeoSV1. As shown in Figure 5.3, the vector contains a neomycin resistance gene inserted into the *U3* region, and carries an ampicillin resistance gene, the pBR322 plasmid origin of replication and the *lac* operator (*lac*O) sequence within its body. Depending on the restriction enzyme used, cellular sequences 5' (*Eco*RI) or 3' (*Bam*HI or *Xho*I) of the provirus can be cloned by plasmid rescue. The *lac*O region permits partial purification of flanking sequence fragments by binding to the Lac repressor protein (Pathak and Temin, 1990; Gossen *et al.*, 1993).

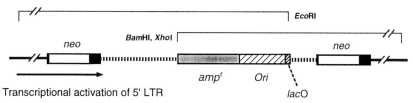

**Figure 5.3.** U3NeoSV1 shuttle vector. U3NeoSV1 is a retrovirus gene trap that allows direct cloning of flanking cellular sequences (solid lines) by plasmid rescue. Circularization of genomic DNA cleaved by *Bam*HI, *Xho*I or *Eco*RI generates plasmids that can be amplified in *E. coli*. Plasmids isolated from AmpR bacterial colonies contain genomic sequences which are either 5' (*Eco*RI) or 3' (*Bam*HI or *Xho*I) of the provirus. For further details, see text.

The vector provides a means for isolating DNA libraries containing promoter regions and 5' exon sequences from the majority of genes expressed in cultured cells. Libraries of cell clones are isolated following gene trap selection and flanking regions adjacent to the integrated proviruses are isolated by plasmid rescue and then sequenced. The process generates sequence tags of expressed genes which we have designated promoter-tagged sites (PTSs). PTSs can be used to identify and map candidate genes in a manner analogous to expressed sequence tags (ESTs; see Chapter 2).

Several features of the strategy are ideally suited for isolating sequence tags of expressed genes. Firstly, most genes, including weakly expressed genes, appear capable of activating *U3* gene expression (Chang *et al.*, 1993). Secondly, whilst the process may favour genes with longer untranslated leaders and possible preferential sites of retrovirus integration or, to a limited extent, highly expressed genes, it is not heavily biased against weakly expressed genes. Therefore, gene representation is likely to be uniform. Thirdly, since the proviruses are positioned in or near transcribed exons, the disrupted genes are relatively easy to identify. In practice, it is only necessary to sequence the 300 nt immediately adjacent to the provirus (by using a single primer complementary to the *U3* gene) to determine if the occupied gene matches transcribed sequences in the databases. To date, 30 out of 450 inserts (approximately 7%) analysed have occurred in previously characterized genes. Fourthly, the transcriptional orientation of the occupied cellular gene is provided by the orientation of the *U3* gene. Finally, relatively large regions of genomic DNA are recovered following plasmid rescue. This will facilitate:

(i) mapping of PTSs by fluorescence *in situ* hybridization (FISH);

(ii) isolation of chromosome-specific PTSs from human–rodent hybrid cells by screening PTS libraries with human repetitive sequence probes; and
(iii) identification of simple sequence repeats (SSRs) or single copy sequences used by PCR- and hybridization-based mapping strategies.

The frequency (7%) with which known genes have been targeted following gene trap selection is similar to the representation of known genes among random cDNAs (i.e. 700 unique genes among 8500 cDNAs; Adams et al., 1993) and presumably reflects the number of vertebrate genes which have been sequenced (Benson et al., 1993; Rice et al., 1993). The nucleic acid databases have doubled in size every 18–24 months over the past 8 years, as have the proportion of flanking sequences that match known genes. Therefore, the percentage of flanking sequences that match known genes (or cDNAs) will probably increase from 7 to 25% within the next 5 years.

We anticipate that genes present in cosmid or YAC clones can be identified by hybridization to flanking sequence libraries (Snell et al., 1993). Conversely, DNA from flanking sequence libraries may be used to probe cosmid or YAC clones. Finally, as discussed above, the flanking region sequences can be used to isolate full-length cDNA clones of the corresponding cellular genes (DeGregori et al., 1994).

PTS databases will complement the use of ESTs for the following reasons:

(i) gene trapping is less biased toward highly expressed genes than is cDNA cloning, and gene representation appears to be more uniform. This is important since most genes are expressed at low levels (e.g. 10 transcripts/cell corresponding to <0.01% of poly(A)-containing RNA; Lewin, 1975) and are underrepresented in cDNA libraries. For example, 10% of ESTs from brain are related to cytoskeletal proteins (Adams et al., 1993), whereas none of the 400 PSTs from a mouse embryonic stem (ES) cell library match cytoskeletal genes. Excluding LINE-1 elements, only four ES cell PSTs (histone H1, P0, hnRNP U and L19 ribosome subunit protein; Table 5.2) were represented among 700 genes that matched brain ESTs (Adams et al., 1993). Only one (P0) was found in a library of ESTs from HepG2 cells (Okubo et al., 1992), and even these four genes constituted only 0.01% of the randomly sequenced cDNAs.
(ii) cDNA probes are unable to distinguish between pseudogenes and functional genes, whereas PTSs are linked to functional genes. For example, gene traps have identified genes encoding ribosomal subunit proteins and expressed LINE-1 elements from a large number of pseudogenes.
(iii) U3 gene traps often insert in or near 5' exon sequences that may be missing from all but full-length cDNAs (von Melchner and Ruley, 1989; Reddy et al., 1991; von Melchner et al., 1992; DeGregori et al., 1994).

## 5.5 Insertional mutagenesis in cultured cells

Chemical mutagenesis has provided a useful means for the analysis of biochemical pathways. However, the affected genes are difficult to isolate.

Gene trap mutagenesis avoids this problem by generating molecular tags that allow the mutant genes to be mapped and cloned.

Retroviruses are well suited for use as insertional mutagens because they integrate widely throughout the genome and cause little if any alteration to the surrounding host DNA (Varmus, 1988; Coffin et al., 1989; Goff, 1990; Sandmeyer et al., 1990; Withers-Ward et al., 1994). However, owing to the large size of mammalian genomes, random mutagenesis requires the screening of large numbers of integrants in order to detect mutations in any specific gene. Therefore, mutation efficiency would be increased if one could select for integrations into gene coding sequences. Gene trap vectors meet this goal since they allow selection for integrations into expressed genes, and thereby increase the frequency of mutations over randomly integrating retroviruses by two to three orders of magnitude (Table 5.3).

**Table 5.3.** Frequency of gene disruptions after random and gene trap mutagenesis

| Gene | Number of integrations required to disrupt gene | | |
|---|---|---|---|
| | Without gene trap selection | With gene trap selection | References |
| HPRT | $10^8$ | – | King et al. (1985) |
| $\beta_2$-Microglobulin | $10^8$ | – | Frankel et al. (1985) |
| src | $\sim 5 \times 10^6$ | – | Varmus et al. (1981) |
| U3Tk(H1) | $10^7$ | $\sim 2 \times 10^4$ | Chang et al. (1993) |
| U3Tk(H4) | $10^7$ | $10^5$ | Chang et al. (1993) |

Unlike chemical mutagens, mutations generated by gene trap retroviruses are generally recessive and consequently a single insertion usually does not result in loss of gene function. To enhance the recovery of null phenotypes, hypodiploid cell lines (i.e. CHO) may be used (Chang et al., 1993; Hubbard et al., 1994) or systems in which gene inactivation leads to a selectable phenotype. For example, growth factor-dependent cell lines may be used to select for gene disruptions in tumour suppressor genes or in genes involved in apoptosis. Selection for growth or survival in the absence of growth factor would allow clones to be isolated in which growth control genes are functionally inactivated. In principle, this enrichment should permit loss of diploid functions, since alleles opposite those disrupted by the provirus may be lost by mechanisms such as gene conversion, non-disjunction or mutation.

## 5.6 Insertional mutagenesis in mice

Gene traps also provide an effective means of analysing gene functions in mice. By using totipotent mouse ES cells as cellular targets, mouse strains with mutationally inactivated gene functions can be constructed. The method exploits the discovery that ES cells derived from the inner cell mass of mouse blastocysts can be cultured *in vitro* for extended periods of time without losing their pluripotent differentiation potential (Doetschman et al., 1985;

Hooper et al., 1987b). Thus, when transferred to pre-implantation mouse blastocysts, ES cells can contribute to all tissue types of the developing embryo, including the germline (Evans and Kaufman, 1981; Martin, 1981). As a result, genes introduced into ES cells can be passaged to transgenic offspring (Bradley et al., 1984; Robertson et al., 1986; Hooper et al., 1987a; Thompson et al., 1989).

Approximately 50% of genes disrupted in ES cell lines after infection/electroporation of gene trap vectors generate recessive phenotypes in mice (Friedrich and Soriano, 1991; von Melchner et al., 1992; Skarnes et al., 1992). This frequency is 10 times greater than that observed following random insertion of retroviruses or microinjected DNA (Jaenisch et al., 1985; Gridley et al., 1987). However, if gene trap vectors always disrupt expressed genes, then why do some fusion genes fail to generate phenotypes?

(i) As has been shown by targeted disruption of some genes, phenotypes may not develop due to compensation by redundant genes (e.g. *MyoD*; Rudnicki et al., 1993).
(ii) Some mutations may be subtle and are thus missed (e.g. *src*; Soriano et al., 1991).
(iii) Gene trap integrations may not always abolish gene function. This could result from splicing around vectors inserted in introns or alternatively spliced exons. In addition, proteins expressed from fusion transcripts may retain some function.

Despite these problems, gene trap vectors work remarkably well as insertional mutagens. To date, only a few genes necessary for normal development and physiology of the mouse have been isolated following unselected insertional mutagenesis. Yet in the last few years, at least seven such genes have been identified and four isolated by gene entrapment (Friedrich and Soriano, 1991; von Melchner et al., 1992; Skarnes et al., 1992; DeGregori et al., 1994). Whilst most mutations recovered by gene trap mutagenesis are recessive, dominant phenotypes have also been observed (Friedrich and Soriano, 1991). These may occur whenever levels of gene expression are insufficient in the heterozygote or disrupted protein coding sequences express dominant-negative activities.

Gene traps induce mutations with such high efficiency that it is feasible to isolate cell lines with integrations in most expressed genes ($1-2 \times 10^4$). However, it is not practical to pass all mutations into the germline, and many mutations will involve genes of lesser significance. Therefore, one would like to pre-screen mutagenized ES cell clones for mutations that affect significant biological processes or genes. Toward this end, two types of screens have been particularly useful.

One approach involves sequencing DNA regions immediately adjacent to a U3 gene trap vector. As described above, comparison of the flanking sequences with nucleic acid databases reveals instances in which the provirus has disrupted known genes. By using the shuttle vector, we were able to analyse 400 clones in less than 6 months (Hicks and Ruley, unpublished observations). The library contained mutations in 21 known genes including: *FUS/TLS*, a gene translocated in malignant liposarcoma (Rabbitts *et al.*, 1993); *PLK*, the mouse homologue of the *POLO* and *CDC5* protein kinase genes of *Drosophila* and *Saccharomyces cerevisiae*, respectively (Clay *et al.*, 1993); *Gas5*, a gene of unknown function, whose expression is induced in growth-arrested cells (Coccia *et al.*, 1992); *GLUT1*, encoding a glucose transporter expressed at blood–tissue boundries (Pessin and Bell, 1992); a gene encoding the p67 kDa high-affinity laminin-binding protein that is regulated in transformation and may function as the Sindbis virus receptor (Wang *et al.*, 1992); *NonO*, encoding an octamer DNA-binding protein related to the *Drosophila nonAdiss* gene (Yang *et al.*, 1993); and *FBP*, encoding a DNA-binding protein that binds to a region upstream of the c-*myc* gene (Duncan *et al.*, 1994). The total effort, even with manual sequencing methods, was similar to that required to target one gene by homologous recombination. Thus, for a fraction of the costs and time of targeting genes by homologous recombination, large numbers of genes can be disrupted and screened by gene trap mutagenesis. Moreover, the process requires neither cDNA sequence nor detailed information on the genomic structure of the genes, thus avoiding analysis of genes without apparent biological (and disease) function.

As described in the following section, a second approach to pre-screen ES cell clones for interesting mutations uses a reporter gene to identify mutations in developmentally regulated genes.

## 5.7 Identification of regulated genes

Gene trap integrations are expected to mimic faithfully the expression pattern of the disrupted endogenous gene and can thus identify genes that are regulated at the transcriptional level by any biological stimulus (i.e. differentiation, hormones, transcription factors).

For studies of differentiation and development, gene traps with reporter genes such as *lacZ* are preferred over the average selectable marker gene since they are more likely to reflect gradients of gene expression. However, to isolate transcriptional targets of hormones and transcription factors, which necessitates the screening of large collections of clones (Figure 5.4), it might be easier to use gene traps with a selectable marker gene. To this end, several vectors have been developed that allow selection for and against activating integrations (Table 5.1). Large collections of clones can be selected: (i) against expression, to obtain clones with integrations in transcriptionally inactive sites of the genome including silent genes, or (ii) for functional gene fusions to obtain clones with integrations in expressed genes.

Each of these cell populations may be cultured under different hormonal conditions or transduced with a transgene expressing a putative transcription factor.

Following selection for changes in gene trap expression, clones may be isolated in which integration occurred next to an active promoter that was subsequently repressed or, conversely, next to an inactive promoter that was subsequently activated (Figure 5.4). In the latter case, however, many more clones are needed since the majority of integrations do not disrupt genes in the

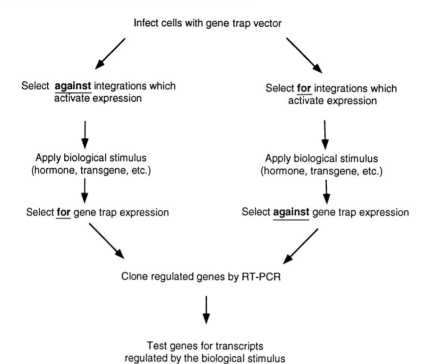

**Figure 5.4.** Strategy to identify genes regulated by a biological stimulus. For further explanation, see text.

initial pool of non-expressed inserts. Finally, cell lines containing regulated fusion genes may be used to isolate genes encoding components of signal transduction pathways and may serve as indicator lines in the mutational analysis of transcription factors.

Several groups have used gene traps to identify and mutate genes that are regulated during differentiation and mouse development (Gossler et al., 1989; Friedrich and Soriano, 1991; Reddy et al., 1992; Rossant and Hopkins, 1992; Skarnes et al., 1992). ES cells differentiate *in vitro* into embryoid bodies that contain derivatives of all three germ layers including muscle, blood, cartilage and other differentiated tissues (Doetschman et al., 1985). In some culture systems, ES cells can be selectively induced to differentiate into haematopoietic cells within a time frame that closely mimics haematopoietic development *in vivo* (Schmitt et al., 1991; Wiles and Keller, 1991; Keller et al., 1993). We are presently using ES cell cultures to identify genes that are repressed during differentiation. Since differentiation seems to require continuous regulation (Blau and Baltimore, 1991), some of these genes could be inhibitors of lineage-specific differentiation programmes. Other genes may act as 'competence' genes that participate in early development without needing to be induced. However, as stem cells differentiate, such genes may be repressed in all but the appropriate cell lineages. Finally, expression of some genes may be loosely regulated in ES cells and become tightly regulated only later in development.

We have used two different vectors to identify mutations in developmentally regulated genes. U3βneo expresses a β-galactosidase–neomycin phosphotransferase fusion protein that allows mutant clones to be selected in G418 and developmentally regulated fusion genes to be identified by changes in X-Gal staining. Of 190 clones tested, 24 expressed developmentally regulated fusion genes (Chen and Ruley, unpublished observations). Another vector using a tkneo fusion protein, proved even more sensitive; 65 out of 134 fusion genes were repressed upon differentiation. Most were inactivated at a time that would correspond *in vivo* to the initial stages of gastrulation (Bruyns, Thorey and von Melchner, unpublished observations). This number appears extremely high considering that only 10–20% of all expressed genes are thought to be unique to any one cell (Lewin, 1975). However, similar results have been obtained by Friedrich and Soriano (1991) in transgenic mice, where 11 out of 24 fusion genes were repressed in early embryos.

Several factors may inflate the number of regulated genes:

(i) regulated genes frequently have longer 5' non-coding exons than widely expressed housekeeping genes (Kozak, 1987), thus providing a larger target capable of activating gene trap expression;
(ii) tagged promoters may be silenced because of *de novo* methylation extending from the provirus into adjacent flanking sequences (Jähner and Jaenisch, 1985). However, *in vivo* studies have shown that most fusion genes that were initially repressed are expressed at later stages of development (Friedrich and Soriano, 1991);
(iii) cultured ES cells may express genes that are not normally expressed *in vivo*. However, in nine cases we have examined, fusion genes expressed in ES cells have also been expressed in pre-implantation embryos (Chen and Ruley, unpublished observations).

Several observations suggest that patterns of reporter gene expression can be used to identify insertion mutations in developmentally regulated genes:

(i) changes in fusion gene expression during *in vitro* differentiation accurately predict changes in expression during *in vivo* development (Reddy *et al.*, 1992; Chen and Ruley, unpublished observations). This agrees with previous observations in which genes regulated *in vitro* were also tightly regulated *in vivo* (Rappolee *et al.*, 1988; Scholer *et al.*, 1990; Sharpe *et al.*, 1990; Rogers *et al.*, 1991; Muthuchamy *et al.*, 1993);
(ii) *lacZ* fusion genes, created by either gene trap selection or homologous recombination, appear to be expressed in a manner identical to that of the occupied cellular gene (Mansour *et al.*, 1990; Le Mouellic *et al.*, 1992; Skarnes *et al.*, 1992; Schneider-Maunoury *et al.*, 1993; Yagi *et al.*, 1993; Tajbakhsh and Buckingham, 1994).

## Acknowledgements

We would like to thank members of our laboratories who have collaborated on work described here. This work has been supported by grants from the

Deutsche Forschungsgemeinschaft (HO-684/2-1) and Deutsche Krebshilfe to H.v.M. and from the National Institutes of Health (R01HG00684, R01GM48688 and R01CA40602) to H.E.R.

## References

Adams MD, Kelley JM, Gocayne JD, Dubnick M, Polymeropoulos MH, Xiao H, Merril CR, Wu A, Olde B, Moreno RF, Kerlavage AR, McCombie WR, Venter JC. (1991) Complementary DNA sequencing: expressed sequence tags and human genome project. *Science* **252**: 1643–1651.

Adams MD, Kerlavage AR, Fields C, Venter JC. (1993) 3400 new expressed sequence tags identify diversity of transcripts in human brain. *Nature Genetics*. **4**: 256–267.

Albritton LM, Tseng L, Scadden D, Cunningham JM. (1989) A putative murine ecotropic retrovirus receptor gene encodes a multiple membrane spanning-protein and confers susceptibility to virus infection. *Cell* **57**: 659–666.

Allen ND, Cran DG, Barton SC, Hettle S, Reik W, Surani MA. (1988) Transgenes as probes for active chromosomal domains in mouse development. *Nature* **333**: 852–855.

Antonarakis SE. (1993) Human chromosome 21: genome mapping and exploration, circa 1993. *Trends Genet.* **9**: 142–148.

Auch D, Reth M. (1990) Exon trap cloning: using PCR to rapidly detect and clone exons from genomic DNA fragments. *Nucleic Acids Res.* **18**: 6743–6744.

Ayane M, Nielsen P, Koehler G. (1989) Cloning and sequencing of mouse ribosomal protein S12 cDNA. *Nucleic Acids Res.* **17**: 6722.

Barnes W. (1994) PCR amplification of up to 35-kb DNA with high fidelity and high yield from lambda bacteriophage templates. *Proc. Natl Acad. Sci. USA* **91**: 2216–2220.

Bellen HJ, O'Kane CJ, Wilson C, Grossniklaus U, Pearson RK, Gehring WL. (1989) P-element-mediated enhancer detection. A versatile method to study development in *Drosophila*. *Genes Dev.* **3**: 1288–1300.

Benson D, Lipman DJ, Ostell J. (1993) GenBank. *Nucleic Acids Res.* **21**: 2963–2965.

Berg C, Berg D, Groisman EA. (1989) Mobile DNA. In: *Transposable Elements and the Genetic Engineering of Bacteria* (eds DE Berg, MM Howe). American Society for Microbiology, Washington, DC, pp. 879–925.

Bhat K, McBurney M, Hamada H. (1988) Functional cloning of mouse chromosomal loci specifically active in embryonal carcinoma stem cells. *Mol. Cell. Biol.* **8**: 3251–3259.

Bird AP. (1986) CpG-rich islands and the function of DNA methylation. *Nature* **321**: 209–213.

Blau HM, Baltimore D. (1991) Differentiation requires continuous regulation. *J. Cell. Biol.* **112**: 781–783.

Bradley A, Evans M, Kaufman MH, Robertson E. (1984) Formation of germ-line chimeras from embryo-derived teratocarcinoma cell lines. *Nature* **309**: 255–256.

Brenner DG, Lin-Chao S, Cohen SN. (1989) Analysis of mammalian cell genetic regulation *in situ* by using retrovirus-derived "portable exons" carrying the *Escherichia coli lacZ* gene. *Proc. Natl Acad. Sci. USA* **86**: 5517–5521.

Breuer B, Fischer J, Alonso A. (1989) Cloning and characterization of the mouse histone H1-0 promoter region. *Gene* **81**: 307–314.

Buckler AJ, Chang DD, Graw SL, Brook JD, Haber DA, Sharp PA, Housman DE. (1991) Exon amplification: a strategy to isolate mammalian genes based on RNA splicing. *Proc. Natl Acad. Sci. USA* **88**: 4005–4009.

Bukrinsky MI, Haggerty S, Dempsey MP, Sharova N, Adzhubei A, Spitz L, Lewis P, Goldfarb D, Emerman M, Stevenson M. (1993) A nuclear localization signal within HIV-1 matrix protein that governs infection of non-dividing cells. *Nature* **365**: 666–669.

Burd C, Swanson M, Gorlach M, Dreyfuss M. (1989) Primary structures of the heterogeneous nuclear ribonucleoprotein A2, B1 and C2 proteins: a diversity of RNA binding proteins is generated by small peptide inserts. *Proc. Natl Acad. Sci. USA* **86**: 9788–9792.

**Burns J, Friedmann T, Driever W, Burrascano M, Yee J.** (1993) Vesicular stomatitis virus G glycoprotein pseudotyped retroviral vectors: concentration to very high titer and efficient gene transfer into mammalian and non-mammalian cells. *Proc. Natl Acad. Sci. USA* **90**: 8033–8037.

**Chang W, Hubbard C, Friedel C, Ruley H.** (1993) Enrichment of insertional mutants following retrovirus gene trap selection. *Virology* **193**: 737–747.

**Chapman VM, Nadeau JH.** (1992) The mouse genome: an overview. *Curr. Opin. Genet. Dev.* **2**: 406–411.

**Clay FJ, McEwen SJ, Bertoncello I, Wilks AF, Dunn AR.** (1993) Identification and cloning of a protein kinase-encoding mouse gene, *Plk*, related to the *polo* gene of *Drosophila*. *Proc. Natl Acad. Sci. USA* **90**: 4882–4886.

**Coccia EM, Cicala C, Charlesworth A, Ciccarelli C, Rossi GB, Philipson L, Sorrentino V.** (1992) Regulation and expression of a growth arrest-specific gene (*gas5*) during growth, differentiation, and development. *Mol. Cell. Biol.* **12**: 3514–3521.

**Coffin JM, Stoye JP, Frankel WN.** (1989) Genetics of endogenous murine leukemia viruses. *Ann. NY Acad. Sci.* **567**: 39–49.

**Cohen J.** (1994) 'Long PCR' leaps into larger DNA sequences. *Science* **263**: 1564–1565.

**Colber DA, Tedeschi MV, Atryzek V, Fausto N.** (1977) Diversity of polyadenylated messenger RNA sequences in normal and 12-hr regenerating liver. *Devel. Biol.* **59**: 111–123.

**Copeland NG, Jenkins NA, Gilbert DJ, Eppig JT, Maltais LJ, Miller JC, Dietrich WF, Weaver A, Lincoln SE, Steen RG.** (1993) A genetic linkage map of the mouse: current applications and future prospects. *Science* **262**: 57–66.

**Danos O, Mulligan RC.** (1988) Safe and efficient generation of recombinant retroviruses with amphotropic and ecotropic host ranges. *Proc. Natl Acad. Sci. USA* **85**: 6460–6464.

**DeGregori J, Russ A, von Melchner H, Rayburn H, Priyaranjan P, Jenkins NA, Copeland NG, Ruley HE.** (1994) A murine homologue of the yeast RNA1 gene is required for postimplantation development. *Genes Dev.* **8**: 265–276.

**Doetschman TC, Eistetter H, Katz M, Schmidt W, Kemler R.** (1985) The *in vitro* development of blastocyst-derived embryonic stem cell lines: formation of visceral yolk sac, blood islands and myocardium. *J. Embryol. Exp. Morphol.* **87**: 27–45.

**Duncan R, Bazar L, Michelotti G, Tomonaga T, Krutzsch H, Avigan M, Levens D.** (1994) A sequence-specific, single-strand binding protein activates the far upstream element of c-*myc* and defines a new DNA-binding motif. *Genes Dev.* **8**: 465–480.

**Evans MJ, Kaufman MH.** (1981) Establishment in culture of pluripotential cells from mouse embryos. *Nature* **292**: 154–156.

**Foote S, Vollrath D, Hilton A, Page DC.** (1992) The human Y chromosome: overlapping DNA clones spanning the euchromatic region. *Science* **258**: 60–66.

**Frankel W, Potter TA, Rosenberg N, Lenz J, Rajan TV.** (1985) Retroviral insertional mutagenesis of a target allele in a heterozygous murine cell line. *Proc. Natl Acad. Sci. USA* **82**: 6600–6604.

**Fried M, Griffith M, Davies B, Bjursell G, Mantia GL, Lania L.** (1983) Isolation of cellular DNA sequences that allow expression of adjacent genes. *Proc. Natl Acad. Sci. USA* **80**: 2117–2121.

**Friedrich G, Soriano P.** (1991) Promoter traps in embryonic stem cells: a genetic screen to identify and mutate developmental genes in mice. *Genes Dev.* **5**: 1513–1523.

**Frohman MA, Dush MK, Martin GR.** (1988) Rapid production of full-length cDNAs from rare transcripts: amplification using a single gene-specific oligonucleotide primer. *Proc. Natl Acad. Sci. USA* **85**: 8998–9002.

**Galau GA, Britten RJ, Davidson EH.** (1974) A measurement of the sequence complexity of polysomal messenger RNA in sea urchin embryos. *Cell* **2**: 920–928.

**Goff SP.** (1990) Integration of retroviral DNA into the genome of the infected cell. *Cancer Cells* **2**: 172–178.

**Gossen JA, de Leeuw WJF, Molijin AC, Vijg J.** (1993) Plasmid rescue from transgenic mouse DNA using LacI repressor protein conjugated to magnetic beads. *Biotechniques* **14**: 624–629.

**Gossler A, Joyner AL, Rossant J, Skarnes WC.** (1989) Mouse embryonic stem cells and reporter constructs to detect developmentally regulated genes. *Science* **244**: 463–465.

**Green MC.** (1989) Catalog of mutant genes and polymorphic loci. In: *Genetic Variants and Strains of the Laboratory Mouse* (eds MF Lyon, AG Searle). Oxford University Press, Oxford, p. 12.

**Gridley T, Soriano O, Jaenisch R.** (1987) Insertional mutagenesis in mice. *Trends Genet.* **3**: 162–166.

**Group NCM.** (1992) A comprehensive genetic linkage map of the human genome. *Science* **258**: 148–162.

**Hamada H.** (1986) Activation of an enhancerless gene by chromosomal integration. *Mol. Cell. Biol.* **6**: 4179–4184.

**Hartmann SC, Mulligan RC.** (1988) Two dominant-acting selectable markers for gene transfer studies. *Proc. Natl Acad. Sci. USA* **85**: 8047–8051.

**Hawkins JD.** (1988) A survey of intron and exon length. *Nucleic Acids Res.* **16**: 9893–9908.

**Hiller S, Hengstler M, Kunze M, Knippers R.** (1988) Insertional activation of a promoterless thymidine kinase gene. *Mol. Cell. Biol.* **8**: 3298–3302.

**Hooper M, Hardy K, Handyside A, Hunter S, Monk M.** (1987a) HPRT-deficient (Lesch-Nyhan) mouse embryos derived from germline colonization by cultured cells. *Nature* **326**: 292–295.

**Hooper M, Hardy K, Handyside A, Hunter S, Monk M.** (1987b) Mouse embryonic stem cells exhibit indefinite proliferative potential. *J. Cell. Physiol.* **133**: 197–201.

**Hosler B, LaRosa G, Grippo J, Gudas L.** (1989) Expression of *REX-1*, a gene containing zinc finger motifs, is rapidly reduced by retinoic acid in F9 teratocarcinoma cells. *Mol. Cell. Biol.* **9**: 5623–5629.

**Hubbard S, Walls L, Ruley H, Muchmore E.** (1994) Generation of Chinese hamster ovary cell glycosylation mutants by retroviral insertional mutagenesis. *J. Biol. Chem.* **269**: 3717–3724.

**Innes CL, Smith PB, Langenbach R, Tindall KR, Boone LR.** (1990) Cationic liposomes (lipofectin) mediate retroviral infection in the absence of specific receptors. *J. Virol.* **64**: 957–961.

**Jaenisch R, Breindl M, Harbers K, Jahner D, Lohler J.** (1985) Retroviruses and insertional mutagenesis. *Cold Spring Harbor Symp. Quant. Biol.* **50**: 439–445.

**Jähner D, Jaenisch R.** (1985) Retrovirus-induced *de novo* methylation of flanking host sequences correlates with gene inactivity. *Nature* **315**: 594–597.

**Keller G, Kennedy M, Papayannopoulou T, Wiles M.** (1993) Hematopoietic commitment during embryonic stem cell differentiation in culture. *Mol. Cell. Biol.* **13**: 473–486.

**King W, Patel M, Lober L, Goff S, Nguyen-Huu M.** (1985) Insertion mutagenesis of embryonal carcinoma cells by retroviruses. *Science* **228**: 554–558.

**Kothary R, Clapoff S, Brown A, Campbell R, Peterson A, Rossant J.** (1988) A transgene containing *lacZ* inserted into the dystonia locus is expressed in the neural tube. *Nature* **335**: 435–437.

**Kozak M.** (1987) An analysis of 5′-noncoding sequences from 699 vertebrate messenger RNAs. *Nucleic Acids Res.* **15**: 8125–8148.

**Kozak M.** (1991) Structural features in eukaryotic mRNAs that modulate the initiation of translation. *J. Biol. Chem.* **266**: 19867–19870.

**Krowczynska A, Coutts M, Makrides S, Brawerman G.** (1989) The mouse homologue of the human acidic ribosomal phosphoprotein PO: a highly conserved polypeptide that is under translational control. *Nucleic Acids Res.* **17**: 6408.

**Landau NR, Littman DR.** (1992) Packaging system for rapid production of murine leukemia virus vectors with variable tropism. *J. Virol.* **66**: 5110–5113.

**Le Mouellic H, Lallemand Y, Brulet P.** (1992) Homeosis in the mouse induced by a null mutation in the *Hox-3.1* gene. *Cell* **17**: 251–264.

**Lewin B.** (1975) Units of transcription and translation: sequence components of heterogeneous nuclear RNA and messenger RNA. *Cell* **4**: 77–93.

**Lynch CM, Miller D.** (1991) Production of high-titer virus-free retroviral vectors by cocultivation of packaging cells with different host ranges. *J. Virol.* **65**: 3887–3890.

**Mandel JL, Monaco AP, Nelson DL, Schlessinger D, Willard H.** (1992) Genome analysis and the human X chromosome. *Science* **258**: 103–109.

**Mansour SL, Thomas KR, Deng C, Capecchi MR.** (1990) Introduction of a *lacZ* reporter gene into the mouse *int-2* locus by homologous recombination. *Proc. Natl Acad. Sci. USA* **87**: 7688–7692.

**Martin GR.** (1981) Isolation of a pluripotent cell line from early mouse embryos cultured in medium conditioned by teratocarcinoma stem cells. *Proc. Natl Acad. Sci. USA* **78**: 7634–7638.

**McKusick VA, Amberger JS.** (1993) The morbid anatomy of the human genome: chromosomal location of mutations causing disease. *J. Med. Genet.* **30**: 1–26.

**McMurray AJ, Shin HS.** (1992) The murine MHC encodes a mammalian homologue of bacterial ribosomal protein S13. Unpublished data; EMBL accession no. M76763.

**von Melchner H, Ruley HE.** (1989) Identification of cellular promoters by using a retrovirus promoter trap. *J. Virol.* **63**: 3227–3233.

**von Melchner H, Reddy S, Ruley HE.** (1990) Isolation of cellular promoters by using a retrovirus promoter trap. *Proc. Natl Acad. Sci. USA* **87**: 3733–3737.

**von Melchner H, DeGregori JV, Rayburn H, Reddy S, Friedel C, Ruley HE.** (1992) Selective disruption of genes expressed in totipotent embryonal stem cells. *Genes Dev.* **6**: 919–927.

**Muthuchamy M, Pajak L, Howles P, Doetschman T, Wieczorek D.** (1993) Developmental analysis of tropomyosin gene expression in embryonic stem cells and mouse embryos. *Mol. Cell. Biol.* **13**: 3311–3323.

**Nakamura T, Onno M.** (1990) Nucleotide sequence of mouse L19 ribosomal protein cDNA isolated in screening with *tre* oncogene probes. *DNA Cell Biol.* **9**: 697–703.

**Nomura N, Miyajima N, Kawarabayashi Y, Tabata S.** (1993) Prediction of new human genes by entire sequencing of randomly sampled cDNA clones. Unpublished data; Genbank accession no. D14812.

**Okamoto K, Okazawa H, Okuda A, Sakai M, Muramatsu M, Hamada H.** (1990) A novel octamer binding transcription factor is differentially expressed in mouse embryonic cells. *Cell* **60**: 461–472.

**Okubo K, Hori N, Matoba R, Niiyama T, Fukushima A, Kojima Y, Matsubara K.** (1992) Large scale cDNA sequencing for analysis of quantitative and qualitative aspects of gene expression. *Nature Genetics.* **2**: 180–185.

**Pathak VK, Temin HM.** (1990) Broad spectrum of *in vivo* forward mutations, hypermutations, and mutational hotspots in a retroviural shuttle vector after a single replication cycle: substitutions frameshifts, and hypermutations. *Proc. Natl Acad. Sci. USA* **87**: 6019–6023.

**Pear W, Nolan G, Scott M, Baltimore D.** (1993) Production of high-titer helper-free retroviruses by transient transfection. *Proc. Natl Acad. Sci. USA* **90**: 8392–8396.

**Pessin JE, Bell GI.** (1992) Mammalian facilitative glucose transporter family: structure and molecular regulation. *Annu. Rev. Physiol.* **54**: 911–930.

**Rabbitts TH, Forster A, Larson R, Nathan P.** (1993) Fusion of the dominant negative transcription regulator *CHOP* with a novel gene *FUS* by translocation t(12;16) in malignant liposarcoma. *Nature Genetics.* **4**: 175–180.

**Rappolee DA, Brenner CA, Schultz R, Mark D, Werb Z.** (1988) Developmental expression of PDGF, TGF-α and TGF-β in preimplantation mouse embryos. *Science* **241**: 1823–1825.

**Reddy S, DeGregori JV, von Melchner H, Ruley HE.** (1991) Retrovirus promoter trap vector to induce *lacZ* gene fusions in mammalian cells. *J. Virol.* **65**: 1507–1515.

**Reddy S, Rayburn H, von Melchner H, Ruley HE.** (1992) Fluorescence activated sorting of totipotent embryonic stem cells expressing developmentally regulated *lacZ* fusion genes. *Proc. Natl Acad. Sci. USA* **89**: 6721–6725.

**Reith AD, Berstein A.** (1991) Molecular basis of mouse developmental mutants. *Genes Dev.* **5**: 1115–1123.

**Rice CM, Fuchs R, Higgins DG, Stoehr PJ, Cameron GN.** (1993) The EMBL data library. *Nucleic Acids Res.* **21**: 2967–2971.

Robertson E, Bradley A, Kuehn M, Evans M. (1986) Germ-line transmission of genes introduced into cultured pluripotential cells by retroviral vector. *Nature* **323:** 445–448.

Rogers MB, Hosler BA, Gudas LJ. (1991) Specific expression of a retinoic acid regulated, zinc-finger gene, *Rex-1*, in preimplanation embryos, trophoblast and spermatocytes. *Development* **113:** 815–824.

Rohdewohld H, Weiher H, Reik W, Jaenisch R, Breindl M. (1987) Retrovirus integration and chromatin structure: Moloney murine leukemia proviral integration sites map near DNase I-hypersensitive sites. *J. Virol.* **61:** 336–343.

Rossant J, Hopkins N. (1992) Of fin and fur: mutational analysis of vertebrate embryonic development. *Genes Dev.* **6:** 1–13.

Rudnicki M, Schnegelsberg P, Stead R, Braun T, Arnold H, Jaenisch R. (1993) *MyoD* or *Myf-5* is required for the formation of skeletal muscle. *Cell* **75:** 1351–1359.

Sablitzky F, Jönsson JI, Cohen BL, Phillips RA. (1993) High frequency expression of integrated proviruses derived from enhancer trap retroviruses. *Cell Growth Differen.* **4:** 451–459.

Sandmeyer SB, Hansen LJ, Chalker DL. (1990) Integration specificity of retrotransposons and retroviruses. *Annu. Rev. Genet.* **24:** 491–518.

Santangelo GM, Tornow J, Moldave K. (1986) Cloning of open reading frames and promotors from the *Saccharomyces cerevisiae* genome: construction of genomic libraries of random small fragments. *Gene* **46:** 181–186.

Santangelo GM, Tornow J, McLaughlin CS, Moldave K. (1988) Properties of promotors cloned randomly from the *Saccharomyces cerevisiae* genome. *Mol. Cell. Biol.* **8:** 4217–4224.

Scherdin U, Rhodes K, Breindl M. (1990) Transcriptionally active genome regions are preferred targets for retrovirus integration. *J. Virol.* **64:** 907–912.

Schmitt R, Bruyns E, Snodgrass H. (1991) Hematopoietic development of embryonic stem cells *in vitro*: cytokine and receptor gene expression. *Genes Dev.* **5:** 728–740.

Schneider-Maunoury S, Topilko P, Seitanidou T, Pournin S, Babinet C, Charney P. (1993) Disruption of *Krox-20* results in alteration of rhombomeres 3 and 5 in the developing hindbrain. *Cell* **75:** 1199–1214.

Scholer HR, Dressler GR, Balling R, Rohdewohld H, Gruss P. (1990) *Oct-4*: a germline-specific transcription factor mapping to the mouse t-complex. *EMBO J.* **9:** 2185–2195.

Shapiro JA. (1983). *Mobile Genetic Elements*. Academic Press, New York.

Sharpe NG, Williams DG, Latchman DS. (1990) Regulated expression of the small nuclear ribonucleoprotein particle protein SmN in embryonic stem cell differentiation. *Mol. Cell. Biol.* **10:** 6817–6820.

Shih CC, Stoye JP, Coffin JM. (1988) Highly preferred targets for retrovirus integration. *Cell* **53:** 531–537.

Skarnes WC, Auerbach BA, Joyner AL. (1992) A gene trap approach in mouse embryonic stem cells: the *lacZ* reporter is activated by splicing, reflects endogenous gene expression, and is mutagenic in mice. *Genes Dev.* **6:** 903–918.

Snell RG, Doucette SL, Gillespie KM, Taylor SA, Riba L, Bates GP, Altherr MR, MacDonald ME, Gusella JF, Wasmuth JJ. (1993) The isolation of cDNAs within the Huntington disease region by hybridisation of yeast artificial chromosomes to a cDNA library. *Hum. Mol. Genet.* **2:** 305–309.

Soriano P, Montgomery C, Geske R, Bradley A. (1991) Targeted disruption of the c-*src* proto-oncogene leads to osteopetrosis in mice. *Cell* **64:** 693–702.

Tajbakhsh S, Buckingham ME. (1994) Mouse limb muscle is determined in the absence of the earliest myogenic factor *myf-5*. *Proc. Natl Acad. Sci. USA* **91:** 747–751.

Thompson S, Clarke AR, Pow AM, Hooper ML, Melton DW. (1989) Germ line transmission and expression of a corrected *HPRT* gene produced by gene targeting in embryonic stem cells. *Cell* **56:** 313–321.

Tsutsui K. (1993) Identification and characterization of a nuclear scaffold protein that binds the matrix attachment region DNA. *J. Biol. Chem.* **268:** 12886–12894.

Varmus H. (1988) Retroviruses. *Science* **240:** 1427–1435.

**Varmus HE, Quintrell N, Ortiy S.** (1981) Retroviruses as mutagens: insertion and excision of a non-transforming provirus alters expression of a resident transforming provirus. *Cell* **25**: 23–36.

**Wang KS, Kuhn RJ, Strauss EG, Ou S, Strauss JH.** (1992) High-affinity laminin receptor is a receptor for Sindbis virus in mammalian cells. *J. Virol.* **66**: 4992–5001.

**Weber F, Villiers J, Schaffner W.** (1984) An SV40 "enhancer trap" incorporates exogenous enhancers or generates enhancers from its own sequences. *Cell* **36**: 983–992.

**Wells D, Hoffman D, Kedes L.** (1987) Unusual structure, evolutionary conservation of non-coding sequences and numerous pseudogenes characterize the human H3.3 histone multigene family. *Nucleic Acids Res.* **15**: 2871–2889.

**Wiles M, Keller G.** (1991) Multiple hematopoietic lineages develop from embryonic stem (ES) cells in culture. *Development* **111**: 259–267.

**Wilson C, Kurth-Pierson R, Bellen HJ, O'Kane CJ, Grossniklaus U, Gehring WJ.** (1989) P-element-mediated enhancer detection: an efficient method for isolating and characterizing developmentally regulated genes in *Drosophila*. *Genes Dev.* **3**: 1301–1313.

**Withers-Ward ES, Kitamura Y, Barnes JP, Coffin JM.** (1994) Distribution of targets for avian retrovirus DNA integration *in vivo*. *Genes Dev.* **8**: 1473–1487.

**Wool I, Chan Y, Paz V, Olvera J.** (1990) The primary structure of rat ribosomal proteins: the amino acid sequences of L27a and L28 and corrections in the sequences of S4 and S12. *Biochim. Biophys. Acta* **1050**: 69–73.

**Yagi T, Shigetani Y, Okado N, Tokunaga T, Ikawa Y, Aizawa S.** (1993) Regional localization of Fyn in adult brain; studies with mice in which *fyn* gene was replaced by *lacZ*. *Oncogene* **8**: 3343–3351.

**Yang YS, Hanke JH, Carayannopoulos L, Craft CM, Capra JD, Tucker PW.** (1993) *NonO*, a non-POU-domain-containing, octamer-binding protein, is the mammalian homolog of *Drosophila nonAdiss*. *Mol. Cell. Biol.* **13**: 5593–5603.

# 6

# Gene transfer studies

David Darling and Marcel Kuiper

## 6.1 Introduction

Transfection studies are virtually obligatory if one is to begin to understand the mode of action of newly cloned genes. This is as true for the human genome as for the genome of any other organism. The reliance on this type of study is particularly acute for human gene analysis, because many types of experiment regularly performed in animal models cannot be done with humans. In this chapter, we shall explore the concept of the transfection of animal cells in the context of a more detailed discussion of the appropriate technology for different experiments. Also included are the choices of methods, plasmids, selection procedure and assays for new protein synthesis.

## 6.2 What is transfection?

Transfection, or gene transfer, covers a popular family of techniques. It involves applying genetic material (in the form of DNA) in such a way that it enters the recipient cell nucleus, is transcribed into mRNA and, in most cases, subsequently translated into protein. For convenience, we can consider this process to consist of a number of separate, although related, events:

(i) the DNA must cross the cell plasma membrane which separates the cytosol from the culture medium;

(ii) the DNA must either cross the membrane of the nuclear envelope, or enter when the nuclear envelope is absent (i.e. in the M phase of the cell cycle);

(iii) having survived this journey, the DNA must be sufficiently intact and contain the correct control sequence(s) in order to allow the cellular transcription apparatus to recognize and transcribe the DNA.

Once all of the above have been achieved, these techniques can be used to examine the factors and conditions involved in the control of transcription and the function/effect of the translated gene products. Additionally, cell lines can be created that have acquired new genes which produce protein in large quantities. Such cell lines may also facilitate the identification of the function of proteins encoded by novel complementary DNA (cDNA) clones. This process immediately presents us with difficult choices and technical problems. For instance, what form should the DNA be in, how can we be sure that it will be transcribed, how can it be maintained in the recipient cells and how can one monitor whether any, or all, of the above has actually taken place?

## 6.3 What form should the DNA be in?

On entering the cell nucleus, the transfected DNA can either integrate (i.e. become incorporated into the genomic DNA) or remain episomal (not integrate into the genome). Whether or not transfected DNA integrates is often merely a function of the time course of the experimental protocol. Short time course experiments involving the transfection of DNA, and the assaying for a gene product within 48–72 h, are often known as *transient assays*. In these cases, the emphasis must be upon optimizing the efficiency with which the DNA will enter the cell nucleus (whilst minimizing the toxicity of the transfection protocol) without regard to the way in which the DNA may integrate.

Since the time course of the culture is short in transient assays, there is no need for selection to be applied. This means that the vector used to carry the gene of interest need not include a selectable marker gene. By contrast, longer term experiments involving integration of the plasmid into the host genome (i.e. stable transfection) require either that the vector contains a selectable marker or, more commonly, that the transfection be done with one plasmid containing the gene of interest in tandem with another containing the selectable marker (co-transfection).

DNA transfected into cells can come in a number of different guises. It can be genomic DNA, that is DNA extracted from cells and cut into fragments with restriction enzymes. This DNA will usually be unidentified, and cells that take up DNA will normally have to be selected on the basis of a change in the phenotype of the transfected cell. This method was used to clone oncogenic DNA sequences from tumour cells (Goldfarb *et al.*, 1982). This DNA will have no independent means of replication, and thus must become integrated into the host genome if it is to be retained in the progeny of the initially transfected cell.

The vast majority of transfections using DNA will, however, be with plasmids (or related vectors called cosmids). These DNA molecules have a number of characteristic inclusions that control different parts of their 'life cycle'. One such inclusion is a gene which confers antibiotic resistance to their bacterial hosts [usually the antibiotics ampicillin (*amp*) or tetracycline (*tet*), although kanamycin (*kan*) and neomycin (*neo*) are also used]. Bacteria containing the plasmid can thus be selected; consistent selection pressure with these antibiotics ensures the maintenance of the plasmid in the bacterial population. In the absence of selection, the bacteria may discard the plasmid. Since these bacteria carry a reduced 'genetic load', they tend to outgrow those which retain such a plasmid.

Plasmids do not need to carry their own replicative genes, but they do need to contain a bacterial origin of replication (e.g. ColE1 *ori*). This small nucleotide sequence allows a plasmid to replicate inside bacterial cells using their own replicative proteins and precursor molecules.

As far as transfections are concerned, cosmids are very much the same as plasmids. The major differences are the presence of bacteriophage λ *cos* sites which increase the efficiency of ligation of inserts, and bacteriophage packaging that allows the efficient transformation of bacterial hosts with vectors containing relatively large inserts (30–50 kb). Many of these vectors also carry sequences which potentiate their replication in eukaryotic cells. This means that in some cases, a whole gene, in its genomic organization complete with *cis*-acting control sequences can be transfected into target cells.

The above are the only requirements as far as the bacterial part of the life cycle is concerned. However, once the DNA has entered the target animal cell nucleus, the requirements are entirely different and are very much dependent on the objective of the study and the type of experiment to be performed.

## 6.4 Generalized requirements for eukaryotic gene transcription

Once foreign DNA has entered the nucleus, how can it replicate in the host cell? As previously pointed out, short-term transient experiments do not tend to pose this problem. For an experimental system involving long-term growth (more than 2 weeks), there are two choices: selection for stable integration or episomal maintenance. Stable integration works on the basis that the DNA, on entering the nucleus, can sometimes integrate into the DNA of the host genome. This is a relatively rare event, but can be selected for on the basis of the transfected DNA containing a gene, the protein product of which confers resistance to a drug to which the cells are otherwise sensitive. The recipient cell with the integration can then be selected for, and identified from, a background of cells that do not have the integrated plasmid (selection procedures and alternatives such as reporter genes will be dealt with later). This works because the transfected plasmids have been engineered to contain sequences which are required to promote the expression of the drug selection gene.

### *6.4.1 Eukaryotic gene transcription*

Even if a plasmid has been designed with an incorporated selection gene (e.g. hygromycin B phosphotransferase, the expression of which confers resistance to hygromycin), it will be of no use unless it can be transcribed and translated by the host/target cell. Thus the plasmid must contain, in the correct position, the basic recognition sequences for the cell to assemble the transcription complexes. In the first instance, a lead has been taken from virus species that infect animal cells. Many of the most commonly used promoters come from viruses such as simian virus 40 (SV40; plasmid pSV2), cytomegalovirus (pCMV), Rous sarcoma virus (pRSV) and murine mammary tumour virus (pMMTV). These viral promoter/enhancer sequences, cloned immediately upstream of the gene of interest, will initiate the transcription complexes required for the adjacent DNA to be transcribed into RNA.

### 6.4.2 SV40-based plasmids

One of the earliest and best characterized SV40-based plasmids is pSV2-neo (Southern and Berg, 1982; Figure 6.1). This plasmid has been based on the structure and life cycle of the SV40 virus and has its root in an evolution culminating in the plasmids pSV2-βG and pSV2-gpt (Mulligan and Berg, 1981). Based on the bacterial plasmid pBR322, it contains an ampicillin resistance gene (*amp$^r$*) and the bacterial origin of replication (ColE1), with three cassette additions from the SV40 virus, flanking the rabbit β-globin gene.

The SV40 promoter/enhancer consists of a complex composed of a 342 bp fragment of SV40 containing the early promoter, a 27 bp palindromic repeat comprising the viral origin of replication (*ori*) and, additionally, three copies of the 21 bp G–C rich sequence and two copies of the 72 bp sequence which together act as enhancer elements and contain binding sites for transcription factors. The orientation is such that the early promoter can direct the transcription of genes cloned downstream (i.e. 3') of the promoter. The rabbit β-globin gene has been included as a cDNA clone containing its own translation initiation codon (ATG) and termination codon (TGA) and placed directly 3' of the SV40 early promoter. Thus mRNA transcripts are initiated in the SV40 *ori* sequence, and read through the β-globin gene and the (translation) stop codon. The intron of the SV40 gene encoding the small t antigen (66 bp in length) has been placed immediately 3' to the above, thereby providing a splice donor and splice acceptor. This intron is thus spliced out in the mature mRNA and was included for the reason that virtually all eukaryotic genes contain introns (the rabbit β-globin cDNA, of course, does not). It has been reasoned that the presence of the intron and the splicing may contribute

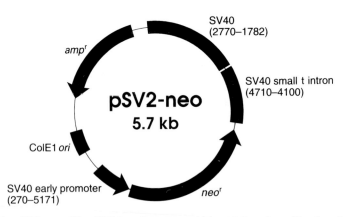

**Figure 6.1.** pSV2-neo. The SV40 early promoter containing the *ori* is placed upstream of the *neo* gene, with the direction of transcription as indicated. The numbers 270–5171 refer to the original designation on the SV40 virus, and the numbers flank the *ori* which was designated as position 1. The SV40 small t intron is only 66 bases within the 610 bp fragment indicated. The SV40 poly(A) signal AATAAA is located within the 988 bp fragment indicated.

to RNA stability. The SV40 poly(A) signal, present within by a 988 bp SV40 DNA fragment, has been placed 3' to the small t intron. This sequence, a 237 bp element, contains a G–T or T rich element followed by the AATAAA motif that ensures the addition of the poly(A) tail to indicate messenger status (m) of the mRNA.

This modular construction contains all the basic requirements to allow most animal cells to transcribe and translate the β-globin gene once the DNA plasmid enters the nucleus of the cell. Cutting this plasmid with the restriction enzymes *Hin*dIII and *Bgl*II linearizes the circular plasmid DNA and liberates the β-globin gene. This allows the replacement of β-globin with other genes which can then be expressed in animal cells. Since low transfection efficiency has always been a problem, the ability to select the cells with the integrated plasmid was an essential further modification. The plasmid pSV2-gpt (Mulligan and Berg, 1981) expressed the bacterial gene guanine phosphoribosyl transferase in place of the β-globin, thus allowing the selection of the transfected cells by virtue of their resistance to mycophenolic acid (see Section 6.8.1). A 1.4 kb fragment of the bacterial transposon 5 (Tn*5*) gene, neomycin phosphotransferase (*neo*) containing its own ATG and some upstream sequences, replaced the β-globin to produce the pSV2-neo plasmid (Southern and Berg, 1982) which confers resistance to the antibiotic geneticin G418 (see Section 6.8.1).

## 6.5 Specialized eukaryotic host cells

Plasmids based on pSV2 have an added advantage in that in certain types of specialized cells they can replicate without a requirement for integration into the host cell genome. The oldest, best characterized, and probably the favourite, of these is the COS (CV-1 origin, SV40) cell system (Gluzman, 1981). These cells are derived from the African green monkey kidney cell line CV-1 transfected with a plasmid containing an *ori*-defective SV40 genome (6 bp deleted from the *ori*, blocking viral DNA replication but not gene transcription). These cells thus contain the normal complement of virus-derived proteins but are unable to promote the replication of the resident SV40 owing to the *ori* sequence deletion. However, when transfected with plasmids possessing an intact SV40 origin of replication, the presence of the SV40 proteins required for replication (specifically large T) allows the replication of the transfected plasmid without the need for its integration into the genome.

In this system, high levels of expression of the encoded genes can be obtained because multiple copies of the plasmid can be maintained in each transfected cell. The presence of the SV40 genome (especially the high level of large T expression) confers high expression levels from the SV40 promoter/enhancer complex. These cell/plasmid combinations are a popular choice for many studies requiring high levels of transcription in a transient assay system. Having once transfected COS cells with an SV40-based plasmid DNA library, clones that contain the desired gene product (e.g. a newly identified gene of interest) can then be processed to recover the episomal plasmid (Hirt, 1967) and used to transform bacteria for further amplification and characterization of the inserts they contain. This particular episomal maintenance only works in COS cells (or other SV40 large T-expressing cells) so it is a little limited, especially since the cells accumulate so many plasmid copies that they die 4–5 days post-transfection (Sambrook *et al.*, 1989).

## 6.6 Specialized plasmids

Plasmids that can only accommodate one insert at a time have restrictions on their use, especially when they are to be used for transfection where the aim is to select stable integrates. Taking pSV2-neo as the example, this plasmid's only ability is to confer resistance to G418 on the recipient cell. What then is the prospect for introducing other genes, for instance those that will not confer a novel resistance, into recipient cells? Genes of interest may only be in the minority in a cDNA library of transfectable plasmids, and they may not result in an easily identifiable change in cell morphology. Additionally they could confer a growth disadvantage to transfected cells over those that did not take up the DNA into their genomes. In many transfection studies, this was overcome by a technique known as co-transfection. In this scenario, the gene of interest was cloned into plasmids similar to pSV2-neo, but with the novel gene in place of *neo*. This plasmid was then transfected into the target cells along with a second plasmid that did contain a selectable marker gene (e.g. pSV2-neo). When the rare integration events did take place, it was also rare for only one plasmid to be involved, and often multiple copies were integrated at the same time. Thus if the non-selected plasmid was in excess (usually 5- to 20-fold), cells resistant to the selectable marker plasmid would more than likely also have integrated the gene of interest, and usually at the same site(s). Selection for resistant cells thus allows the potential gene of interest to 'piggy-back' on the selection.

The inherent advantage of this approach is that the same selection plasmids can be used for most transfections, and that plasmids with other genes can usually be used without further modification. The major disadvantage is that although the selection plasmid can be retained, the unselected plasmid is often lost during the subsequent replication of the cell, particularly if it confers no growth advantage upon the host. Other disadvantages include the amount of DNA required, the fact that the co-transfection does not always work, and that DNA quality cannot be guaranteed for different plasmid preparations so that the ratio of DNA added may not always reflect the ratio of high quality DNA.

### 6.6.1 Double insert plasmids

Plasmids are now readily available that allow more than one insert to be accommodated at the same time. These plasmids represent a technological development using knowledge gleaned from earlier generations of plasmids. Simply explained, they have all the requirements for their expression in the target cells *twice* over. Thus the selected gene will have its own promoter (e.g. SV40) with the requisite ATG (translation start codon), TGA (translation stop codon) and poly(A) signal whilst a second promoter (e.g. CMV) will be present at another location in the same plasmid with an additional ATG, TGA and poly(A) signal. Examples of such plasmids are given in the catalogues of various commercial companies [e.g. Invitrogen pRc/RSV, which has the *neo* gene under the control of the same SV40 sequences as in pSV2-neo, with an additional empty cloning site 3' to the RSV long terminal repeat (LTR) which acts as the promoter. The multiple cloning site (MCS) is then coupled with the bovine growth hormone transcription termination sequences and poly(A) signal]. Also available, from Clontech, is pMAMneo, supplied with SV40 early promoter-driven *neo* and an empty MCS positioned 3' to the MMTV LTR, which acts as a promoter and is situated 5' to a second SV40 splice and poly(A) signal. Since the transcriptional activity of the MMTV LTR is regulated by glucocorticoids, this construct acts as a dexamethasone-inducible promoter (Lee *et al.*, 1981). Plasmids such as these allow genes of interest to be selected for indirectly, whilst only one plasmid need be transfected.

The major disadvantages of these plasmids is their increased size (pMAMneo 8.4 kb, pRc/RSV 5.2 kb; remember that the gene of interest has yet to be added!), which tends to reduce both their replication in bacteria and their efficiency of transfection into eukaryotic hosts. These constructs are becoming more commonly used since the selection of drug-resistant cells almost always allows the selection of the second insert. Transfection protocols can now be optimized to generate single inserts rather than the multiple array inserts required in the past.

### 6.6.2 Inducible expression

What can be done if the gene of interest is suspected of being cytotoxic, or inhibitory to growth? These genes would not be fully represented in a transfected library since the cells would not grow in culture. One way around this is to use inducible promoters which are normally silent but can be activated under the appropriate conditions. The archetypal promoter of this type is the MMTV LTR (Lee *et al.*, 1981). When this promoter/enhancer sequence was cloned into a plasmid, it was determined that transcription of the downstream insert could be induced 5- to 10-fold over that of the background level of expression by the addition of the glucocorticoid agonist dexamethasone. This construct, either alone in a plasmid (for co-transfection) or in a double insert plasmid [as in pMAMneo or pMSG (Pharmacia)] allows one to switch on (and switch off) the transcription of the gene under its control at will. Thus, genes that may be potentially inhibitory to growth can be better represented in the cell transfection. One obligatory requirement for the MMTV LTR is that the target cells be glucocorticoid responsive (i.e. they must have the correct receptor) which may be a problem with some cells. Another potential pitfall is the presence of phenol red (a weak glucocorticoid agonist) in the medium, or glucocorticoids in the serum. These may well contribute to the 'leaky' nature of this promoter, which is weakly active even in the absence of additional dexamethosone. These types of problems, which also affect other promoters such those containing retinoic acid (RA)-responsive elements, have led to the search for alternatives. One which has proved popular is the mouse metallothionein promoter (Stuart et al., 1985) which can be induced by heavy metals such as cadmium. Another, which is claimed to have a basal level of expression as low as only 10–20 molecules/cell, is based on the repressor of the bacterial *lac* operon (Stratagene LacSwitch). In this case, the *lac* repressor protein has been cloned into a plasmid (p3'SS) under the control of the F9-1 polyoma promoter, and selection is present as hygromycin resistance ($hygro^r$). This plasmid, when integrated, stably expresses the *lac* repressor protein. The gene of interest is cloned into another plasmid (pOPRSVI CAT or pOP13 CAT) under the control of the RSV promoter with *lac* operator sequences intervening between promoter and insert. The plasmid selection in this case is $neo^r$. This plasmid is then transfected into cells expressing the plasmid-borne repressor protein. Transfectants can be selected for by resistance to G418, but will not express the gene under the control of the *lac* operator-interrupted RSV promoter. After addition of isopropyl thiogalactosidase (IPTG), the *lac* operator protein is competed for by its binding to the IPTG molecule, and transcription from the RSV promoter can then proceed. The concentration of IPTG that allows transcription of this RSV promoter is not itself toxic to the eukaryotic transfected cells. Of course this is always a potential problem and one must ensure (as with any inducible vector) that the effect of the novel (transfected) gene is not confused with phenotypic changes induced in non-transfected cells either by induced expression of endogenous genes, or by non-specific cytotoxic events.

### 6.6.3 Epstein–Barr virus-based plasmids

Episomal replication based on the SV40/COS1 system is an extremely useful tool (Gluzman, 1981). However, it is limited to the cells that express SV40 large T antigen. More flexible methods that allow episomal replication in other cells types are now available. Epstein–Barr virus (EBV)-based plasmids were initially constructed in order to allow episomal replication and expression of foreign genes in human B lymphocytes. The advantages of these plasmids were that the episomal replication allowed an increase in the transfection efficiency and that the plasmids possessed a natural ability to infect and replicate in human B lymphocytes. The original construct pHebo (Sugden et al., 1985) was only capable of replicating episomally in cells expressing the EBV nuclear antigen 1 (EBNA-1). This was because the origin of replication (*oriP*) within pHebo required trans-activation by the EBNA-1 protein, which limited recipient cells to those that were EBNA-1 positive (e.g. DAUDI Burkitt lymphoma cells). It was also found that the plasmid was maintained efficiently in these cells with a copy number of 1–60 per cell, which meant that these plasmids tended not to kill the cell (unlike the

COS1/SV40 system). Subsequent vectors included the *EBNA-1* gene; these plasmids could therefore trans-activate their own replication and thus be used in EBNA-1-negative cells (Yates *et al.*, 1985). These vectors can be used in a variety of cells from species as diverse as human, monkey and dog, but *not* rodent cells (Yates *et al.*, 1985). Vectors based on these constructs are now commercially available from suppliers such as Invitrogen (pREP10, Figure 6.2). These can contain a truncated *EBNA-1* gene which can still trans-activate the *ori*P. They also contain a dominant selectable marker driven by the thymidine kinase (TK) promoter, and an empty MCS under the control of either CMV or RSV promoters. These plasmids combine episomal replication with the advantage of double inserts.

Bovine papilloma virus (BPV)-based plasmids also replicate episomally in most mammalian cell lines (Lowry *et al.*, 1980). In one construct (pBPV; Pharmacia), BPV sequences are included that not only allow episomal replication but also result in a transformed phenotype (focus formation) which can be used as a dominant selectable marker. An MCS, flanked by the inducible mouse metallothionein I promoter, is also included. One problem with this system is that the morphological changes induced by the BPV may mask the effect of the gene of interest. This plasmid tends to be inefficiently maintained in human cells although it can be used in some rodent cell lines.

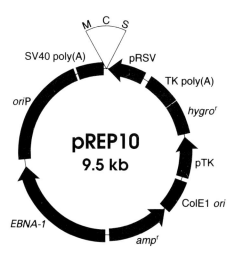

**Figure 6.2.** pREP10. pREP10 includes the *ori*P, the EBV origin of replication, which is trans-activated by the *EBNA-1* gene also present; this pairing allows episomal replication of the plasmid. The requirements for bacterial replication and selection are present ($amp^r$ and ColE1 *ori*). Hygromycin resistance is promoted by the TK promoter and polyadenylated by the TK gene poly(A) signal; direction of transcription is as indicated. A cloning site is available between the RSV promoter and the SV40 poly(A) signal. This plasmid can thus contain the selection gene and the gene of interest and be maintained episomally.

### 6.6.4 Shuttle vectors

A shuttle vector is a vector that can be used to carry an insert between different organisms. All the plasmids so far discussed contain some of the features required of a true shuttle vector. The inclusion of the bacterial origin of replication allows their propagation in bacteria, and the $amp^r$ allows selection for bacterial transformation. The inclusion of sequences that allow episomal replication and expression in eukaryotic cells allows these plasmids to be replicated in these cells whilst at the same time expressing their encoded gene. The episomal replication allows intact circular plasmids to be recovered from the transfected cells by Hirt extraction (Hirt, 1967) so that they can then be shuttled back into bacteria for further characterization. Thus a shuttle vector is represented by an SV40-based plasmid in COS cells.

### 6.6.5 Multifunctional plasmids

More and more functions can now be included in a single vector but, as yet, no single vector incorporates all the functions that would ideally be required. Cosmids for instance can be packaged in bacteriophages, propagated in bacteria, and expressed in eukaryotic cells or episomally maintained in COS1 cells (i.e. Stratagene pWE15), but they have no eukaryotic expression promoters other than those that may be present in the genomic insert. Modified λ phage (such as DR2) cDNA libraries are a useful alternative; these can be plated as normal phage plaques, and screened as a standard library. In addition, an insert can be rescued from the phage as a eukaryotic expression plasmid containing the selected gene sequence under the control of the RSV LTR alongside dominant selection under the herpes simplex virus thymidine kinase (HSV-TK) promoter. These features are coupled with the EBV *oriP*, allowing its episomal replication in EBNA-1-expressing cells. The ColE1 *ori* and $amp^r$ also allow this plasmid to replicate and be selected for in *E. coli*. Note that Invitrogen's EBV-*cre/lox* system does essentially the same job by means of a P1 recombinase-induced recombination of two *lox*P sites to produce a circular plasmid pPOP6 that also includes the *EBNA-1* gene.

Other plasmids include all that is required for the bacterial and eukaryotic 'life cycle' and in addition contain the F1 *ori* (or the M13 *ori*) for single-stranded DNA rescue, and either T3, T7 or SP6 RNA polymerase promoters for the *in vitro* production of RNA for use as single-stranded RNA probes.

All these modifications are designed to reduce the labour involved in moving a DNA sequence into a host that optimizes further analysis. As yet, no shuttle vector contains all the desirable features but the technology is moving towards that position. In the meantime, probably the least efficient part of the whole process is the actual transfection of the naked DNA into the recipient cells.

## 6.7 Transfection procedures

The number of available techniques for the introduction of foreign genes into eukaryotic cells has steadily increased and grown in diversity. There are, however, still a number of basic techniques which remain the most commonly used. Added to these are some more recently described techniques that are very promising and have advantages in some cases. For all these methods, however, there are still some basic rules to follow.

The only major requirements for the DNA to be transfected are that it should be clean and of good quality, with an optical density ratio at 260:280 nm of 1.6–1.8. The presence of contaminants such as protein, phenol, bacterial genomic DNA and endotoxins will tend to reduce the effectiveness of transfection (Ehlert et al., 1993). In addition, if the DNA is degraded then it will not be able to integrate, or be transcribed, in recipient cells. Many workers find that with plasmid DNA, a high degree of supercoiling increases the effectiveness of transfection. This could be for a number of reasons; for instance the volume occupied by a supercoiled plasmid is smaller than that required by a nicked plasmid which means the supercoiled version may present a smaller complex to traverse the membrane. Additionally, tightly coiled DNA may be more resistant to residual endonuclease activity, and may also travel better within the cell. Alternatively, in electroporation, it has been determined that linearization of the plasmid can increase the efficiency of transient transfection (Toneguzzo et al., 1986). However, if episomal plasmids were linearized, they would be non-functional unless they could recircularize within the cell.

In general, as the size of the plasmid increases, the potential efficiency of transfection decreases, and for a long time this has imposed a limit on the size of the DNA fragment that can be transfected; cosmids, for example, can be difficult to transfect. The health and density of the recipient cells are also of critical importance to the transfection, and the cells must be proliferating. There are many reasons why this is the case but the most critical is probably that the DNA can best gain access to the nucleus when the nuclear envelope is absent (i.e. during mitosis).

### 6.7.1 Calcium phosphate co-precipitation

The calcium phosphate method of transfecting DNA was published in 1973 by Graham and Van der Eb and is the oldest method of experimental gene transfer. This method depends on the formation of a DNA/calcium phosphate co-precipitate which then settles on, and is taken up by, the target cells. The mechanisms by which this takes place are not well understood, but for a long time this was the only reliable method for transfecting adherent cells to produce stable integrants. The suspicion is that precipitation of the DNA can protect it from degradation by endogenous nucleases so that the DNA can access the nucleus intact. With the relatively low intracellular $Ca^{2+}$ concentration, the co-precipitate breaks down and the DNA can then be transcribed. Modifications of this technique have been adapted for suspension cells (Chu and Sharp, 1981) but without spectacular results. Other additions to the method include a glycerol (Parker and Stark, 1979), dimethyl sulphoxide (DMSO; Lopata et al., 1984) or chloroquine (Luthman and Magnusson, 1983) shock. The first two are thought to work because the chemicals are trans-membranal and are assumed to carry the DNA across the membrane. Chloroquine is thought to work because it increases the pH of lysosomes in the cell, thereby reducing the intracellular DNA degradation.

The efficiency of the transfection is based largely on the quality of the precipitates. The finer the precipitate the better, since it is less cytotoxic and is taken up more effectively. This in turn is to some extent dependent on the pH at which the transfection is performed. The HEPES [$N$-(2hydroxyethyl)piperazine-$N'$-(2-ethanesulphonic acid)] buffer (Graham and Van der Eb, 1973), can be replaced by BES [$N,N$-bis(2-hydroxyethyl)-2-aminoethanesulphonic acid; Chen and Okayama, 1987] in a protocol which is both labour saving and often more efficient. The precipitation in this case takes place in the medium in which the cells are cultured over a period of 18–24 h and provides a very fine precipitate. With the HEPES method, the precipitate was made before addition to the cells and it can be very difficult to get the balance right between the precipitate forming reasonably quickly and it not being too crude. The only caveat with the BES method is that the protocol takes account of the volume of medium that is added. Therefore, if the volume of medium in the culture is either decreased or increased, then the amount of DNA, BES and calcium chloride must be altered correspondingly so that the DNA does not come out of solution. Attention must also be paid to the calcium concentration in the medium because, in low calcium, the DNA may go back into solution.

### 6.7.2 DEAE–dextran

Whereas calcium phosphate-mediated transfection is a favoured method for adherent cells, DEAE (diethylaminoethyl)–dextran transfection (Somapayrac and Danna, 1981) provides a technique often used for the transient transfection of suspension cells. There are now many modifications to the original protocol (Sambrook et al., 1989; Ausubel et al., 1994) which was first published as a method for improving the efficiency of SV40 infection (McCutchan and Pagano, 1968). This method is usually accompanied by a DMSO or glycerol shock (Lopata et al., 1984).

Although the method of uptake is poorly understood, the transfection depends on the association of DNA with high molecular weight DEAE–dextran (a long chain polycation). The DEAE molecular weights most commonly used range from $5 \times 10^5$ up to $2 \times 10^6$. This molecule can be cytotoxic, and pilot studies on its effects on target cells should be carried out around an optimum of 200 µg ml$^{-1}$. During the course of incubation, the DNA/DEAE–dextran complex is thought to adhere to the cell surface, thus associating the DNA with the cell. A modification introduced to reduce toxicity involved the pre-incubation of cells with DEAE–dextran followed by a washing step, with subsequent addition of DNA to the culture medium (Holler et al., 1989). This presumably works by DEAE–dextran adhering to the surface of the cell with the DNA in the medium then adhering to the DEAE–dextran on the cell surface. Taken together with the demonstration that efficiency can be substantially enhanced by removing adherent cells from the substratum (Golub et al., 1989), it may be that future modifications will make this method even more efficient.

## 6.7.3 Electroporation

Electroporation is amongst the favourite methods for those who find their suspension cells difficult to transfect, but it can be performed on both adherent and suspension cells (Neumann *et al.*, 1982; Potter *et al.*, 1984). Cells are suspended in a buffer containing DNA, and the voltage applied opens pores in the cell membrane (the membrane also loses its polarity) which then enables the DNA to enter the cell. The design of the electroporation equipment is constantly evolving. Initially a spike pulse was delivered, although in subsequent models this tended to be modified with the delivered pulse as a square wave discharge. In the latest models, some have resumed the square wave technique with the further option of two shocks, one short with a high voltage pulse and the second, a long low voltage. In theory, the first creates the pores, and the second moves the DNA via the pores into the cells.

Opinion is fluid regarding the optimum buffer preparation. Recommendations include growth medium (with or without serum), phosphate-buffered saline (PBS) and mannitol. In order to reduce toxicity, buffers were originally designed to mimic the conditions found inside the cells along with non-conductive buffers. The emphasis is now a little different because this no longer seems so crucial. It is often the case that a survival of 20–30% is found 24 h after the transfection, and reduced toxicity often tends to accompany reduced transfection efficiency. Electroporation has not been found to be particularly effective in the production of stable integrants. Most workers find that optimum transfection is more easily achieved using linear DNA. This may mean that electroporation of episomal plasmids may not be particularly effective.

## 6.7.4 Liposomes and lipid-based transfection

Lipid, originally phosphatidyl-L-serine, can be treated such that when it is mixed with the DNA of interest it forms a vesicle around the DNA (Itani *et al.*, 1987). These lipid vesicles (liposomes) were mixed with the target cells, and the lipid was then thought to fuse with the lipid in the membrane of the cells and the DNA was released into the cytoplasm. Originally this method required the synthesis of one's own liposomes (Itani *et al.*, 1987). These were difficult and complicated to make; the lipid had to be protected from oxidation by storage under argon and the resulting liposomes tended to be rather unreliable. It was first thought that the DNA was required to be inside the lipid envelope, but now it appears that association with the lipid is itself sufficient. This has become a popular technique for both adherent and suspension cells, and a number of companies now sell lipofection kits. The reagents used in this method are relatively non-toxic and it appears that the longer the cells are in contact with the lipid, the better the transfection (up to a maximum of 8 h). Long exposure can lead to a high percentage of cells taking up the DNA, and relatively high expression levels can be obtained. Since few cells die after this treatment, it has become a popular choice for transient expression experiments, although it tends to be less efficient at producing cells with stable integrations.

### 6.7.5 Adenovirus and poly-L-lysine-conjugated complexes

Adenovirus has recently been introduced as one of the few methods capable of transfecting cells which are not actively proliferating. As such it has huge potential for transfection because of its high infection efficiency (Levrero et al., 1991). The main problem, however, is that the virus must be engineered to contain the gene of interest, and defective virus (adenovirus E1a gene defective) must then be transfected into, and propagated in, special helper cells (293 cells). This method would not be applicable for the introduction of many different genes (as in a library), or large inserts, into target cells of interest.

Receptor-mediated transfer of DNA using protamine or poly-L-lysine conjugates is a new technique that utilizes the features of the internalization of occupied cell surface receptors into endosomes. Poly-L-lysine can be covalently conjugated to a variety of ligands (Wagner et al., 1990; Wu et al., 1991; Citro et al., 1994), for example transferrin. Receptors for transferrin are found on all cells and, when mixed with DNA, the poly-L-lysine component of the conjugate both condenses and becomes closely (but non-covalently) associated with the DNA. This complex can then bind to the cell surface receptor for transferrin, the whole receptor/transferrin/poly-L-lysine–DNA complex is then endocytosed into the cells, and captured within the endosomes. This method can introduce DNA ranging in size from 18 bases (Citro et al., 1994) up to over 20 kb (Cotten et al., 1992).

One limit to this process was thought to be the efficiency with which the DNA escapes from the endosome, which resulted in the proposal that co-cultivation of the complexes in the presence of adenovirus would improve this escape. Proteins on the surface of adenovirus particles have an endosome disruption activity that is manifested when the internal endosome pH drops; the particles are thus released into the cytoplasm. Co-addition of DNA–transferrin/poly-L-lysine complexes with adenovirus was found to give an enhancement of more than two orders of magnitude in detectable product from the transfected gene. Furthermore, the destruction of the adenoviral genome, although inactivating adenoviral gene expression (i.e. the addition of dead virus) had no corresponding effect on endosome disruption activity. This method was capable of transfection of up to 48 kb constructs into target cells with little or no diminution of efficiency over that noted with an 8 kb construct (Cotten et al., 1992). The full potential of this method is however limited to those cells expressing the adenoviral receptor; cells from a blood origin, such as human myeloid/erythroid leukaemia cells K562, were relatively inefficiently transfected (Wagner et al., 1992). However, a further modification (Figure 6.3) involving the direct coupling of inactivated virus to the DNA–transferrin/poly-L-lysine complex however removes this limitation (Wagner et al., 1992). Manipulation of different ligand constructs is now also being used to target these complexes to particular cells *in vitro* and *in vivo* (Gottschalk et al., 1994).

### 6.7.6 Alternative transfection procedures

Alternative transfection procedures are still available, and all have their place. Scrape loading (Feccheimer *et al.*, 1987) is very easy to perform and can load cells with protein as well as DNA. Polybrene/DMSO (Kawai and Nishizawa, 1984) works well on some suspension cells. Protoplast fusion (Yokoyama and Imamoto, 1987) cuts out many steps in the purification of DNA and works with large plasmids in cases where many other methods have failed. Microinjection (Capecchi, 1980) is very efficient but hard to perform and best used when the transfer of DNA to only a few target cells is required. Biolistic methods (Klein *et al.*, 1987; Sanford *et al.*, 1993) load microprojectiles with DNA and then shoot them at the target cells. Using this technique, the authors were even able to punch DNA through a plant cellulose cell wall, an approach which has since been adapted to animal cells.

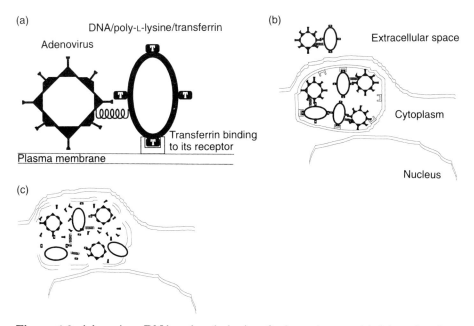

**Figure 6.3.** Adenovirus–DNA–poly-L-lysine/tranferrin conjugates. (a) Adenovirus is covalently linked to poly-L-lysine by transglutaminase (linkage represented by coiled motif). This complex is then added to DNA, neutralizing 25% of its negative charge. A further conjugate of poly-L-lysine/transferrin is then added to neutralize the remainder of the DNA, resulting in the complex shown. This complex then binds in a specific manner to the cell transferrin receptor. (b) Occupation of the receptor allows the whole complex to be internalized within the endosome and to enter the cell's cytoplasm. (c) As the endosomal pH drops, the disruption activity situated on the stalks of the adenovirus protein coat breaks open the endosome and allows the DNA to escape with only minimal degradation.

## 6.8 Assays for new protein synthesis

Once the cells have been transfected, it is usually necessary to determine whether or not the transfection has worked efficiently. In some cases, the transfected DNA will confer a phenotypic difference upon the cells that can be readily visualized without any further manipulations, as is the case with the BPV plasmids discussed earlier (Lowry et al., 1980). However, in the majority of cases, the cells must either be selected by their newly acquired resistance to drug treatment or reporter gene expression must be detected.

### 6.8.1 Dominant selectable marker genes

Dominant selectable markers confer upon the cells a new ability to resist the effect of cytotoxic drugs. It is important that resistance to these genes cannot occur spontaneously, or at least occurs with a *very* low frequency. The first genes of this type allowed cells to express the herpes virus-derived TK protein (Wigler et al., 1977), or the bacterial gene, xanthine–guanine phosphoribosyl-transferase (*XGPRT*; Mulligan and Berg, 1981), or the eukaryote genes, adenine phosphoribosyltransferase (*APRT*; Wigler et al., 1979) and dihydrofolate reductase, (*DHFR*; Crouse et al., 1983). All of the above can allow metabolic pathway-deficient mutant cell lines to proliferate in specialized selective medium. In the case of *tk*, *APRT* and *XGPRT*, the transfected cells will survive and prosper in HAT (hypoxanthine–aminopterin–thymidine) medium whereas untransfected cells will die. The only problem is that these genes replace an endogenous gene activity; thus specialized enzyme-deficient mutant cells are often required. Therefore these are not truly dominant selectable markers, and this can seriously limit the utility of these selections. However, a modification to the selection for *XGPRT* is the addition of aminopterin and mycophenolic acid which inhibit the activity of the endogenous enzyme (dihydrofolate reductase) synthesizing inosine monophosphate (IMP) and the IMP dehydrogenase respectively. This allows selection in cells that do not have a metabolic deficiency since the bacterial *XGPRT* enables the utilization of xanthine (an activity with no eukaryotic equivalent) and hypoxanthine. Under these circumstances, *XGPRT* does act as a dominant selectable marker.

True dominant selectable markers include the hygromycin phosphotransferase (*hygro*) gene (Blochlinger and Diggleman, 1984) conferring resistance to the antibiotic hygromycin B; and the neomycin phosphotransferase (*neo*) gene conferring resistance to the antibiotic geneticin (G418; Southern and Berg, 1982). These are bacterial genes for which there is no eukaryotic equivalent and can thus be used for any cells. Sensitivity to these drugs can vary between cell types and must be determined for each new application. More recently, resistance genes to puromycin (*puro*; Hartman and Mulligan, 1988) and bleomycin/phleomycin/zeomycin (*bleo*; de la Luna and Ortin, 1982; Mulsant et al., 1988; Semon et al., 1987) have become available. None of the above (*hygro*, *neo*, *puro* or *bleo*) appear to cross-react significantly. Thus multiple plasmids can be transfected independently into a recipient cell and the selection then applied either sequentially or in combination. We find that suitably transfected cells have the ability to grow in hygromycin, G418 and puromycin simultaneously. Further resistance genes include histidinol dehydrogenase (*HisD*; Hartman and Mulligan, 1988) which allows cells to grow in the absence of exogenous L-histidine, and confers resistance to high levels of L-histidinol. Another dominant selectable marker is blasticine S deaminase (blasticine resistance, *bsr*) conferring resistance to blasticine S by its deamination to the less toxic deaminohydroxy derivative (Izumi *et al.*, 1991).

Finally, interesting examples of specialized uses include the herpes *tk* gene, and the *DHFR* gene which confers resistance to methotrexate. It has been shown that amplification of the *DHFR* gene, and associated DNA, can be selected for by step-wise increases in the concentration of methotrexate (Milbrandt *et al.*, 1981). This can sometimes allow the selection of high level expression of the associated marker gene (Kaufmann and Sharp, 1982). The HSV-TK promoter is a popular choice for initiating eukaryotic expression. The *tk* gene confers a greatly increased sensitivity to the effects of the toxic thymidine analogue gancyclovir. Endogenous (i.e. eukaryotic) TK has a low affinity for gancyclovir, whilst HSV-TK, being more promiscuous in its usage of precursors, has a high affinity. Using this selection system, cells that have taken up the gene can be selectively killed, a crucial requirement for the generation of insertions by homologous recombination, since the design of the vector ensures that cells having undergone homologous recombination will not express HSV-TK (Capecchi, 1989).

### 6.8.2 Reporter genes

There are genes that do not confer resistance to a drug, but are quick and easy to assay and the amount of protein produced can be quantitated. These genes are often used to give an idea of the efficiency of the transfection, and also to investigate the modulation of promoter usage. Commonly used reporter genes include chloramphenicol acetyltransferase (CAT; Gorman *et al.*, 1982), β-galactosidase (Edlund *et al.*, 1985) and firefly luciferase (Brasier *et al.*, 1989). These genes code for proteins which have no eukaryotic counterparts. This prevents high background readings produced by endogenous enzymes. However, extracts for CAT assays should be incubated at 65°C for 10 min to inactivate any de-acetylating enzymes (Gorman *et al.*, 1982).

The protein products of these genes are readily detectable in cytoplasmic extracts. In the case of CAT activity, its ability to acetylate chloramphenicol in the presence of acetyl coenzyme A makes it readily detectable. For this assay, the chloramphenicol is labelled and changes in mobility resulting from one, two or three additional acetyl groups are detected by thin layer chromatography. Alternatively, radiolabelled acetyl coenzyme A is used and the radioactive acetyl groups transferred to the chloramphenicol are measured after extraction with ethyl acetate, thereby indicating the amount of CAT protein present. β-Galactosidase is assayed in cell extracts or in paraformaldehyde-fixed cells by the conversion of colourless substrate (i.e. *o*-nitrophenyl-β-D-galactopyranoside) to a blue product that can be visualized or quantitated spectrophotometrically.

Another reporter system, 10- to 100-fold more sensitive than CAT, is the firefly luciferase assay. Cell lysates containing the luciferase gene product can be assayed quantitatively using luciferin as a substrate and ATP as the energy source. Light output resulting from the ATP-dependent oxidation of luciferin is then measured in a luminometer (Brasier *et al.*, 1989).

## 6.9 Analysis of cloned genes

Putting all of the above techniques together, we now have the ability to clone novel genes into vectors, introduce them into cells of interest, monitor the efficiency of this uptake and investigate the transcription and translation of the new gene. All that is now required is a way to analyse the expressed genes, and assign functions to their protein products. In many cases, it is essential to have at least some idea of a possible function for the gene of interest. For this reason, the databases of gene sequences (many of which have a defined function) are an invaluable aid. Comparing the DNA sequence of the translated portion of a new gene can give a good idea of a possible function. In this section, some of the more common methods for analysis of function will be described, by examining two examples from the literature that demonstrate the power of these techniques.

### 6.9.1 Identification of ligands for novel receptors

The genes for a number of steroid hormone receptors have recently been cloned and characterized. These include the receptors for oestrogen (Green *et al.*, 1986), progesterone (Misrahi *et al.*, 1987) and glucocorticoids (Weinberger *et al.*, 1985). All show significant intermolecular homology (Evans, 1988). In addition, the modular structure of these proteins allows the excision and fusion of sections from different receptors, generating chimeric products which retain their function as ligand-dependent nuclear receptor/transcription factors (Evans, 1988). Using sequences from one of these receptors as a probe on a low stringency cDNA library screen, a further putative member of the gene family was isolated, but its ligand was unknown (Petkovich *et al.*, 1987).

One candidate ligand for the new gene (based on binding studies) was RA, but the challenge was to demonstrate this when there was no known genomic DNA target (DNA-binding site) for the putative RA–receptor complex. The strategy that proved successful was firstly to take the cDNA from the well characterized oestrogen receptor, excise its DNA-binding domain (domain C) and use this to replace the equivalent domain in the novel receptor. This hybrid gene was placed in a plasmid under the control of the SV40 early promoter (similar to the construction of pSV2-neo), and the whole expression vector construct termed RAR-ER.CAS. Theoretically this engineered protein product would remain sensitive to its as yet unknown ligand, and still transactivate transcription of a reporter gene in constructs possessing the DNA target (response element) for the oestrogen-activated oestrogen receptor. A second construct (the reporter plasmid *vit-tk*-CAT) was then used. This contained the oestrogen-responsive element from the *Xenopus* vitellogen in A2 gene (*vit*, acting as an inducible enhancer) cloned 5' to the HSV-TK promoter which promotes transcription of the CAT gene. Transient co-transfection experiments with these two plasmid constructs (Figure 6.4) in HeLa cells showed that, in the absence of any ligand, only a low 'basal' level of CAT activity could be detected. In the presence of $10^{-8}$M RA, however, the detectable level of CAT activity was induced more than 10-fold, demonstrating increased receptor-mediated transcription of the CAT gene after RA treatment. This was not seen with the addition of molar equivalents of oestrogen, testosterone, retinol, vitamin D3 or triiodothyronine. This effect was also absent after RA treatment if the RAR-ER.CAS construct was replaced with an alternative CAT reporter construct driven by the glucocorticoid-inducible MMTV LTR.

These data (along with their respective control experiments) demonstrated that, of the molecules tested, the preferred ligand for the novel gene was RA. This gene was thus designated RARα, a RA-dependent, sequence-specific transcription factor (Petkovich *et al.*, 1987). Another group (Giguere *et al.*, 1987) reported essentially the same result with the same gene, independently isolated, using virtually the same technology in COS cells coupled with a chimeric molecule with elements from the glucocorticoid receptor and its relevant response element. These methods were then also used to designate a second gene (de The *et al.*, 1987) as the RA receptor-β (RARβ; Brand *et al.*, 1988).

### *6.9.2 Identification of transcription factors*

A novel gene may contain elements that make it a potential transcription factor, but the location of its response element, and the genes which it regulates (if they exist) are usually unknown. How then can its function as a transcription factor be confirmed or refuted? One elegant technique uses a protein cloned from the yeast *Saccharomyces cerevisiae* (the Gal4 protein). When expressed in eukaryotic cells, this protein will interact with its DNA element (sequence $UAS_G$) to promote transcription of a downstream reporter. Deletion studies have shown that the first 147 amino acids of this protein constitute the DNA-binding domain (GAL4-DBD), but with much reduced transcription activation ability. If this 147 amino acid portion is fused to a transcription factor (or part thereof), it can then fully regain its function and activate transcription in a sequence-specific manner from the $UAS_G$.

The DNA sequences of putative transcription factors, or transcriptional domains of novel proteins can then be cloned in-frame (cloned in such a

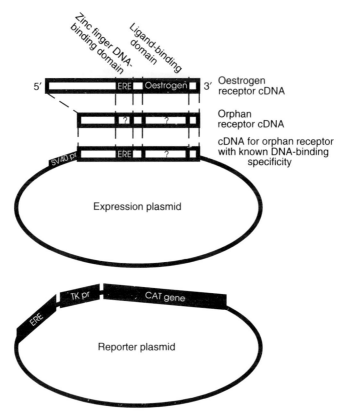

**Figure 6.4.** Chimeric finger swap constructs. As detailed in the text, the expression plasmid has been engineered with the zinc-finger DNA-binding domain of the oestrogen receptor replacing (or exchanged with) the same region in the unknown (or orphan) receptor. Transcription and translation of this chimera results in a protein that, on activation, will trans-activate genes containing the oestrogen-response element (ERE) enhancer. The reporter plasmid containing the minimal TK promoter is weakly active in the absence of stimulation via the ERE, but is strongly enhanced on trans-activation by ligand (i.e. all-*trans*-RA) activation of the orphan receptor chimeric protein.

way that transcription of GAL4 is immediately followed by the novel sequences) to yield a hybrid fusion protein. Co-transfection of these plasmids with $UAS_G$-controlled reporter plasmids can then allow the investigation of the transcriptional regulatory activity of novel proteins without the need for information about their cognate DNA response elements (Figure 6.5).

This method has been used to map the transcriptionally active domains HOB1 and HOB2 of Fos and Jun (Sutherland *et al.*, 1992) and the activator and suppressor functions of the Visna virus Tat protein (Carruth *et al.*, 1994).

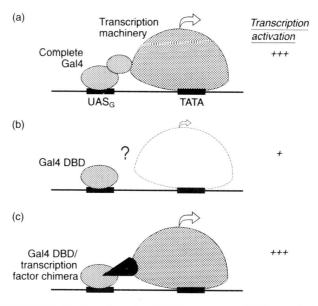

**Figure 6.5.** Gal4 DNA-binding domain. (a) The complete Gal4 protein is composed of the DNA-binding domain (DBD), shown bound to the $UAS_G$, along with protein determinants that recruit the cells transcription machinery for TATA-dependent RNA transcription. (b) The DBD alone allows only minimal TATA-dependent transcription. (c) DBD fusion proteins (dark segment) allow the determination of the ability of the new determinant to trans-activate TATA-dependent RNA transcription. Since the DBD alone has some residual activity, this can also be used to investigate the repressor function of chimeric determinants.

## References

**Ausubel FM, Brent R, Kingston RE, Moore DD, Seidman JG, Smith JA, Struhl K.** (1994) (eds) *Current Protocols in Molecular Biology.* Wiley, New York.

**Blochlinger K, Diggleman H.** (1984) Hygromycin B phosphotransferase as a selectable marker for DNA transfer experiments with higher eukaryotic cells. *Mol. Cell. Biol.* **4:** 2929–2931.

**Brand N, Petkovich M, Krust A, Chambon P, de The H, Marchio A, Tiollais P, Dejean A.** (1988) Identification of a second human retinoic acid receptor. *Nature* **332:** 850–853.

**Brasier AR, Tate JE, Habener JF.** (1989) Optimized use of the firefly luciferase assay as a reporter gene in mammalian cell lines. *Biotechniques* **7,** 1116-1122.

**Capecchi MR.** (1989) Altering the genome by homologous recombination. *Science* **244,** 1288-1292

**Capecchi MR.** (1980) High efficiency transformation by direct microinjection of DNA into cultured mammalian cells. *Cell* **22,** 479-488.

**Carruth LM, Hardwick JM, Morse BA, Clements JE.** (1994) Visna virus Tat protein: a potent transcription factor with both activator and suppressor domains. *J. Virol.* **68,** 6137-6146.

**Chen C, Okayama H.** (1987) High-efficiency transformation of mammalian cells by plasmid DNA. *Mol. Cell. Biol.* **7,** 2745-2752.

**Chu G, Sharp PA.** (1981) SV40 DNA transfection of cells in suspension: analysis of the efficiency of transcription and translation of T-antigen. *Gene* **13,** 197-202.

Citro G, Szczylik P, Ginobbi P, Zupi G, Calabretta B. (1994) Inhibition of leukaemia cells proliferation by folic acid–polylysine-mediated introduction of c-*myb* antisense oligodeoxynucleotides into HL-60 cells. *Br. J. Cancer* **69**, 463–467.

Cotten M, Wagner E, Zatlouka K, Phillips S, Curiel D, Birnstiel M. (1992) High efficiency receptor mediated delivery of small and large <48 kb gene constructs using the endosome-disruption activity of defective chemically inactivated adenovirus particles. *Proc. Natl Acad. Sci. USA* **89**: 6094–6098.

Crouse GF, McEwan RN, Pearson ML. (1983) Expression and amplification of engineered mouse dihydrofolate reductase minigenes. *Mol. Cell. Biol.* **3**, 257–266

Edlund T, Walker M, Barr P, Rutter W. (1985) Cell-specific expression of the rat insulin gene – evidence for role of two distinct-5′ flanking elements. *Science* **230**: 912–916.

Ehlert F, Bierbaum P, Schorr J. (1993) Importance of DNA quality for transfection efficiency. *Biotechniques* **14**: 546.

Evans RM (1988) The steroid and thyroid hormone receptor superfamily. *Science* **240**: 889–895.

Fechheimer M, Boylan JF, Parker S, Siskens JE, Patel GL, Zimmer SG. (1987) Transfection of mammalian cells with plasmid DNA by scrape loading and sonication loading. *Proc. Natl Acad. Sci. USA* **84**: 8463–8467.

Giguère V, Ong ES, Segui P, Evans R. (1987) Identification of a receptor for the morphogen retinoic acid. *Nature* **330**: 624–629.

Gluzman Y (1981) SV40-transformed simian cells support the replication of early SV40 mutants. *Cell* **23**: 175–182.

Goldfarb M, Shimizu K, Perucho M, Wigler M. (1982) Isolation and preliminary characterization of a human transforming gene from T24 bladder carcinoma cells. *Nature* **296**: 404–409.

Golub EI, Kim H, Volsky DJ. (1989) Transfection of DNA into adherent cells by DEAE–dextran DMSO method increases drastically if the cells are removed from surface and treated in suspension. *Nucleic Acids Res.* **17**: 4902.

Gorman CM, Moffat LF, Howard BH. (1982) Recombinant genomes which express chloramphenicol acetyltransferase in mammalian cells. *Mol. Cell. Biol.* **2**: 1044–1051.

Gottschalk S, Cristiano RJ, Smith LC, Woo SLC. (1994) Folate receptor mediated DNA delivery into tumor cells: protosomal disruption results in enhanced gene expression. *Gene Ther.* **1**: 185–191.

Graham FL, Van der Eb AJ. (1973) A new technique for the assay of infectivity of human adenovirus 5 DNA. *Virology* **52**: 456–461.

Green S, Walter P, Kumar V, Krust A, Bonert J-M, Argos P, Chambon P. (1986) Human oestrogen receptor cDNA: sequence, expression and homology to v-*erb-A*. *Nature* **320**: 134–139.

Hartmann SC, Mulligan RC. (1988) Two dominantly acting selectable markers for gene transfer studies in mammalian cells. *Proc. Natl Acad. Sci. USA* **85**: 8047–8051.

Hirt B. (1967) Selective extraction of polyoma DNA from infected mouse cell cultures. *J. Mol. Biol.* **26**: 365–369.

Holler W, Fordis CM, Howard BH. (1989) Efficient gene transfer by sequential treatment of mammalian cells with DEAE–dextran and deoxyribonucleic acid. *Exp. Cell Res.* **184**: 546–551.

Itani T, Ariga H, Yamaguchi N, Tadakuma T, Yasuda T. (1987) A simple and efficient liposome method for transfection of DNA into mammalian cells grown in suspension. *Gene* **56**: 267–276.

Izumi M, Miyazawa H, Kamakura T, Yamaguchi I, Endo T, Hanaoka F. (1991) Blasticidin S-resistance gene (*bsr*): a novel selectable marker for mammalian cells. *Exp. Cell Res.* **197**: 229–233.

Kaufmann RJ, Sharp PA. (1982) Amplification and expression of sequences cotransfected with a modular dihydrofolate reductase complementary DNA gene. *J. Mol. Biol.* **159**: 601–621.

Kawai S, Nishizawa M. (1984) New procedure for DNA transfection with polycation and dimethyl sulfoxide. *Mol. Cell. Biol.* **4**: 1172–1174.

Klein TM, Wolf ED, Wu R, Sanford JC. (1987) High velocity microprojectiles for delivering nucleic acids into living cells. *Nature* **327**: 70–73.

**Lee F, Mulligan R, Berg P, Ringold G.** (1981) Glucocorticoids regulate expression of dihydrofolate reductase cDNA in mouse mammary tumour virus chimeric plasmids. *Nature* **214:** 228–232.

**Levrero M, Barban V, Manteca S, Ballary A, Balsamo C, Avanitaggiat ML, Natoli G, Skellckens H, Tiollas P, Perricaudet M.** (1991) Defective and nondefective adenovirus vectors for expressing foreign genes *in vitro* and *in vivo*. *Gene* **101:** 195–202.

**Lopata MA, Cleveland DW, Sollner-Webb B.** (1984) High-level expression of a chloramphenicol acetyltransferase gene by DEAE–dextran-mediated DNA transfection coupled with a dimethylsulfoxide or glycerol shock treatment. *Nucleic Acids Res.* **12:** 5707–5717.

**Lowry DR, Dvoretzky I, Shober R, Law M-F, Engel L, Howley PM.** (1980) *In vitro* tumorigenic transformation by a defined sub-genomic fragment of bovine papilloma virus DNA. *Nature* **287:** 72–74.

**de la Luna S, Ortin J.** (1982) (33) *pac* gene as efficient dominant marker and reporter gene in mammalian cells. *Methods Enzymol.* **216,** 376–385.

**Luthman H, Magnusson G.** (1983) High efficiency polyoma DNA transfection of chloroquine treated cells. *Nucleic Acids Res.* **11:** 1295

**McCutchan JH, Pagano JS.** (1968) Enhancement of the infectivity of simian virus 40 deoxyribonucleic acid with diethyl-aminoethyl-dextran. *J. Natl Cancer Inst.* **41:** 351–357.

**Milbrandt JD, Heintz NH, White WC, Rothman SM, Hamlin JL.** (1981) Methotrexate resistant chinese hamster ovary cells have amplified a 135-kilobase-pair region that includes the dihydrofolate reductase gene. *Proc. Natl Acad. Sci. USA* **78:** 6043–6047.

**Misrahi M, Atger M, Dauriol L, Loosfelt H, Meriel C, Fridlansky F, Guiochonmantel A, Galibert F, Milgrom E.** (1987) Complete amino acid sequence of the human progesterone receptor deduced from cloned cDNA. *Biochem. Biophys. Res. Commun.* **143:** 740–748.

**Mulligan RC, Berg P.** (1981) Selection for animal cells that express the *E. coli* gene coding for xanthine–guanine phosphoribosyltransferase. *Proc. Natl Acad. Sci. USA* **78:** 2072–2076.

**Mulsant P, Gatignol A, Dalens M, Tiraby G.** (1988) Phleomycin resistance as a dominant selectable marker in CHO cells. *Somatic Cell Mol. Genet.* **14:** 243–252.

**Neumann E, Schaefner-Ridder M, Wang Y, Hofschneider PH.** (1982) Gene transfer into mouse myeloma cells by electroporation in high electric fields. *EMBO J.* **1:** 841–845.

**Parker BA, Stark GR.** (1979) Regulation of simian virus 40 transcription: sensitive analysis of the RNA species present early in infections by virus or viral DNA. *J. Virol.* **31:** 360–369.

**Petkovich M, Brand NJ, Krust A, Chambon P.** (1987) A human retinoic acid receptor which belongs to the family of nuclear receptors. *Nature* **330:** 444–450.

**Potter HL, Weir U, Leder P.** (1984) Enhancer-dependent expression of human kappa immunoglobulin genes introduced into mouse pre-B lymphocytes by electroporation. *Proc. Natl Acad. Sci. USA* **81:** 7161–7165.

**Sambrook J, Fritsch EF, Maniatis T.** (1989) *Molecular Cloning: A Laboratory Manual*, 2nd edn. Cold Spring Harbor Laboratory Press, Cold Spring Harbor, NY.

**Sanford JC, Smith FD, Russell JA.** (1993) Optimising the biolistic process for different biological applications. *Methods Enzymol.* **271:** 483–509.

**Semon D, Movva NR, Smith TF.** (1987) Plasmid-determined bleomycin resistance in *Staphylococcus aureus*. *Plasmid* **17:** 46–53.

**Somapayrac LM, Danna KJ.** (1981) Efficient infection of monkey cells with DNA from simian virus 40. *Proc. Natl Acad. Sci. USA* **78:** 7575–7578.

**Southern PJ, Berg P.** (1982) Transformation of mammalian cells to antibiotic resistance with a bacterial gene under control of the SV40 early gene promoter. *J. Mol. Appl. Genet.* **1:** 327–341.

**Stuart GW, Searle PF, Palmiter RD.** (1985) Identification of multiple metal regulatory elements in mouse metallothionein-II promoter by assaying synthetic sequences. *Nature* **317:** 828–831.

**Sugden M, Marsh K, Yates J.** (1985) A vector that replicates as a plasmid and can be efficiently selected in B-lymphoblasts transformed by Epstein–Barr virus. *Mol. Cell. Biol.* **5:** 410–413.

**Sutherland JA, Cook A, Bannister AJ, Kouzarides T.** (1992) Conserved motifs in Fos and Jun define a new class of activation domain. *Genes Dev.* **6:** 1810–1819.

de The H, Marchio A, Tiollais P, Dejean A. (1987) A novel steroid hormone receptor-related gene inappropriately expressed in human hepatocelluar carcinoma. *Nature* **330:** 667–670.

Toneguzzo F, Hayday AC, Keating A. (1986) Electric field-mediated DNA transfer: transient and stable gene expression in human and mouse lymphoid cells. *Mol. Cell. Biol.* **6:** 703–706.

Wagner E, Zenke M, Cotten M, Beug H, Birnstiel ML. (1990) Transferrin–polycation conjugates as carriers for DNA uptake into cells. *Proc. Natl Acad. Sci. USA* **87:** 3410–3414.

Wagner E, Zatlouka K, Cotten M, Kirlappos H, Mechtler K, Curiel D, Birnstiel M. (1992) Coupling of adenovirus to transferrin–polylysine/DNA complexes greatly enhances receptor mediated gene delivery and expression of transfected genes. *Proc. Natl Acad. Sci. USA* **89:** 6099–6103.

Weinberger C, Hollenberg SM, Ong ES, Harmon JM, Brower ST, Cidlowski J, Thompson EB, Rosenfeld MG, Evans RM. (1985) Identification of human glucocorticoid receptor complementary DNA clones by epitope mapping. *Science* **228:** 740–742.

Wigler M, Silverstein S, Lee LS, Pellicer A, Cheng YC, Axel R. (1977) Transfer of purified herpes virus thymidine kinase gene into cultured mouse cells. *Cell* **11:** 223–232.

Wigler M, Pellicer A, Silverstein S, Axel R, Urlaub G, Chasin L. (1979) DNA-mediated transfer of the adenine phosphoribosyltransferase locus into mammalian cells. *Proc. Natl Acad. Sci. USA* **76:** 1373–1376.

Wu GY, Wilson JM, Shalaby F, Grossman M, Shafritz DA, Wu CH. (1991) Receptor-mediated gene delivery *in vivo*. Partial correction of genetic analbuminemia in Nagase rats. *J. Biol. Chem.* **266:** 14338–14342.

Yates JL, Warren N, Sugden B. (1985) Stable replication of plasmids derived from Epstein–Barr virus in various mammalian cells. *Nature* **313:** 812–815.

Yokoyama K, Imamoto F. (1987) Transcriptional control of the endogenous *myc* protooncogene by antisense RNA. *Proc. Natl Acad. Sci. USA* **84:** 7363–7367.

# 7

# Foreign DNA integration and DNA methylation patterns

Walter Doerfler

## 7.1 Introduction

In spite of a great deal of experimental work in different biological systems, the functional meaning of DNA methylation in mammalian genomes is still only partly understood. The available evidence suggests that each gene or DNA segment exhibits a distinct pattern of DNA methylation which is dependent on cell type in an organism or stage of development. These patterns appear to be subject to change in, for example, different periods of development or when growth or culture conditions of a given cell type are altered. In this context, we have been interested in alterations in DNA methylation in tumour cells, particularly in adenovirus type 12 (Ad12)-induced hamster tumour cells. It is not known whether these changes in cellular DNA methylation patterns are causally related to the mechanism of oncogenic transformation or whether they are, instead, one of its consequences, or both.

For more than a decade, the author's laboratory has contributed detailed experimental evidence to support the notion that the sequence-specific methylation of eukaryotic promoter sequences plays an important role in the long-term inactivation of genes and DNA segments. Obviously, additional factors at the level of DNA–protein interactions participate in this inactivation. The binding of specific proteins to DNA motifs and the formation of complex structures mediated by protein–protein interactions have been heralded as decisive factors in the regulation of gene activity. The unexpected complexity of these interactions has, however, compromised hopes for a plausible model based on the analyses of a multitude of proteins that bind to DNA and/or to each other.

When foreign DNA is inserted into an established genome, the newly added sequences are methylated *de novo* in specific sequences. The author has proposed that this *de novo* methylation is part of an ancient cellular defence mechanism directed against the activity of foreign DNA in an evolved cellular or organismic system (Doerfler, 1991b). The molecular mechanism of this *de novo* methylation and the recognition of sites of initiation of *de novo* methylation are not understood and deserve intensive investigation. We have tried to make a contribution toward the elucidation of these problems with analyses of the adenovirus system.

*De novo* methylation is intrinsically coupled to the integration of foreign DNA into an established genome. In this chapter experiments from our own laboratory are summarized that have aimed to improve our understanding of the mechanism of foreign DNA integration. We have tried to imitate at least certain aspects of the integration reaction in a highly purified cell-free system by using extracts of hamster cell nuclei. In *in vitro* reactions, a known pre-insertion site of Ad12 DNA in hamster cells and the nucleotide (nt) 20 885–24 053 fragment of Ad12 DNA have been shown to recombine. The hamster nuclear extracts have been highly purified to a small number of proteins. Some of these proteins have been microsequenced with surprising results. A future challenge will be to investigate whether and how foreign DNA integration and methylation are related.

Foreign DNA ingested by mammals with the food supply can survive the digestive regimes of the gastrointestinal tract, and at least a small proportion of this DNA in fragmented form can be retrieved from the animals' faeces. An even smaller fraction of the ingested DNA has been detected in the animals' white blood cell population. In these experiments, DNA from bacteriophage M13 has been fed to mice. The effects of a (presumably) constant flow of foreign DNA from food to the blood of an organism are unknown.

We shall further pursue the concept that foreign DNA integration into a cell's genome and the associated alterations in the patterns of methylation in the foreign and host target DNAs can contribute to mutagenesis and oncogenesis in the host organism. Owing to changes in the methylation of cellular DNA sequences remote from the site of foreign DNA insertion, the concept of insertional mutagenesis has to be interpreted in a wider context.

## 7.2 The adenovirus system as a model

With the notion that practically any foreign DNA can be incorporated into pre-existing mammalian genomes, it should be explained why we have chosen the adenovirus system as a model. The prime motivation for this choice was the well established oncogenic nature of the virus in rodents, particularly in newborn hamsters. In this regard, adenovirus work has been intimately connected to the biological questions of cell transformation by oncogenic viruses and to the underlying mechanisms. Moreover, after several decades of intensive research, adenoviruses have proven themselves to be impressive tools for the study of the molecular biology of mammalian cells. It was with the help of this viral system that many fundamental mechanisms in the molecular biology of mammalian cells have been recognized and, at least partly, unravelled. One may also argue that, in studies on the fate of foreign DNA in mammalian cells, any choice of DNA will prove somewhat arbitrary. Thus, the selection of a well characterized viral genome, whose biochemistry and molecular biology are understood in considerable detail, has obvious advantages. Of course, the use of DNA from an oncogenic virus will involve selection of cells that have been transformed by this virus and which exhibit very specific biological properties. This choice was intentional, and aimed at contributing to our understanding of the transformation mechanism of cells by oncogenic viruses. Nevertheless, in the more general interpretation of the data from the adenovirus system, it will be prudent to keep this selection and its limitations in mind. On the other hand, any DNA molecule chosen for similar studies might have entailed different yet similarly selective mechanisms which would have been less apparent for a less well characterized DNA molecule. All the more general conclusions in this chapter should be prefaced by the thought that we have made a deliberate choice which will reflect upon all data obtained.

A detailed description of the molecular biology of adenoviruses cannot be provided within the scope of this chapter. However, several books on this subject have been published (Doerfler, 1983b, 1984b; Ginsberg, 1985; Doerfler and Böhm, 1995). The reader is referred to these volumes for a basic introduction to the adenovirus system.

### 7.2.1 Site selection in the integration of adenovirus DNA

With the possibility that the insertion of adenovirus DNA into the mammalian genome could fundamentally alter the transcriptional programme of infected and transformed cells, it was important to investigate whether the foreign DNA was inserted randomly at many different, or at highly specific sites in the genomes of the affected cells. We have spent considerable time and effort in analysing the sites of adenovirus DNA integration in a large number of virus-transformed or Ad12-induced tumour cells from rodents, mainly from hamsters (for reviews, see Doerfler, 1982, 1991a; Doerfler *et al.*, 1983). We have studied integrated genomes in rodent cells transformed in cell culture by infection with adenovirus (adenovirus-transformed cells), in Ad12-induced tumours or in cells cultured from these Ad12-induced tumours. It is not known whether these different types of cells can be considered as similar with respect to their tumorigenic phenotype.

Three types of analytical approaches were chosen to investigate and to prove the integrated state of the adenovirus genomes within the cellular genomes.

(i) The DNA from transformed or tumour cells was cleaved with different restriction endonucleases, the fragments transferred to membranes, and the distribution of adenovirus-specific DNA fragments determined by DNA–DNA hybridization to adenovirus DNA or to the cloned terminal fragments of virion DNA. In this way, the internal and terminal viral DNA fragments could be localized in the cellular genome relative to cellular DNA fragments. In many instances, the viral DNA, particularly Ad12 DNA, was found to be integrated in an orientation co-linear with that in the virion genome (i.e. the viral genome inside purified virus particles). Consequently, the terminal viral DNA segments flanked by cellular DNA sequences and excised together with them, did not co-migrate upon gel electrophoresis with any of the known virion DNA fragments, but appeared instead in 'off-size' positions. From the analysis of at least 80 different transformed and tumour cell lines performed since 1976 in the author's laboratory, it appeared that adenovirus DNA was never found free in any of these transformed or tumour cells, but always integrated into the host genome. In several instances, the 'off-size' fragments also contained rearranged viral DNA sequences (Sutter et al., 1978; Stabel et al., 1980; Kuhlmann and Doerfler, 1982; Kuhlmann et al., 1982; Orend et al., 1991, 1994, 1995). When analysed at the nucleotide sequence level (see below), these integration patterns have proved to be very complicated in individual instances. Additional complexities of the system will be described below.

(ii) In order to prove the covalent linkage (integration) of adenoviral to cellular DNA sequences, it was necessary to molecularly clone and determine the nucleotide sequence of some of the junction sites between viral and cellular DNAs. From the data available so far, all the cellular DNA sequences flanking integrated viral DNA sequences are different from each other (Deuring et al., 1981a; Gahlmann et al., 1982; Stabel and Doerfler, 1982; Gahlmann and Doerfler, 1983; Deuring and Doerfler, 1983; Schulz and Doerfler, 1984; Lichtenberg et al., 1987; Jessberger et al., 1989a). These data provide no evidence for the notion that adenovirus DNA had integrated at highly specific cellular DNA sequences. However, only a relatively small number of such junction sites have so far been investigated. Moreover, the extent of DNA sequences determined in the flanking cellular DNA have been limited. In addition, many of the cells in which junction sequences between adenovirus DNA and cellular DNA were analysed, were cloned cell lines which had been maintained in culture for many years. It was thus possible that the integrated viral genomes could have become rearranged, transposed or altered in other ways after the original integration event. In a few instances (see below), evidence was obtained for the occurrence of selective sites of viral DNA integration in different tumour cell lines.

(iii) Recently, we have initiated studies on Ad12-transformed cell lines and on Ad12-induced tumour cell lines to determine the chromosomal locations of the integrated viral genomes. The data collated so far on cell lines T637, HA12/7, H191 and H281 using fluorescence *in situ* hybridization (FISH) methods with biotinylated Ad12 probes (P. Wilgenbus, G. Meyer zu Altenschildesche, M. Lutze, S.T. Tjia, and W. Doerfler, unpublished observations) indicate that the Ad12 genomes are indeed chromosomally located and that the bulk of the integrated viral genomes lie at a single chromosomal site.

### 7.2.2 Modes of adenovirus DNA integration – a synopsis of data

Under the premise that we have necessarily studied only a limited set of adenovirus-transformed cell lines or Ad12-induced tumour cell lines, various conclusions can presently be drawn, as outlined in Table 7.1.

**Table 7.1.** Elements of adenovirus DNA integration into the mammalian genome: a pliable mechanism

- Chromosomal location of adenovirus integrates
- Intact genomes and/or fragments can be integrated
- Rearranged parts of the viral genome can be integrated in the form of fragments close to intact viral DNA molecules
- Sites of linkage exhibit frequently, but not always, patchy sequence homologies between recombination partners
- Integrated adenovirus DNA can be intact, rearranged or carry deletions
- Mode of viral DNA persistence depend on the permissivity of the virus–host system used
- At sites of linkage, viral nucleotides can be deleted
- Adenovirus DNA integration can entail deletions of cellular DNA at the site of insertion or can be effected without the loss of a single cellular nucleotide
- Cellular DNA sequences at sites of insertion have been found to be transcriptionally active
- Integrated adenovirus DNA can be partly or completely lost from the cellular genome: morphological revertants of transformed cells can arise
- Upon cultivation of Ad12-induced tumour cells, cells can be selected that carry integrated viral genomes at selective cellular sites
- Integrated adenovirus genomes become methylated *de novo*
- Patterns of *de novo* methylation depend on the site of adenovirus DNA integration, perhaps also on other factors
- Integration of foreign (adenovirus) DNA into established mammalian genomes can be associated with extensive changes in the methylation patterns of cellular genes
- Integration/recombination of adenovirus DNA into/with cellular DNA can be imitated in a cell-free system with purified components from hamster nuclear extracts
- The *in vitro* generated recombinants exhibit similarities to the *in vivo* observed integrates: patch homologies

In evaluating the outcome of an infection with adenoviruses with respect to cellular transformation, the biology of the virus–host cell system has to be considered. By contrast, transfection of viral DNA fragments presents a very different situation which must affect the outcome of the transformation event. Lastly, although almost any experimental system can be criticized as having non-natural properties, the induction of tumours in living animals can be considered to be as close to the full range of complexities as possibly attainable. Nevertheless, one can rightfully argue that Ad12 in reality might not often have had the opportunity to infect *Mesocricetus auratus*, the Syrian gold hamster, before Trentin and colleagues performed their pioneering experiments in 1961/1962 (Trentin *et al.*, 1962).

Ad12 infects hamster cells non-productively; the infection is completely abortive, with failure of Ad12 virion production, Ad12 DNA replication and late gene transcription. Newly assembled viral particles can never be found. The basis for this non-permissive interaction is complex and multi-tiered (for reviews, see Doerfler, 1991a; Zock and Doerfler, 1993). We have been able to show that the major late promoter of Ad12 DNA carries a mitigator element in the downstream sequence which, at least in part, appears responsible for the inability of this promoter to function in hamster cells (Zock and Doerfler, 1990; Zock et al., 1993). Upon the experimental removal of this mitigator element, the major late promoter of Ad12 DNA becomes functional in hamster cells, and its activity in the permissive human cells is enhanced. As a consequence of the non-permissive interaction of Ad12 with hamster cells, there is no selective pressure against the persistence of the entire Ad12 genome in hamster cells, and we have thus frequently found that Ad12 DNA can be integrated into the hamster cellular genome almost intact and co-linear with the arrangement of Ad12 DNA sequences as found in the DNA extracted from the virus particle. Realistically, it must be added that, in some Ad12-transformed cell lines which carry multiple Ad12 DNA copies (e.g. cell line T637), certain parts of the integrated Ad12 genomes can be very markedly rearranged (Eick and Doerfler, 1982; Orend et al., 1995).

By contrast, adenovirus type 2 (Ad2) infects hamster cells productively and can grow to appreciable titres. As a consequence, the persistence of the intact viral genome in hamster cells appears to be selected against or else the intact Ad2 genome would be replicated, the infected cells would all be killed and transformed cells would probably not arise. Consistent with this line of reasoning, we have found fragments of integrated Ad2 sequences or integrated Ad2 genomes with varying lengths of internal deletions in Ad2-transformed hamster cells. Thus, patterns of persistence and integration can be decisively influenced by the biology of the virus–host system. At least, that is one possible interpretation of the data. In some of the cell lines investigated, the Ad2 genome, usually with internal deletions, is integrated in an orientation co-linear with the arrangement of the Ad2 DNA sequence in the virion. In some instances, the orientation has been rearranged (Vardimon and Doerfler, 1981).

In most cases, certainly in cells carrying Ad12 genomes in an integrated form, linkage of the viral sequences to the adjacent cellular DNA sequences was via the terminal viral DNA sequences. At most junction sites between viral and cellular DNAs analysed so far, a number of viral nucleotides were deleted in the process. This number of deleted nucleotides ranged from 0 to 174 in different cell lines. At the left end of integrated Ad12 DNA in the Ad12-transformed hamster cell line HA12/7, for example, not a single viral nucleotide was found to be deleted. At the other end of the spectrum, viral DNA integration could proceed without the deletion of a single cellular nucleotide at the site of linkage, or large segments of cellular DNA could be lost in the process of inserting foreign (viral) DNA. Most frequently, multiple copies of viral DNA molecules became integrated upon infection of cells or hamsters with adenoviruses. We were able to demonstrate that these multiple copies were not integrated in true tandem fashion but that other nucleotide sequences, cellular DNA or rearranged viral DNA sequences, had become interspersed between adjacent viral DNA molecules. Viral DNA termini were found to be rearranged or partly inverted in some of the integrated DNA molecules, particularly when multiple copies of viral DNA were integrated.

When cell lines carrying integrated viral DNA molecules were passaged in culture for longer periods of time, the loss of all or part of the viral genomes from the transformed cell lines was occasionally observed (Groneberg et al., 1978; Eick et al., 1980; Kuhlmann et al., 1982). This loss was accompanied by changes in cell morphology. However, in a few instances, the loss of Ad12 DNA sequences, including the left terminal segment, which was considered to be essential in the transformation of cells by adenoviruses, did not affect the tumorigenic phenotype of these revertants (Kuhlmann et al., 1982). Thus, at least in these cell lines, persistence of the viral genome could not be considered an absolute prerequisite for the maintenance of the tumorigenic cell phenotype. It was conceivable that, as a consequence of viral infection and/or integration of Ad12 DNA into the cellular genome and the subsequent loss of the integrated foreign DNA, the organization of the cellular genome or the expression patterns of cellular genes were altered such that the affected cells were transformed to malignant cells. In cells that had lost the Ad12 genomes, Ad12 DNA could no longer be detected by the very sensitive Southern blotting technique (Southern, 1975) and subsequent hybridization to $^{32}$P-labelled cloned fragments of Ad12 DNA. We are currently re-examining whether traces of Ad12 DNA can still be detected in these revertants by the polymerase chain reaction (PCR). For this purpose, the revertant cell lines will have to be rigorously recloned to ascertain absence of a few cells from the original Ad12-induced tumour cells.

Junction sequences between adenoviral and cellular DNA sequences were cloned and their nucleotide sequences were determined from the following cell lines: the Ad12-transformed hamster cell lines T637 (M. Lutze, B. Schmitz, and W. Doerfler, unpublished observations) and HA12/7 (Jessberger et al., 1989a), from the Ad2-transformed hamster cell line HE5 (Gahlmann et al., 1982; Gahlmann and Doerfler, 1983), the Ad12-induced hamster tumour cell lines CLAC1 (Stabel and Doerfler, 1982), CLAC3 (Deuring et al., 1981a), T1111/2 (Lichtenberg et al., 1987) and H191 (M. Lutze, B. Schmitz and W. Doerfler, unpublished observations), from the Ad12-induced mouse tumour line CBA12/T1 (Schulz and Doerfler, 1984) and from the symmetric recombinant (SYREC2) of Ad12 virus (Deuring and Doerfler, 1983).

In the Ad12-encapsidated SYREC2 DNA molecule, the DNA consisted of the left terminal 2081 nucleotide pairs of Ad12 DNA positioned at both SYREC termini, and a large palindrome of human cellular DNA of partly unique and partly repetitive DNA sequences to make up a molecule that had about the length of Ad12 DNA. This recombinant DNA molecule generated in cell culture could be packaged into Ad12 virions owing to the presence of terminal Ad12 sequences that might be akin to known packaging sequences identified at the left terminus of other adenoviruses. Upon denaturation and reannealing of SYREC2 DNA, molecules were generated with about half the length of Ad12 DNA, attesting to the proposed structure of palindromic DNA molecules (Deuring et al., 1981b). In addition, restriction analyses of these DNA molecules revealed that they contained the left terminus of Ad12 DNA at both termini. These recombinant molecules of course required wild-type Ad12 as a helper for their replication in cell culture. It will be interesting to evaluate these SYREC molecules for their potential as adenovirus vectors for the encapsidation and transfer of large segments of foreign DNA. With an original length similar to that of Ad12 DNA with 34 125 nucleotide pairs (Sprengel et al., 1994), the SYREC DNA molecule might accommodate up to 30 kbp (kilobase pairs) of foreign DNA.

Another aspect raised by the structure and composition of the SYREC2 DNA molecule was that of recombination between Ad12 DNA and human cellular DNA in productively infected human cells. After the infection of permissive human cells with Ad12 or Ad2 virions, transformed human cells were never obtained. The few adenovirus-transformed human cell lines available were generated by transfecting viral DNA fragments into human cells. The existence of SYREC2 DNA molecules proved that, even in human cells productively infected with Ad12, recombination could proceed between Ad12 DNA and cellular DNA. It could not be decided whether the SYREC molecules were generated as a consequence of viral DNA integration and excision or in the course of recombination between Ad12 DNA and fragments of cellular DNA produced during viral infection. By cytogenetic and *in situ* hybridization methods, the preferential association of Ad12 DNA with human chromosome 1 in Ad12-infected cells was demonstrated early and late after Ad12 infection of permissive human cells (McDougall et al., 1972, 1973; Rosahl and Doerfler, 1988).

Although there was no evidence for site-specific integration in many cell lines in which the junctions between Ad12 DNA and cellular DNA were analysed there were a few examples, suggesting selectivity in certain integration events. Our conclusions on the absence of site-specific integration of viral DNA in adenovirus-transformed or Ad12-induced tumour cells were based on results obtained with cloned cell lines that had been kept in culture for long periods. It was unknown to what extent these experimental parameters might have selected for cells with the foreign viral DNA integrated in a particular manner, whether the integrated viral DNA could have been rearranged during cell culture, or whether the results obtained were actually representative of the primary integrative behaviour of adenovirus genomes.

We recently isolated and characterized a set of five Ad12-induced hamster tumour cell lines from Ad12-induced tumours by maintaining uncloned tumour cells in culture for longer periods of time. Subsequently, the patterns of Ad12 DNA integration were analysed by using five different restriction endonucleases and the Southern blotting procedure (Orend et al., 1994). Terminal Ad12 DNA fragments on these blots were identified by hybridizing the DNA fragments to the cloned terminal Ad12 DNA fragments. In these experiments, the patterns of Ad12 DNA integration appeared very similar, if not identical, for all five hamster tumour cell lines which were derived from different oncogenesis experiments. We do not yet understand what selective procedure during cell culture might have led to the isolation of these five cell lines, one of which was isolated about 10 years prior to the isolation of the other four (Kuhlmann and Doerfler, 1982). Integration patterns of Ad12 DNA had also been determined at early passage levels after explantation of the cells from the tumour-bearing animals and, at that time, the integration patterns were different in these different cell lines. Upon prolonged cultivation, cells might have been selected that had growth advantages and integration patterns which could have changed under the selective conditions of cell culture employed. Such changes might not occur very frequently because, in most cloned cell lines analysed for integration patterns over a long period of time, such changes were not apparent. It was probably significant that the cell lines with similar integration sites of Ad12 DNA originated from uncloned populations of Ad12-induced hamster tumour cells. In cell lines generated from cloned tumour cells, the selection of cells with preferred integration sites of Ad12 DNA might have been prevented by excluding a cell population capable of rearranging integrated viral DNA during cell culture.

## 7.3 On the mechanism of integrative recombination

### 7.3.1 Insertion of foreign DNA by a versatile mechanism

The results of studies on the model of adenovirus DNA integration in mammalian cells suggested that the insertion mechanism had to be rather pliable. Usually, many copies of viral DNA were integrated, often not in true tandem fashion; individual viral DNA molecules could be interspersed with cellular or rearranged viral DNA segments. Intact or nearly intact viral genomes or additional viral DNA fragments were integrated, sometimes in proximity to the intact viral genomes. At the sites of junction to cellular DNA, viral nucleotides could be deleted; occasionally the terminal viral nucleotide sequence was preserved. At several of the nucleotide sequences linking viral and cellular DNAs, patchy homologies were observed either between the linked viral and cellular nucleotide sequences or between the deleted terminal viral sequences and the cellular segments replacing them. However, junctions devoid of such patchy sequence homologies were also noted. The mechanism of integrative recombination operating in mammalian cells could apparently take advantage of patchy homologies but did not have to depend upon their availability. Recombinants with patchy homologies at the sites of junction were perhaps more frequently found than those without patchy homologies. When taking into account the flexibility of the integrative recombination mechanism with respect to the necessity for nucleotide sequence homologies between the reaction partners, one appreciated that it was logically difficult to categorize this mechanism as being akin to homologous or to non-homologous recombination.

The sometimes frustrating desire for neat, simple classifications disregards the fact that integrative recombination in mammalian cells tends to utilize pliable mechanisms that cannot easily be subsumed under idealized categories. The mechanism can work under a variety of molecular constellations and, perhaps for that reason, has proved to be quite efficient and successful. This mechanism may have evolved over long periods and plays a major role in permitting cells to incorporate foreign DNA very efficiently, and subsequently to select or counterselect for the persistence and the continued expression of certain foreign DNA sequences or for their persistence and permanent silencing by sequence-specific DNA methylation, respectively.

There is an additional element to consider in investigations into the mechanism of insertional recombination in mammalian cells. We have reported that all the cellular DNA sequences at the junction sites between adenovirus DNA and cellular DNA, that we have examined in detail, are transcriptionally active (Gahlmann et al., 1984; Lichtenberg et al., 1987; Schulz et al., 1987). Transcriptional activity for these sequences was found in hamster, mouse or human cells that had never been exposed to adenoviruses, in the Ad12-induced tumour cells and in adenovirus-transformed cells. The transcription products derived from these cellular sites vary in length and quality. They may constitute short RNAs, probably without coding capacity, or represent transcription products exhibiting open reading frames. I propose that the cellular DNA sequences which are actively transcribed assume an appropriate chromatin configuration and are bound by cellular proteins that render them particularly amenable to the machinery for integrative recombination with foreign DNA. For foreign genes to be transcribed, it would be advantageous to have access to active transcription processes. Although this argument is merely one of plausibility, it is still supported by the fact that integrative recombination has also been found to be directed towards cellular sites of transcriptional activity in other viral and cellular systems (Mooslehner et al., 1990; Scherdin et al., 1990).

In the adenovirus integration system, the question is still unresolved as to whether cellular factors alone suffice to expedite adenovirus DNA integration, whether viral gene products participate as essential components of the integrative recombination machinery or whether they merely modify cellular factors. Since adenovirus DNA, like any other foreign DNA transfected into mammalian cells, can be integrated into the cellular genome in the absence of viral infection, the bias lies in favour of a mechanism of integrative recombination supplied by the recipient cell. However, it is likely that, upon adenovirus infection and subsequent adenovirus DNA integration, viral gene products can exert an auxiliary or modifying role in integrative recombination between viral and cellular DNAs. Integrative recombination may not be dependent on adenoviral functions but they could render the event more efficient or alter essential parameters to render the reaction more specific. It is prudent to keep in mind that the mechanism under investigation has a high degree of flexibility. A number of years ago, we began to study this mechanism further by using a cell-free system from nuclear extracts of uninfected hamster cells (Jessberger et al., 1989b; Tatzelt et al., 1992, 1993; Fechteler et al., 1995).

## 7.3.2 Studies on the mechanism of integrative recombination in a cell-free system

High-salt extraction of isolated nuclei from BHK21 hamster cells generates nuclear, cell-free extracts that facilitate *in vitro* recombination between fragments of Ad12 DNA and a pre-insertion sequence from hamster cells (Jessberger *et al.*, 1989b). In designing this experimental approach, it has been reasoned that, in the absence of information about the actual requirements for the integrative recombination reaction, a cellular DNA sequence that had already previously served as an integration target in a living organism should be chosen as recombination partner with Ad12 DNA. We have cloned and sequenced the pre-insertion sequence from BHK21 hamster cells that corresponds to the insertion sequence of Ad12 DNA in the Ad12-induced hamster tumour cell line CLAC1 (Stabel and Doerfler, 1982). This cellular pre-insertion sequence has been designated p7. Of course, it is impossible to predict what nucleotide sequence requirements a cellular DNA segment must exhibit to qualify as an integrative recombination partner with foreign DNA. Thus, a previously identified cellular DNA segment as the target of Ad12 DNA integration appeared to be a qualified choice.

In a series of such cell-free recombination experiments, we have observed that the segment of Ad12 DNA comprising nucleotides 20 885–24 053 in the complete Ad12 DNA sequence (Sprengel *et al.*, 1994) recombines *in vitro* more frequently with the p7 hamster pre-insertion DNA sequence than do other Ad12 DNA segments (Jessberger *et al.*, 1989b). At present, this phenomenon cannot be explained, but it is not due to obvious nucleotide sequence homologies between p7 hamster DNA and the nt 20 885–24 053 fragment of Ad12 DNA. This and other Ad12 DNA segments recombine much more frequently with the p7 hamster DNA sequence than with the adjacent pBR322 vector sequence in the construct used. With a second Ad12 pre-insertion DNA segment (p16) from hamster cells (Lichtenberg *et al.*, 1987), similar results of enhanced recombination with the Ad12 DNA fragment have been obtained. We have also investigated the possibility that randomly selected hamster cell DNA sequences would recombine with Ad12 DNA fragments in the cell-free system, but have so far not found recombinants (Jessberger *et al.*, 1989b; Tatzelt *et al.*, 1992). It is, therefore, likely that the p7 pre-insertion sequence contains motifs that are preferentially recognized by the recombination machinery and are used as recombination targets for Ad12 DNA. Non-pre-insertion hamster DNA sequences may thus be utilized much less efficiently.

By using standard gel filtration and chromatographic procedures, we have been able to purify components from the nuclear extracts to a considerable extent. The most highly purified preparations that still catalyse *in vitro* recombination of the selected p7 and Ad12 DNA (nt 20 885–24 053 fragment) partners, exhibit a limited number of proteins when analysed by sodium dodecyl sulphate–polyacrylamide gel electrophoresis (SDS–PAGE) followed by silver staining of the polypeptides (Tatzelt *et al.*, 1993). More recent experimental work has led to further purification of the components. The most highly purified fractions exhibit four or five major protein bands when analysed as described (Fechteler *et al.*, 1995). The functional characterization of these components has been initiated. It may be surmised that the recombination machinery consists of several proteins, perhaps combined in a complex. Thus, further purification may, at some stage, lead to the disruption of the necessary ensemble of cellular proteins. The fact that we have been able to demonstrate recombination between p7 hamster cell DNA and Ad12 DNA in fractionated nuclear extracts from uninfected hamster cells demonstrates that, at least for the imitation reaction described here, cellular factors do suffice. Of course, our model system may still not catalyse the true integration reaction of Ad12 DNA. We have yet to analyse whether viral-encoded functions participate in the reaction qualitatively or quantitatively.

We have applied an *Escherichia coli* transfection test using recA$^-$ bacterial strains as an assay system for the identification of recombinants, and have documented in a large number of control experiments that recombination between p7 hamster DNA and Ad12 DNA fragments has not occurred after transfection into *E. coli*, but in the cell-free extracts (Jessberger *et al.*, 1989b; Tatzelt *et al.*, 1992). This interpretation has been corroborated by the finding that a completely different assay system also successfully identified recombinants generated in a cell-free system: PCR (Saiki *et al.*, 1988) reveals recombinants when used with DNA re-extracted directly from the fractionated nuclear mixture, and these recombinants resemble those identified by the transfection method (Tatzelt *et al.*, 1993). This assay obviates the involvement of *E. coli* transfection in the isolation of recombinants. Again, the results of a large number of control experiments have verified our interpretations and ruled out the possibility of PCR artefacts. In particular, when the reaction partners were incubated separately with fractionated nuclear extracts, subsequently re-extracted, mixed and then subjected to PCR, recombinants were not found. With the improved purification of the system, it may become feasible in the future to identify recombinants directly by Southern blotting without the use of additional procedures.

An appreciable number of *in vitro*-generated recombinants have been analysed for their nucleotide sequences at the junctions between Ad12 DNA and the p7 hamster DNA sequence. It is striking that, as described for the junctions from adenovirus-transformed cells or Ad12-induced tumour cells, patchy homologies are apparent at the junction sites of the recombinants generated in the cell-free system as well (Jessberger *et al.*, 1989b; Tatzelt *et al.*, 1992). This finding encourages expectations that this cell-free system simulates, to some extent, elements of the *in vivo* integration reaction.

The data obtained so far should be complemented in the future by work examining the activity of extracts from nuclei of Ad12-infected hamster cells. Moreover, we have initiated experiments utilizing cosmid constructs with cellular pre-insertion sequences and the intact Ad12 genome, either as isolated DNA or complexed with authentic viral core proteins, in the hope of approximating the actual integration reaction more closely. We have also explored a cell-free system from nuclei of *Spodoptera frugiperda* insect cells which catalyse the *in vitro* recombination between adenovirus DNA and *Autographa californica* nuclear polyhedrosis virus DNA fragments (Schorr and Doerfler, 1993). Apparently, the capacity for this type of non-homologous recombination is inherent also in insect cells.

## 7.4 *De novo* DNA methylation of integrated foreign DNA

### 7.4.1 De novo *methylation of integrated foreign DNA: a cellular defence mechanism?*

In our investigations on the structure of integrated Ad12 genomes in transformed hamster cells, we found that these genomes became methylated *de novo* in very specific patterns (Sutter *et al.*, 1978; Sutter and Doerfler, 1979, 1980; Vardimon *et al.*, 1980; Kuhlmann and Doerfler, 1982; Orend *et al.*, 1991, 1995). We had previously shown that the virion DNA from purified Ad12 particles was not detectably methylated (Günthert *et al.*, 1976; Wienhues and Doerfler, 1985; C. Kämmer and W. Doerfler, unpublished observations). Thus, it was possible to demonstrate unequivocally that the previously unmethylated Ad12 DNA was methylated *de novo* after integration into the cellular genome.

An inverse correlation was observed in many parts of integrated adenovirus genomes between the extent of DNA methylation and the level of transcription (Sutter and Doerfler, 1979, 1980; Vardimon *et al.*, 1980). This correlation has subsequently been refined for the promoter regions of integrated adenovirus genes (Kruczek and Doerfler, 1982). These observations initiated a decade of research on the role that sequence-specific promoter methylation can play in the long-term silencing of eukaryotic genes. In these studies (for reviews, see Doerfler, 1981, 1983a, 1984a, 1989, 1992, 1993; Doerfler *et al.*, 1993), we have mainly, but not exclusively, used viral promoters. The contents of earlier reviews are not duplicated here. DNA methylation has more recently been recognized for its importance in long-term gene inactivation both in the sphere of developmental biology and in human genetics. The earlier work on eukaryotic viral and cellular promoters has conceptually opened the path for studies on complex genetic phenomena (e.g. genomic imprinting in mammalian genomes; Surani *et al.*, 1984; Sapienza *et al.*, 1987; Reik *et al.*, 1987; Swain *et al.*, 1987; Hu *et al.*, 1993). The observation that integrated foreign DNA molecules, such as the Ad12 genome, can be methylated *de novo* and consequently partly or completely inactivated, has not been restricted to adenovirus genomes. Other integrated viral genomes or, for that matter, any foreign DNA integrated into established genomes [e.g. after transfection and selection in mice (Lettmann *et al.*, 1991) or in plants (Linn *et al.*, 1990)], have been subjected to the same, apparently ubiquitous control mechanism and have become extensively methylated *de novo* and inactivated. There are also examples in which *de novo* methylation has not ensued.

The genome of Epstein–Barr virus (EBV), a member of the herpesvirus group, which can persist in virus-transformed cells predominantly in a non-integrated, circular, episomally free form, can also become *de novo* methylated in specific patterns (Ernberg *et al.*, 1989; Hu *et al.*, 1991). Thus *de novo* methylation of foreign genomes in mammalian cells is not solely associated with the integrated state of the newly acquired DNA. Persisting EBV genomes continue to be replicated in synchrony with the cycle of the cellular genome, presumably by cellular DNA polymerase systems. It is challenging to consider the possibility that the cellular replication machinery may be intimately associated with the apparatus for the *de novo* methylation of DNA which would then be responsible for the methylation of the EBV DNA. By contrast, free adenovirus DNA replication is self-sufficient and provides its own replication system which lacks, however, a DNA methyltransferase system. Perhaps, for that reason, intracellular, free adenovirus DNA has never been found to become methylated *de novo* (Wienhues and Doerfler, 1985; C. Kämmer and W. Doerfler, unpublished observations).

Since the insertion of foreign DNA into established genomes and its continued transcription constitutes a major goal of many, though not all, strategies in gene therapy, the mechanism of *de novo* methylation and subsequent long-term inactivation of integrated foreign genomes requires serious consideration and detailed investigation. *De novo* methylation may represent a major obstacle on this frequently considered path towards the successful repair of genetic defects in mammalian cells. Alternative approaches, such as the presentation of foreign genes in free non-integrated form (e.g. in free adenovirus genomes; Ragot *et al.*, 1993), under conditions in which they do not predominantly integrate, may have a better chance of providing the means for long-term, non-obstructed expression of foreign genes designed to substitute for missing genetic functions in a cell or an organism. Even if a defective gene could be replaced precisely by the wild-type allele, the question remains as to whether this replaced gene or DNA segment would also be subject to the defence mechanism of *de novo* methylation, because, owing to the lack of an authentic cellular methylation pattern, it might be recognized as foreign by the *de novo* methylation system of the cell.

As mentioned earlier, the *de novo* methylation of integrated foreign DNA in established genomes can be viewed as an ultimate cellular defence mechanism which apparently can operate selectively. Possibly by survival and selection of cells with an optimized pattern of *de novo* methylated (and inactivated) or non-methylated (and continually expressing) foreign genes, specific patterns of *de novo* methylated genes persist and contribute to the constellation of newly introduced genes in an altered genome. Adenovirus-transformed cells provide an example of this mechanism. Frequently, the early viral genes, specifically the E1 and E4 genes, do not succumb to this cellular defence mechanism Sutter and Doerfler, 1980; Orend *et al.*, 1991, 1995), probably because, by selection, they can escape inactivation and contribute to the transformed state of those cells in which they continue to be expressed.

It appears that cells have developed different mechanisms for their defence against the insertion of foreign DNA and genes. Under experimental conditions, a variety of options have become available for the introduction of foreign DNA into cells in culture. It is not known how frequently cells of an intact organism are exposed to, take up, and chromosomally integrate foreign DNA. It is likely that the cytoplasmic membrane is a first barrier against the penetration of foreign DNA molecules. Once that barrier is penetrated, foreign DNA can be nucleolytically degraded in the cytoplasm or in its organelles. Nevertheless, foreign DNA can be transported to the nucleus and become integrated. Such integrated genomes can be lost again from the cellular genome, as exemplified by the existence of morphological revertants of Ad12-transformed cells in which viral genomes, in part or *in toto*, have been excised (Groneberg *et al.*, 1978; Eick *et al.*, 1980). Finally, should all these possible mechanisms for the elimination of foreign genes fail, the genes could be *de novo* methylated and thus become inactivated.

### 7.4.2 Initiation of de novo methylation in mammalian cells is not predominantly dependent upon the nucleotide sequence of foreign DNA

Integrated Ad12 DNA in hamster tumour cells or Ad12 DNA fixed in the hamster cell genome by transfection and selection is not immediately methylated *de novo*. It requires an unknown number of cell generations – and other unknown factors – to initiate *de novo* methylation. We have investigated where in the co-linearly integrated Ad12 genome *de novo* methylation is initiated. It commences in two paracentrally located regions of Ad12 DNA and not, for example, at the termini of Ad12 DNA which are contiguous with cellular DNA sequences that probably present an established methylation pattern (Orend *et al.*, 1991, 1995).

There is evidence from several different systems in which *de novo* methylation has been studied that certain nucleotide sequences may be preferentially methylated *de novo* (Szyf *et al.*, 1990; Mummaneni *et al.*, 1993; Hasse and Schulz, 1994). While nucleotide sequence may play a role in selecting sites for the initiation of *de novo* methylation, the results from investigations of *de novo* methylation in the adenovirus system argue that nucleotide sequence cannot by itself be the sole determinant of the sites of initiation of *de novo* methylation.

(i) In integrated Ad12 genomes, *de novo* methylation is initiated in the paracentrally located nt 20 885–24 053 fragment of the viral genome, more precisely in an internal segment of this region (Orend *et al.*, 1995). When the same viral DNA segment is transposed (e.g. to the left end of the integrated Ad12 genomes in cell line T637), the transposed Ad12 DNA segment with the same nucleotide sequence is not methylated, or, at least, not to the same extent as the internally located segment.

(ii) When fragments of Ad12 DNA, such as the nt 1–5574 (*Eco*RI-C) fragment or the nt 20 885–24 053 (*Pst*I-D) fragment are transfected into and fixed by integration in the genomes of mammalian cells, these DNA fragments become methylated in some cell lines, but remain unmethylated in others, possibly dependent on the site of foreign DNA integration (Orend et al., 1995). By contrast, in Ad12-transformed cells, which have been transformed by infection with Ad12 virions and carry the entire Ad12 genomes in an integrated form, the 1–5574 nt fragment remains hypomethylated, the internal *Pst*I-D fragment becomes heavily methylated.

(iii) In the Ad2-transformed hamster cell line HE1 (Cook and Lewis, 1979), the late E2A promoter in the Ad2 genome is completely methylated at all 5'-CG-3' sequences, as determined by genomic sequencing. Exactly the same nucleotide sequence is completely unmethylated in another Ad2-transformed hamster cell line, HE2 (Toth et al., 1989, 1990). In HE1, the E2A gene has been silenced; in HE2, it is transcribed and translated into DNA-binding protein (Johannsson et al., 1978).

(iv) The large segment of human cellular DNA sequences in the symmetric Ad12 DNA recombinant SYREC2 (Deuring et al., 1981b; Deuring and Doerfler, 1983) is not methylated in its 5'-CCGG-3' sequences in the SYREC2 genome isolated from purified virions. The same cellular nucleotide sequences are, however, very heavily methylated in the 5'-CCGG-3' sequences within the human cellular genome in cells growing in culture.

It may therefore be tentatively concluded that the *de novo* methylation mechanism is not predominantly regulated simply by a specific nucleotide sequence, but that additional parameters such as location in different intranuclear compartments, DNA structure, the type of proteins bound at such structures, the site of foreign DNA integration and the replicative state of the cell may all have an important influence. For future experimental designs in related projects, it is important to recognize that one must not select for the same foreign DNA segments whose *de novo* methylation one intends to follow.

Unfortunately, next to nothing is known about the enzymatic mechanism of *de novo* methylation. It is not clear whether *de novo* and maintenance methylation are effected by the same or by different enzymes, or by one enzyme in conjunction with different co-factors. We still do not know how many DNA methyltransferases exist in mammalian cells (Li et al., 1993).

In our own work on DNA methyltransferases, we have turned to studies on frog virus 3 (FV3), a member of the iridovirus group (Willis et al., 1989). The FV3 genome in the virion is heavily methylated (Willis and Granoff, 1980). Using the genomic sequencing technique, we have demonstrated that, in the viral DNA segments investigated, all 5'-CG-3' sequences are methylated and that 5-methylcytosine (5mC) occurs exclusively in these dinucleotide sequences (Schetter et al., 1993). From the results of experiments using several methylation-sensitive restriction enzymes, it appears likely that all

5'-CG-3' sequences in the genome are methylated (Willis and Granoff, 1980; Schetter *et al.*, 1993). There is evidence that, after infection of cells with FV3, the newly synthesized viral DNA is not immediately methylated after replication, but becomes rapidly and completely methylated *de novo* shortly after. This system thus offers the possibility to study characteristics of the enzyme(s) involved in *de novo* methylation. Such studies have been initiated (Schetter *et al.*, 1993).

### 7.4.3 Methylation of triplet repeat amplifications in the human genome: manifestation of the cellular defence mechanism

Several human genetic diseases are known in which the apparently autonomous amplification of naturally occurring triplet repeats occurs [e.g. myotonic dystrophy, the fragile X syndrome (*FRAXA*), Kennedy disease, Huntington's disease, mental retardation associated with the fragile site *FRAXE* on the human X chromosome, spinocerebellar ataxia type I and hereditary dentatorubralpallidoluysian ataxia (Caskey *et al.*, 1992; Richards and Sutherland, 1992; Riggins *et al.*, 1992; The Huntington's Disease Collaborative Research Group, 1993; Knight *et al.*, 1993; Orr *et al.*, 1993; Koide *et al.*, 1994; Nagafuchi *et al.*, 1994)]. These sequence amplifications either occur in the coding sequence of genes or in their 3' or 5' non-coding regions. It is not yet understood in any detail how these 'dynamic mutations' are related to the pathogenetic mechanisms of these disorders. Moreover, the mechanism by which triplet repeat expansion takes place is still unknown. We have observed that synthetic oligodeoxyribonucleotides such as $(CGG)_{17}$, $(GCC)_{17}$, $(CG)_{25}$ [but not $(TAA)_{17}$ or $(CAGG)_{13}$] can be expanded autonomously *in vitro* by *Taq* DNA polymerase (PCR conditions) or by Klenow polymerase (without cycling) to oligomers of up to 2000 bp in length. This *in vitro* amplification, which apparently requires a certain nucleotide sequence (and hence a specific structure), can be inhibited, though not abolished, by the methylation of the C residues in the oligodeoxyribonucleotides; much shorter chains are then synthesized *in vitro* (Behn-Krappa and Doerfler, 1994).

In some of the amplified sequences associated with human genetic diseases (e.g. the *FMR1* gene in the fragile X syndrome), an increase of DNA methylation in these sequences has been observed (Oberlé *et al.*, 1991). We have suggested (Behn-Krappa and Doerfler, 1994) that this *de novo* methylation may represent another manifestation of the cellular defence mechanism against foreign DNA mentioned above. According to this reasoning, the amplified triplet repeats containing many 5'-CG-3' dinucleotides could be recognized as foreign by the DNA methyltransferase system of the host cell and could thus become methylated *de novo* as a barrier against further expansion. This interpretation would be consistent with the results of our *in vitro* amplification studies and the effect of 5'-CG-3' methylation which inhibits amplification.

## 7.5 Alterations in patterns of cellular DNA methylation and gene expression as consequences of foreign DNA insertions into mammalian genomes?

We have investigated the possibility that the insertion of foreign DNA into an established mammalian genome can lead to far-reaching alterations in patterns of cellular DNA methylation and gene expression. These alterations might be as significant to the oncogenic transformation of cells as some of the viral gene products thought to be involved in this process. In the pursuit of this concept, we have initially been able to demonstrate that the methylation state of a hamster cell DNA sequence of about 1 kbp in length that immediately abuts the site of insertion of Ad12 DNA in the Ad12-induced tumour T1111/2 has been altered in that all the sequences that are completely methylated in normal hamster DNA have lost the methyl group (Lichtenberg et al., 1988).

More recently, we screened the Syrian hamster cell genome with several different randomly selected genomic DNA or cDNA probes by Southern blot hybridization after cutting the hamster cell DNA with *Hpa*II, *Msp*I or *Hha*I. Cellular DNA was extracted from the following cell types:

(i) a number of Ad12-transformed hamster cell lines;
(ii) Ad12-induced tumour cell lines;
(iii) BHK21 cells that carried integrated Ad12 genomes fixed by neomycin (*neo*) gene co-transfection and *neo* selection in the hamster genome, but which did not show the Ad12-transformed phenotype. In these cells, the E1 region of the Ad12 genome was not detectably transcribed;
(iv) normal BHK21 hamster cells;
(v) Ad12-infected BHK21 hamster cells at 30 h and several weeks post-infection;
(vi) BHK21 cells with integrated plasmid DNA.

Among the different cellular hybridization probes, several showed very striking increases in DNA methylation in cellular genes in some of the Ad12-transformed cells, in some of the Ad12-induced tumour cells, in BHK21 cells with integrated Ad12 DNA lacking the Ad12-transformed phenotype or in BHK21 cells carrying integrated plasmid DNA, whereas others showed no changes (Heller et al., 1995).

It was shown by FISH that the hybridization probes used in these experiments were located on different hamster chromosomes (Heller et al., 1995). One of the cellular DNA hybridization probes used in these studies, the intracisternal A particle (IAP) DNA, which exhibited a very striking increase in DNA methylation in cell line T637 as compared to BHK21 hamster cells, was distributed on most hamster chromosomes.

Several possible interpretations can be offered to account for these changes in patterns of cellular gene methylation:

(i) transformation by Ad12;
(ii) insertion of foreign DNA into the established hamster cell genome;
(iii) the action of early Ad12 gene products synthesized in Ad12-transformed or Ad12-induced tumour cells.

We have tried to distinguish between these possibilities and analysed cellular DNA isolated from BHK21 cells at 30 h or several weeks after infection with Ad12. In these cells, none of the aforementioned alterations in cellular gene methylation have been observed. Two BHK21 cell lines, HAd12-neo2 and HAd12-neo5 (Orend et al., 1995), which contained multiple copies of integrated Ad12 DNA, or BHK21 cells carrying integrated plasmid DNA only, were also investigated. These cell lines did not show the transformed phenotype typical of Ad12 but were indistinguishable from BHK21 cells. An increase in DNA methylation for the IAP probe was also observed in these cells.

We therefore favour the interpretation that, perhaps in conjunction with the transformed state of the cells, the insertion of the Ad12 genomes or plasmid DNA into the established hamster genome contributes to the increase in methylation in many regions of the cellular genomes. The mechanism, by which these changes are effected, is not known. Since the patterns of early Ad12 DNA expression are similar in Ad12-infected and in Ad12-transformed hamster cells, it is unlikely that Ad12 gene products play a decisive role in rapidly changing the patterns of methylation in cellular genes of transformed cells. It should be emphasized that there are considerable differences in the extent and locations of these changes in different Ad12-transformed cells or Ad12-induced hamster tumour cells. Moreover, only a subset of genes seems to be affected. The cellular genes with altered methylation patterns are located on hamster chromosomes distinct from the chromosome on which Ad12 DNA is integrated. Since we have used only a relatively small number of randomly selected hamster gene probes and found changes in DNA methylation in a high proportion of them, these alterations must be extensive and widely distributed.

It has been shown in many different systems that changes in DNA methylation are associated with changes in patterns of gene expression. In Ad12-transformed cells, we demonstrated that among 40 different genes tested, the expression of five genes was altered in comparison to non-Ad12-transformed BHK21 cells (Rosahl and Doerfler, 1992). Although the ratio of the number of genes with alterations in expression to the number of genes tested argues for frequent changes, more work will be required to support this interpretation. We continue to pursue the possibility that the integration of foreign DNA into the hamster genome is associated with widespread changes in DNA methylation and consequently in expression patterns among hamster cellular genes. These findings will have significance for the mechanism of viral oncology, for gene therapy and for the interpretation of results obtained from transgenic organisms.

## 7.6 DNA methylation and gene activity

### 7.6.1 A fully 5'-CG-3' but not a 5'-CCGG-3' methylated late FV3 promoter retains activity

Several lines of evidence demonstrate that the DNA of the iridovirus FV3 is methylated in all 5'-CG-3' sequences both in virion DNA and in the intracellular viral DNA at late times after infection. The 5mC residues in this viral DNA occur exclusively in 5'-CG-3' dinucleotide positions. We have cloned and sequenced the *L1140* gene and its promoter from FV3 DNA. The gene encodes a 40 kDa protein. The results of transcriptional pattern analyses for this gene in fat head minnow (FHM) fish cells document that this gene is transcribed exclusively late after FV3 infection. The *L1140* gene and its promoter are fully methylated at late times after infection. We have been interested in resolving the apparent paradox that the methylated *L1140* promoter is methylated and active late in FV3-infected cells. Of course, the possibility cannot be excluded that one or a few 5'-CG-3' sequences outside restriction endonuclease sites might have escaped *de novo* methylation after FV3 DNA replication. A construct has been devised that places the chloramphenicol acetyltransferase (CAT) gene construct under the control of the *L1140* promoter. Upon transfection, this construct exhibits activity only in FV3-infected BHK21 hamster cells, and not in uninfected BHK21 cells. For technical reasons, FHM cells have proved less suitable for transfection experiments. The fully 5'-CG-3' or 5'-GCGC-3' (*Hha*I) methylated, *Hpa*II-mock-methylated or unmethylated *L1140* promoter–*CAT* gene construct is active in FV3-infected BHK21 cells, whereas the same construct 5'-CCGG-3' (*Hpa*II) methylated has lost all activity. Apparently, complete methylation of the late *L1140* promoter in FV3 DNA is compatible with activity. However, a very specific 5'-CCGG-3' methylation pattern, that does not naturally occur in authentic FV3 DNA in infected cells, abrogates promoter function. These results further support the notion that very specific patterns of methylation are required to inhibit or inactivate eukaryotic and viral promoters (Munnes *et al.*, 1995).

### 7.6.2 Topology of the promoter of RNA polymerase II- and III-transcribed genes is modified by the methylation of 5'-CG-3' dinucleotides

In eukaryotic cells, RNA polymerase II- and III-transcribed promoters can be inactivated by sequence-specific methylation. For some promoter motifs, the introduction of 5mC residues has been shown to alter specific promoter motif–protein interactions. To what extent does the presence of 5mC in promoter or regulatory sequences affect the structure of DNA itself? We have investigated changes in DNA bending in three naturally occurring DNA elements, the late E2A promoter of Ad2 DNA, one of our main model systems, the VAI (virus-associated) RNA gene of Ad2 DNA, and an *Alu* element associated with the human angiogenin gene. Alterations in electrophoretic mobility of differently permuted promoter segments in non-denaturing polyacrylamide gels have been used as the assay system. In the late E2A promoter of Ad2 DNA, a major, and possibly some minor, DNA-bending motifs exist which cause deviations in electrophoretic mobility in comparison to co-electrophoresed marker DNA fragments devoid of DNA-bending motifs. DNA elements have been specifically methylated *in vitro* by the *Hpa*II (M-*Hpa*II; 5'-CCGG-3'), the *Fnu*DII (M-*Fnu*DII; 5'-CGCG-3') or the CpG DNA methyltransferase from *Spiroplasma* species (M-*Sss*I; 5'-CG-3').

Methylation by one of these DNA methyltransferases influences the electrophoretic mobility of the three tested promoter elements significantly, although to different extents. It cannot be predicted whether sequence-specific promoter methylation increases or decreases electrophoretic mobility; these changes must be determined experimentally. Methylation of the *E. coli* dcm (5'-CCATGG-3') sites in some of the DNA constructs does not elicit mobility changes. It is concluded that sequence-specific methylation of promoter or regulatory DNA elements can alter the bending of DNA very markedly. This parameter may contribute significantly to the silencing of promoters, probably via the alteration of spatial relationships among DNA-bound transcription factors (Muiznieks and Doerfler, 1994a).

### 7.6.3 Impact of 5'-CG-3' methylation on the activity of different eukaryotic promoters: a comparison

The inhibitory or inactivating effects of sequence-specific promoter methylation in different viral or human cellular promoters (Ad2 E2AL, SV40, LTR-MMTV, HSV-tk, TNFα) have been compared by *in vitro* methylation with the M-*Hpa*II or the M-*Sss*I DNA methyltransferase. In most promoters, 5'-CG-3' methylation reduces activity to a few percent of that of mock-methylated promoters. The number of 5'-CG-3' dinucleotides in a promoter does not strictly correlate with the extent of methylation inhibition. The long terminal repeat of the mouse mammary tumour virus (LTR-MMTV) promoter, which lacks 5'-CG-3' dinucleotides, is not affected by methylation. The late E2A promoter of Ad2 DNA cannot be inactivated by 5'-CCGG-3' methylation when the construct carries the strong cytomegalovirus (CMV) enhancer devoid of this sequence. By contrast, 5'-CG-3' methylation shuts this promoter off even in the presence of the CMV enhancer which contains 5'-CG-3' sites (Muiznieks and Doerfler, 1994b).

## 7.7 Uptake of foreign DNA through the gastrointestinal tract

We have set out to explore the possibility that traces of foreign DNA, that are constantly ingested with the routine food intake, might be taken up by the cells of an organism and become integrated at random into the cellular genome. By subsequently eliciting alterations in the methylation and expression patterns of the affected cells, targets might thus be generated in which these changes lead to the oncogenic transformation of individual cells. This, as yet hypothetical, mechanism of oncogenic transformation is related to the daily exposure of the animals' gastrointestinal tract to foreign DNA.

We investigated whether foreign DNA taken up by mammals with the food supply could, at least in part, survive the digestive regimen of the gastrointestinal tract and eventually enter the bloodstream (Schubbert *et al.*, 1994). In early model experiments, 3- to 6-month-old mice were fed bacteriophage M13 DNA (Hofschneider, 1963) in amounts between 10 and 50 μg. The DNA was supplied in double-stranded supercoiled circular or

*Eco*RI-linearized form directly by pipette-feeding to the animals' oral cavity or with food pellets. M13 DNA was chosen as a traceable food additive because we failed to find any homologies between this phage DNA and DNA repurified from the faeces or blood of control mice that had never come into contact with this DNA. Moreover, the entire nucleotide sequence of this viral DNA had been determined (Yanisch-Perron, 1985). At various times after feeding M13 DNA, DNA was extracted: (i) from the faeces, either extracorporeally or taken from the animal's rectum, or (ii) from whole blood, from isolated blood cells or from the serum. These preparations were subsequently analysed for the presence of M13 DNA sequences by electrophoresis and Southern blot hybridization (Southern, 1975), by dot–blot hybridization or by PCR (Saiki *et al.*, 1988).

The results of these analyses (Schubbert *et al.*, 1994) demonstrated that:

(i) M13 DNA sequences were detected in the animals' faeces between 1 and 7 h after feeding, and
(ii) M13 DNA sequences were present in the bloodstream 2–8 h after feeding.

From the faeces, a few percent of the ingested DNA were recovered, in the bloodstream less than or equal to 0.1%. The bulk of the faeces-excreted M13 DNA was in the size range between 100 and 400 nucleotide pairs. Using PCR, M13 DNA molecules of up to a length of 1692 nucleotide pairs were detected. The PCR-amplified M13 DNA was resequenced and, apart from occasional, non-systematic deviations, was found to be identical to the published nucleotide sequence. The results of these studies were identical regardless of whether DNA was extracted from extracorporeally deposited faeces or from faeces removed from the terminal gut of the animals. The latter precaution precluded the possibility that the faeces extracorporeally deposited might have been externally contaminated by unnoticed oral contacts by the animals.

Similarly, and surprisingly, M13 DNA sequences were also detected both in DNA extracted from total blood, or from leukocytes, but not from the serum of M13-fed animals. The maximal lengths of M13 DNA fragments observed were 976 nucleotide pairs. Upon resequencing this PCR-amplified DNA, authentic M13 DNA was found with occasional rare deviations in nucleotide sequence. DNA isolated from the bloodstream of animals that had not been fed M13 DNA was consistently found to be free of M13 DNA by any of the analytical techniques applied.

These data were confirmed with DNA preparations from 50 different animals (faecal samples) plus 16 buffer-fed (0.01 M Tris–HCl, pH 7.5, 1 mM EDTA) controls, and from 105 different animals (blood samples) plus 30 buffer-fed controls with essentially identical results (Schubbert *et al.*, 1994). We are currently investigating whether integrated M13 DNA sequences exist and can be cloned by suitable vector systems from the DNA of organs (leukocytes, spleen, liver) taken from M13-fed animals.

The results described (Schubbert *et al.*, 1994) verify that food-ingested foreign DNA is not completely degraded in the gastrointestinal tract and can reach the bloodstream, although in minute amounts and in fragmented form.

Of course, it is known that fragments of DNA are highly recombinogenic. Since the exposure of many organ systems to recombinogenic DNA fragments is continuous over the entire life span of an organism, it will be very interesting to consider their possible contributions to mutagenic and oncogenic events which are likely to be cumulative over the duration of an individual's life span.

## 7.8 A concept of oncogenesis – implications for gene therapy and research on transgenic organisms

The classical concept of insertional mutagenesis relates to damage caused to cellular functions or genes that are encoded at the sites of foreign (adenoviral) DNA insertion into the host genome. Since a very sizeable part of the mammalian genome consists of repetitive sequences with essentially unknown functions, insertion might frequently be non-consequential to the repertoire of cellular functions. However, we have adduced evidence that the insertion of adenovirus DNA or plasmid DNA into the hamster cell genome, possibly in conjunction with the cells' transformed phenotype, can be associated with extensive changes in the methylation patterns of cellular genes and of their expression. In that way, foreign DNA insertion at a restricted region on one chromosome could have important *sequelae* for the expression profile of the afflicted cell involving a large but so far unknown number of genes at remote cellular locations. It is likely that each insertion event generates a different pattern of changes and, in that sense, a unique disturbance in the recipient nucleus. The overall consequences for cellular survival will probably range over a wide continuum from cell death to the absence of detectable functional changes. For oncogenic transformation to ensue with derailed growth control, very specific subsets of alterations in methylation and expression patterns may be required.

By linking the observation of apparently frequent alterations in patterns of DNA methylation in cellular genes upon insertion of, and transformation by, the adenovirus genome to the discovery that food-ingested DNA reaches the bloodstream of mammals with the potential of dissemination to many organs of the animal, a concept for oncogenic transformation emerges. How frequently do food-ingested DNA fragments obtain access to cells of the organism and become integrated into their genomes? How specific or variable are the changes in DNA methylation and expression patterns, and do they cause cell transformation or various stages of loss of growth control? One can envisage a very wide gamut of possibilities that will be difficult to prove or disprove in an individual tumour in which it will be impossible to differentiate between primary and secondary events (e.g. changes in DNA methylation and expression patterns), which may be consequences rather than causes of oncogenesis.

In the future, we shall extend research to the basic mechanisms of foreign DNA insertion in animals, to its frequency and sites of insertion, and to changes in patterns of DNA methylation. For this latter aspect, it would be advisable to use foreign DNA as a model that does not have coding capacity or cannot be expressed in eukaryotic cells, in order to circumvent the difficulty of having to differentiate between the effects of the insertion event and of gene products of biologically active DNA, such as Ad12 DNA. A wide field of research will have to be addressed here. These projects will be relevant for oncogenesis, but also for gene therapy and the interpretation of experiments in which transgenic animals or plants are utilized. In gene therapy and in transgenic organisms, foreign DNA can possibly affect and alter many parts of an established genome at sites remote from the targeted site of insertion via changes in DNA methylation. Although these considerations might complicate the interpretation of some experiments, these concepts will have to be carefully weighed in the design of future projects.

## Acknowledgements

I am indebted to Petra Böhm for expert editorial work. Research in the author's laboratory was supported by the Deutsche Forschungsgemeinschaft (SFB274-A1 and Schwerpunkt Virulenzfaktoren und Wirtstropismus animaler Viren), the Bundesministerium für Forschung und Technologie (project 2.03) (Genzentrum Köln), and the Fritz Thyssen-Stiftung.

## References

Behn-Krappa A, Doerfler W. (1994) Enzymatic amplification of synthetic oligodeoxyribonucleotides: implication for triplet repeat expansions in the human genome. *Hum. Mut.* **3**: 19–24.

Caskey CT, Pizutti A, Fu Y-H, Fenwick RG Jr, Nelson DL. (1992) Triplet repeat mutations in human disease. *Science* **256**: 784–789.

Cook JL, Lewis AM Jr. (1979) Host response to adenovirus 2-transformed hamster embryo cells. *Cancer Res.* **39**: 1455–1461.

Deuring R, Doerfler W. (1983) Proof of recombination between viral and cellular genomes in human KB cells productively infected by adenovirus type 12: structure of the junction site in a symmetric recombinant (SYREC). *Gene* **26**: 283–289.

Deuring R, Winterhoff U, Tamanoi F, Stabel S, Doerfler W. (1981a) Site of linkage between adenovirus type 12 and cell DNAs in hamster tumour line CLAC3. *Nature* **293**: 81–84.

Deuring R, Klotz G, Doerfler W. (1981b) An unusual symmetric recombinant between adenovirus type 12 DNA and human cell DNA. *Proc. Natl Acad. Sci. USA* **78**: 3142–3146.

Doerfler W. (1981) DNA methylation – a regulatory signal in eukaryotic gene expression. *J. Gen. Virol.* **57**: 1–20.

Doerfler W. (1982) Uptake, fixation, and expression of foreign DNA in mammalian cells: the organization of integrated adenovirus DNA sequences. *Curr. Top. Microbiol. Immunol.* **101**: 127–194.

Doerfler W. (1983a) DNA methylation and gene activity. *Annu. Rev. Biochem.* **52**: 93–124.

Doerfler W. (ed.) (1983b) The molecular biology of adenoviruses. *Curr. Top. Microbiol. Immunol.*, vols 109 and 110.

Doerfler W. (1984a) DNA methylation: role in viral transformation and persistence. *Adv. Viral Oncol.* **4**: 217–247.

Doerfler W. (ed.) (1984b) The molecular biology of adenoviruses. *Curr. Top. Microbiol. Immunol.*, vol. 111.

Doerfler W. (1989) Complexities in gene regulation by promoter methylation. *Nucleic Acids Mol. Biol.* **3**: 92–119.

**Doerfler W.** (1991a) The abortive infection and malignant transformation by adenoviruses: integration of viral DNA and control of viral gene expression by specific patterns of DNA methylation. *Adv. Virus Res.* **39**: 89–128.

**Doerfler W.** (1991b) Patterns of DNA methylation – evolutionary vestiges of foreign DNA inactivation as a host defense mechanism – a proposal. *Biol. Chem. Hoppe-Seyler* **372**: 557–564.

**Doerfler W.** (1992) Transformation of cells by adenoviruses: less frequently discussed mechanisms. In: *Malignant Transformation by DNA Viruses. Molecular Mechanisms* (eds W Doerfler, P Böhm). VCH, Weinheim, pp. 87–109.

**Doerfler W.** (1993) Adenoviral DNA integration and changes in DNA methylation patterns: a different view of insertional mutagenesis. *Progr. Nucleic Acid Res. Mol. Biol.* **46**: 1–36.

**Doerfler W, Böhm, P.** (eds) (1995). The molecular repertoire of adenoviruses. *Curr. Top. Microbiol. Immunol*, vol. 199, I–III.

**Doerfler W, Gahlmann R, Stabel S, Deuring R, Lichtenberg U, Schulz M, Eick D, Leisten R.** (1983) *Curr. Top. Microbiol. Immunol.* **109**: 193–228.

**Doerfler W, Tatzelt J, Orend G, Schorr J, Rosahl T, Fechteler K, Zock C.** (1993) Viral and cellular strategies on integrated adenovirus genomes. In: *Virus Strategies* (eds W Doerfler, P Böhm). VCH, Weinheim, pp. 369–400.

**Eick D, Doerfler W.** (1982) Integrated adenovirus type 12 DNA in the transformed hamster cell line T637: sequence arrangements at the termini of viral DNA and mode of amplification. *J. Virol.* **42**: 317–321.

**Eick D, Stabel S, Doerfler W.** (1980) Revertants of adenovirus type 12-transformed hamster cell line T637 as tools in the analysis of integration patterns. *J. Virol.* **37**: 887–892.

**Ernberg I, Falk K, Minarowitz J, Busson P, Rursz T, Masucci MG, Klein G.** (1989) The role of methylation in the phenotype-dependent modulation of Epstein–Barr nuclear antigen 2 and latent membrane protein genes in cells latently infected with Epstein–Barr virus. *J. Gen. Virol.* **70**: 2989–3002.

**Fechteler K, Tatzelt J, Huppertz S, Wilgenbus P, Doerfler W.** (1995) The mechanism of adenovirus DNA integration: studies in a cell-free system. *Curr. Top. Microbiol. Immunol.* **199**/II, 109–137.

**Gahlmann R, Doerfler W.** (1983) Integration of viral DNA into the genome of the adenovirus type 2-transformed hamster cell line HE5 without loss or alteration of cellular nucleotides. *Nucleic Acids Res.* **11**: 7347–7361.

**Gahlmann R, Leisten R, Vardimon L, Doerfler W.** (1982) Patch homologies and the integration of adenovirus DNA in mammalian cells. *EMBO J.* **1**: 1101–1104.

**Gahlmann R, Schulz M, Doerfler W.** (1984) Low molecular weight RNAs with homologies to cellular DNA at sites of adenovirus DNA insertion in hamster or mouse cells. EMBO J. **3**: 3263–3269.

**Ginsberg H.** (ed) (1985) *Adenoviruses*. Plenum Press, New York.

**Groneberg J, Sutter D, Soboll H, Doerfler W.** (1978) Morphological revertants of adenovirus type 12-transformed hamster cells. *J. Gen. Virol.* **40**: 635–645.

**Günthert U, Schweiger M, Stupp M., Doerfler, W.** (1976) DNA methylation in adenovirus, adenovirus-transformed cells, and host cells. *Proc. Natl Acad. Sci. USA* **73**: 3923–3927.

**Hasse A, Schulz WA.** (1994) Enhancement of reporter gene *de novo* methylation by DNA fragments from the α-fetoprotein control region. *J. Biol. Chem.* **269**: 1–6.

**Heller H, Kammer C, Wilgenbus P, Doerfler W.** (1995) The chromosomal insertion of foreign (adenovirus type 12 or bacteriophage lambda) DNA is associated with enhanced methylation of cellular DNA segments. *Proc. Natl Acad. Sci. USA*, in press.

**Hofschneider PH.** (1963) Untersuchungen Über "kleine" *E. coli* K12 Bakteriophagen. *Z. Naturforsch.* **18b**: 203–210.

**Hu L-F, Minarovits J, Cao S-L, Contreras-Salazar B, Rymo L, Falk K, Klein G, Ernberg I.** (1991) Variable expression of latent membrane protein in nasopharyngeal carcinoma can be related to methylation status of the Epstein–Barr virus BNLF-1 5′-flanking region. *J. Virol.* **65**: 1558–1567.

**Huntington's Disease Collaborative Research Group.** (1993) A novel gene containing a trinucleotide repeat that is expanded and unstable on Huntington's disease chromosomes. *Cell* **72**: 971–983.

**Jessberger R, Weisshaar B, Stabel S, Doerfler W.** (1989a) Arrangement and expression of integrated adenovirus type 12 DNA in the transformed hamster cell line HA12/7: amplification of Ad12 and c-*myc* DNAs and evidence for hybrid viral–cellular transcripts. *Virus Res.* **13**: 113–128.

**Jessberger R, Heuss D, Doerfler W.** (1989b) Recombination in hamster cell nuclear extracts between adenovirus type 12 DNA and two hamster preinsertion sequences. *EMBO J.* **8**: 869-878.

**Johansson K, Persson H, Lewis AM, Pettersson U, Tibbetts C, Philipson L.** (1978) Viral DNA sequences and gene products in hamster cells transformed by adenovirus type 2. *J. Virol.* **27**: 628–639.

**Knight SJL, Flannery AV, Hirst MC, Campbell L, Christodoulou Z, Phelps SR, Pointon J, Middleton-Price HR, Barnicoat A, Pembrey ME, Holland J, Oostra BA, Bobrow M, Davies KE.** (1993) Trinucleotide repeat amplification and hypermethylation of a CpG island in *FRAXE* mental retardation. *Cell* **74**: 127–134.

**Koide R, Ikeuchi T, Onodera O, Tanaka H, Igarashi S, Endo K, Takahashi H, Kondo R, Ishikawa A, Hayashi T, Saito M, Tomoda A, Miike T, Naito H, Ikuta F, Tsuji S.** (1994) Unstable expansion of CAG repeat in hereditary dentatorubral-pallidoluysian atrophy (*DRPLA*). *Nature Genetics.* **6**: 9–13.

**Kruczek I, Doerfler W.** (1982) The unmethylated state of the promoter/leader and 5'-regions of integrated adenovirus genes correlates with gene expression. *EMBO J.* **1**: 409–414.

**Kuhlmann I, Doerfler W.** (1982) Shifts in the extent and patterns of DNA methylation upon explantation and subcultivation of adenovirus type 12-induced hamster tumor cells. *Virology* **118**: 169–180.

**Kuhlmann I, Achten S, Rudolph R, Doerfler W.** (1982) Tumor induction by human adenovirus type 12 in hamsters: loss of the viral genome from adenovirus type 12-induced tumor cells is compatible with tumor formation. *EMBO J.* **1**: 79–86.

**Lettmann C, Schmitz B, Doerfler W.** (1991) Persistence or loss of preimposed methylation patterns and *de novo* methylation of foreign DNA integrated in transgenic mice. *Nucleic Acids Res.* **19**: 7131–7137.

**Li E, Beard C, Jaenisch R.** (1993) Role of DNA methylation in genomic imprinting. *Nature* **366**: 362–365.

**Lichtenberg U, Zock C, Doerfler W.** (1987) Insertion of adenovirus type 12 DNA in the vicinity of an intracisternal A particle genome in Syrian hamster tumor cells. *J. Virol.* **61**: 2719–2726.

**Lichtenberg U, Zock C, Doerfler W.** (1988) Integration of foreign DNA into mammalian genome can be associated with hypomethylation at site of insertion. *Virus Res.* **11**: 335–342.

**Linn F, Heidmann H, Saedler H, Meyer P.** (1990) Epigenetic changes in the expression of the maize A1 gene in *Petunia hybrida*: role of numbers of integrated gene copies and state of methylation. *Mol. Gen. Genet.* **222**: 329–336.

**McDougall JK, Dunn AR, Jones KW.** (1972) *In situ* hybridization of adenovirus RNA and DNA. *Nature* **236**: 346–348.

**McDougall JK, Kucherlapati R, Ruddle FH.** (1973) Localization and induction of the human thymidine kinase gene by adenovirus 12. *Nature* **245**: 172–175.

**Mooslehner K, Karls U, Harbers K.** (1990) Retroviral integration sites in transgenic *Mov* mice frequently map in the vicinity of transcribed DNA regions. *J. Virol.* **64**: 3056–3058.

**Muiznieks I, Doerfler W.** (1994a) The topology of the promoter of RNA polymerase II- and III-transcribed genes is modified by the methylation of 5'-CG-3' dinucleotides. *Nucleic Acids Res.* **22**: 2568–2575.

**Muiznieks I, Doerfler W.** (1994b) The impact of 5'- CG-3' methylation on the activity of different eukaryotic promoters: a comparative study. *FEBS Lett.* **344**: 251–254.

Mummaneni P, Bishop PL, Turker MS. (1993) A *cis*-acting element accounts for a conserved methylation pattern upstream of the mouse adenine phosphoribosyltransferase gene. *J. Biol. Chem.* **268**: 552–558.

Munnes M, Schetter C, Hölker I, Doerfler W. (1995) A fully 5'-CG-3' but not a 5'-CCGG-3' methylated late frog virus 3 promoter retains activity. *J. Virol.* **69**: in press.

Nagafuchi S, Yanagisawa H, Sato K, Shirayama T, Ohsaki E, Bundo M, Takeda T, Tadokoro K, Kondo I, Murayama N, Tanaka Y, Kikushima H, Umino K, Kurosawa H, Furukawa T, Nihei K, Inoue T, Sano A, Komure O, Takahashi M, Yoshizawa T, Kanazawa I, Yamada M. (1994). Dentatorubal and pallidoluysian atrophy expansion of an unstable CAG trinucleotide on chromosome 12p. *Nature Genetics* **6**: 14–18.

Oberlé I, Rousseau F, Heitz D, Kretz C, Devys D, Hanauer A, Boue J, Bertheas MF, Mandel JL. (1991) Instability of a 550-base pair DNA segment and abnormal methylation in fragile X syndrome. *Science* **252**: 1097–1102.

Orend G, Kuhlmann I, Doerfler W. (1991) Spreading of DNA methylation across integrated foreign (adenovirus type 12) genomes in mammalian cells. *J. Virol.* **65**: 4301–4308.

Orend G, Linkwitz A, Doerfler W. (1994) Selective sites of adenovirus (foreign) DNA integration into the hamster genome: changes in integration patterns. *J. Virol.* **68**: 187–194.

Orend G, Knoblauch M, Kämmer C, Tjia ST, Schmitz B, Linkwitz A, Meyer zu Altenschildesche G, Maas J, Doerfler W. (1995) The initation of *de novo* methylation of foreign DNA integrated into a mammalian genome is not exclusively targeted by nucleotide sequence. *J. Virol.* **69**: 1226–1242.

Orr HT, Chung M-Y, Banfi S, Kwiatkowski TJ Jr, Servadio A, Beaudet AL, McCall AE, Duvick LA, Ranum LPW, Zoghbi HY. (1993) Expansion of an unstable trinucleotide CAG repeat in spinocerebellar ataxia. *Nature Genetics* **4**: 221–226.

Ragot T, Vincent N, Chafey P, Vigne E, Gilgenkrantz H, Couton D, Cartaud J, Briand P, Kaplan J-C, Perricaudet M, Kahn A. (1993) Efficient adenovirus-mediated transfer of a human minidystrophin gene to skeletal muscle of *mdx* mice. *Nature* **361**: 647–650.

Reik W, Collick A, Norris ML, Barton SC, Surani MA. (1987) Genomic imprinting determines methylation of parental alleles in transgenic mice. *Nature* **328**: 248–251.

Richards RI, Sutherland GR. (1992) Dynamic mutations: a new class of mutations causing human disease. *Cell* **70**: 709–712.

Riggins GJ, Lokey LK, Chastain JL, Leiner HA, Sherman SL, Wilkinson KD, Warren ST. (1992) Human genes containing polymorphic trinucleotide repeats. *Nature Genetics* **2**: 186–191.

Rosahl T, Doerfler W. (1988) Predominant association of adenovirus type 12 DNA with human chromosome 1 early in productive infection. *Virology* **162**: 494–497.

Rosahl T, Doerfler W. (1992) Alterations in the levels of expression of specific cellular genes in adenovirus-infected and -transformed cells. *Virus Res.* **26**: 71–90.

Saiki RK, Gelfand DH, Stoffel S, Scharff SJ, Higuchi R, Horn GT, Mullis KB, Erlich H. (1988) Primer-directed enzymatic amplification of DNA with a thermostable DNA polymerase. *Science* **239**: 487–491.

Sapienza C, Peterson AC, Rossant J, Balling R. (1987) Degree of methylation of transgenes is dependent on gamete of origin. *Nature* **328**: 251–254.

Scherdin U, Rhodes K, Breindl M. (1990) Transcriptionally active genome regions are preferred targets for retrovirus integration. *J. Virol.* **64**: 907–912.

Schetter C, Grünemann B, Hölker I, Doerfler W. (1993) Frog virus 3 DNA methylation and DNA methyltransferase activity from nuclei of infected cells. *J. Virol.* **67**: 6973–6978.

Schorr J, Doerfler W. (1993) Non-homologous recombination between adenovirus and AcNPV DNA fragments in cell-free extracts from insect *Spodoptera frugiperda* nuclei. *Virus Res.* **28**: 153–170.

Schubbert R, Lettmann CM, Doerfler W. (1994) Ingested foreign DNA (phage M13) survives transiently in the gastrointestinal tract and enters the bloodstream of mice. *Mol. Gen. Genet.* **242**: 495–504.

Schulz M, Doerfler W. (1984) Deletion of cellular DNA at site of viral DNA insertion in the adenovirus type 12-induced mouse tumor CBA-12-1-T. *Nucleic Acids Res.* **12**: 4959–4976.

Schulz M, Freisem-Rabien U, Jessberger R, Doerfler W. (1987) Transcriptional activities of mammalian genomes at sites of recombination with foreign DNA. *J. Virol.* **61**: 344–353.

Southern EM. (1975) Detection of specific sequences among DNA fragments separated by gel electrophoresis. *J. Mol. Biol.* **98**: 503–517.

Sprengel J, Schmitz B, Heuss-Neitzel D, Zock C, Doerfler W. (1994) The nucleotide sequence of human adenovirus type 12 DNA: a comparative functional evaluation. *J. Virol.* **68**: 379–389.

Stabel S, Doerfler W. (1982) Nucleotide sequence at the site of junction between adenovirus type 12 DNA and repetitive hamster cell DNA in transformed cell line CLAC1. *Nucleic Acids Res.* **10**: 8807–8023.

Stabel S, Doerfler W, Friis RR. (1980) Integration sites of adenovirus type 12 DNA in transformed hamster cells and hamster tumor cells. *J. Virol.* **36**: 22–40.

Surani MAH, Barton SC, Norris ML. (1984) Development of reconstituted mouse eggs suggests imprinting of the genome during gametogenesis. *Nature* **308**: 548–550.

Sutter D, Doerfler W. (1979) Methylation of integrated viral DNA sequences in hamster cells transformed by adenovirus 12. *Cold Spring Harbor Symp. Quant. Biol.* **44**: 565–568.

Sutter D, Doerfler W. (1980) Methylation of integrated adenovirus type 12 DNA sequences in transformed cells is inversely correlated with viral gene expresion. *Proc. Natl Acad. Sci. USA* **77**: 253–256.

Sutter D, Westphal M, Doerfler W. (1978) Patterns of integration of viral DNA sequences in the genomes of adenovirus type 12-transformed hamster cells. *Cell* **14**: 569–585.

Swain JL, Stewart TA, Leder P. (1987) Parental legacy determines methylation and expression of an autosomal transgene: a molecular mechanism for parental imprinting. *Cell* **50**: 719–727.

Szyf M, Tanigawa G, McCarthy PL Jr. (1990) A DNA signal from the *Thy-1* gene defined *de novo* methylation patterns in embryonic stem cells. *Mol. Cell. Biol.* **10**: 4396–4400.

Tatzelt J, Scholz B, Fechteler K, Jessberger R, Doerfler W. (1992) Recombination between adenovirus type 12 DNA and a hamster preinsertion sequence in a cell-free system. Patch homologies and fractionation of nuclear extracts. *J. Mol. Biol.* **226**: 117–126.

Tatzelt J, Fechteler K, Langenbach P, Doerfler W. (1993) Fractionated nuclear extracts from hamster cells catalyze cell-free recombination at selective sequences between adenovirus DNA and a hamster preinsertion site. *Proc. Natl Acad. Sci. USA* **90**: 7356–7360.

Toth M, Lichtenberg U, Doerfler W. (1989) Genomic sequencing reveals a 5-methyl deoxycytosine-free domain in active promoters and the spreading of preimposed methylation patterns. *Proc. Natl Acad. Sci. USA* **86**: 3728–3732.

Toth M, Müller U, Doerfler W. (1990) Establishment of *de novo* DNA methylation patterns. Transcription factor binding and deoxycytidine methylation at CpG and non-CpG sequences in an integrated adenovirus promoter. *J. Mol. Biol.* **214**: 673–683.

Trentin JJ, Yabe Y, Taylor G. (1962) The quest for human cancer viruses. A new approach to an old problem reveals cancer induction in hamsters by human adenovirus. *Science* **137**: 835–841.

Vardimon L, Doerfler W. (1981) Patterns of integration of viral DNA in adenovirus type 2-transformed hamster cells. *J. Mol. Biol.* **147**: 227–246.

Vardimon L, Neumann R, Kuhlmann I, Sutter D, Doerfler W. (1980) DNA methylation and viral gene expression in adenovirus-transformed and -infected cells. *Nucleic Acids Res.* **8**: 2461–2473.

Willis DB, Granoff A. (1980) Frog virus 3 DNA is heavily methylated at CpG sequences. *Virology* **107**: 250–257.

Willis DB, Thompson JP, Essani K, Goorha R. (1989) Transcription of methylated viral DNA by eukaryotic RNA polymerase II. *Cell Biophys.* **15**: 97–111.

Wienhues U, Doerfler W. (1985) Lack of evidence for methylation of parental and newly synthesized adenovirus type 2 DNA in productive infections. *J. Virol.* **56**: 320–324.

**Yanisch-Perron C, Vieira J, Messing J.** (1985) Improved M13 phage cloning vectors and host strains: nucleotide sequences of the M13mp18 and pUC19 vectors. *Gene* **33**: 103–119.

**Zock C, Doerfler W.** (1990) A mitigator sequence in the downstream region of the major late promoter of adenovirus type 12 DNA. *EMBO J.* **9**: 1615–1623.

**Zock C, Doerfler W.** (1993) Role of the mitigator sequence in the species-specific expression of the major late promoter in adenovirus type 12 DNA. In: *Virus Strategies* (eds W Doerfler, P Böhm). VCH, Weinheim, pp. 305–317.

**Zock C, Iselt A, Doerfler W.** (1993) A unique mitigator sequence determines the species specificity of the major late promoter in adenovirus type 12 DNA. *J. Virol.* **67**: 682–693.

# 8

# Transgenic animals in human gene analysis

Franz Theuring

## 8.1 Introduction

The ability to carry out a comprehensive molecular genetic analysis of an organism becomes more difficult and more limited as the complexity of that organism increases, and more effort is required to analyse a given developmental process, mutation or disease. Progress in the genetic analysis of mammalian development can never be as rapid as with *Caenorhabditis* and *Drosophila*. Not only is the mammalian genome size larger and the generation time longer, but the embryo also develops much more slowly. In addition, experimental manipulation of embryos within their uterine environment is more difficult.

The remarkably large and varied collection of known mouse mutations has been obtained principally by relying upon serendipity to expose visible variation in behaviour and/or morphology (e.g. studies on the inheritance of coat colours in a variety of domestic animals which can be traced back to experiments carried out in the early 1900s). Traditionally, the biological questions posed were determined mainly by the nature of the mutants available. Thus, although phenotypic consequences of mutated genes could be analysed, the nature of the genes themselves remained unknown. It is here that 'reverse genetics' (i.e. recombinant DNA technology) provided an important experimental approach which has changed this situation dramatically. The expression of cloned genes in cells and intact animals has allowed us to both explore and understand the relationship between gene structure, gene regulation and gene function by analysing the activities of normal and mutant genes. Furthermore, we can now evaluate both naturally occurring mutant alleles and experimentally constructed mutants.

The ability to introduce foreign genes into the mammalian germline has proved to be one of the most powerful tools in mammalian developmental genetics. This technology arose from the knowledge, obtained through gene transfer experiments in tissue culture, that novel DNA species could be inserted into the genome of a host cell and subsequently efficiently expressed (Graessmann et al., 1979; Capecchi, 1980). At the time when efforts were first being made to insert DNA into an intact mammal, the key unknown factors were the frequency of gene insertion and the question of whether an embryo (which, by contrast to cultured cells, must complete a complex developmental programme) could tolerate the insertion of novel genes without suffering lethal consequences. Developmental biologists wanted to determine the consequences of altering the genetic programme of a cell or whole organism in order to study the effects on the developmental potential of individual cells. The limited cell types available for the analysis of gene expression during mammalian development and differentiation in tissue culture stimulated the search for methods to introduce cloned genes into the whole animal, and many important advances have been achieved.

Little was known about the ability of developing embryos to tolerate integration and expression of new genetic material, and for that DNA to be transmitted as a Mendelian trait to offspring. However, in 1974, the first 'transgenic' animal was produced by injecting SV40 DNA into the blastocoel cavity of mouse embryos (Jaenisch and Mintz, 1974). Two years later, Jaenisch (1976) found that Moloney murine leukaemia virus (MMLV) could infect mouse embryos, insert a proviral copy of its DNA into the genome, and transmit that DNA in Mendelian fashion. These findings paved the way for the insertion of DNA from any source directly into the mouse, a feat first demonstrated by Gordon et al. (1980), employing the strategy of pronuclear microinjection. Within the next few months several groups reported the stable integration, expression and germline transmission of genes in both fetuses and adult mice (Brinster et al., 1981; Costantini and Lacy, 1981; Gordon and Ruddle, 1981; Harbers et al., 1981; Wagner et al., 1981a,b). Mice transformed in this way were subsequently called 'transgenic' by Gordon and Ruddle (1981), a term which is now accepted and used to describe all forms of gene transfer into either animal or plant.

In parallel, a stem cell system for efficient gene transfer studies was developed. The successful establishment of pluripotent embryonic stem (ES) cells was reported by two groups (Evans and Kaufman, 1981; Martin, 1981). These cells, derived directly from embryos, not only resemble embryonic carcinoma cells in their biological properties but are better cells for exploring their *in vivo* potential. This was subsequently demonstrated by the efficient formation of germline chimeras (Bradley et al., 1984; see Chapter 9).

The experimental power of both these systems for the analysis of gene regulation and gene function in mammals is well documented by the numerous reports and reviews published over the past 14 years (see Figure 1 in Aguzzi et al., 1994). Because of the rapidly expanding nature of this research area, I have decided not to attempt an exhaustive description of all published studies, but rather to select those which, in my opinion, are relevant to the different topics which provide entry points into this technology.

## 8.2 Methodology

There are three basic methods in use for the generation of transgenic animals:

(i) microinjection of linearized DNA fragments into the pronucleus of one-cell embryos, the most widely used method for transferring genetic information;
(ii) retroviral infection of pre- and post-implantation embryos, and
(iii) injection of genetically modified embryonic stem cells into host blastocysts to generate chimeric animals (see Chapter 9).

The procedure for pronuclear DNA microinjection is essentially the same today as it was in the pioneering experiments of the early 1980s (for detailed descriptions, see Hogan *et al.*, 1986; Gordon, 1993): one-cell embryos are obtained for microinjection from superovulated pregnant females at a stage when the two pronuclei are visible. The two most critical parameters affecting transgene integration are the form and concentration of the DNA (Brinster *et al.*, 1985). It has been shown that integration of linear molecules is much more efficient than that of circular ones. The DNA concentration should be 1–2 mg ml$^{-1}$, equivalent to several hundred copies of DNA injected into the pronucleus. The size of the transgenes (ranging from 0.7 to 50 kb) does not seem to exert any influence upon transgene integration.

Micromanipulated embryos can then be transferred directly into the oviduct of a pseudopregnant foster mother or cultured overnight and transferred the following day at the two-cell stage to a pseudopregnant recipient. Random integration into a single chromosomal locus normally occurs and the number of integrated copies varies from one to over 50. Rarely, two sites of integration are observed in a single transgenic animal. Multiple copies are arranged primarily in head-to-tail arrays.

Evidence from retroviral infection studies suggests that preferred viral integration into regions accessible to digestion with DNase I occurs (Rohdewohld *et al.*, 1987). The results demonstrated that all integrations occurred into chromosomal loci containing several DNase-hypersensitive sites, and invariably the proviral integration sites mapped to within a few hundred base-pairs of a DNase-hypersensitive site.

In most cases, integration of the transgene occurs at the one-cell stage so that every cell of the resulting transgenic animal (founder) contains the foreign DNA and the transgene can be passed to the next generation in Mendelian fashion. If integration of the DNA is delayed until after several cell divisions, a genetic mosaic founder will develop in which only a few cells contain the transgene. It might turn out that this founder is unable to pass the transgene to the next generation because the foreign DNA is not represented in the germline. Transgene-bearing animals are most commonly identified by standard procedures such as Southern blot, dot blot or polymerase chain reaction (PCR). Once identified, founder animals are usually bred to wild-type animals to provide positive F1 offspring for the establishment of a transgenic line (Figure 8.1) and expression analysis, including mRNA and protein detection.

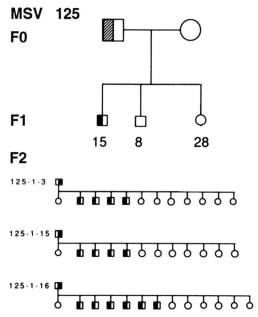

**Figure 8.1.** Example of a transgenic pedigree. In this transgenic line (MSV–SV40T, line 125), 15 males out of 23 from the F1 generation were transgene-positive, whereas none of the 28 females carried the transgene. The expected Y-integration of the transgene was confirmed by analysing the F2 generation, in which only males were found to be transgene-positive.

The rat is a particularly attractive animal model for physiological, pharmacological and behavioural studies, especially in cardiovascular and neuropsychiatric research. Although mice are easier to maintain, less expensive and are much more thoroughly characterized genetically, the application of transgenic mice to biomedical research may be difficult since the physical size of the mouse, its behavioural repertoire and capability for instrumentalization are limiting. Therefore, it would be highly desirable to modify the rat genome in a way which is equivalent to that achievable in the mouse. In general, transgenic rats are generated in the same way as outlined for mice, but there are several technical details which have to be modified. Since superovulation tended to be more difficult to achieve in rats, both Hammer *et al.* (1990) and Mullins *et al.* (1990) made considerable use of the observations of Armstrong and Opavsky (1988) relating to the induction of superovulation. By employing an osmotic mini-pump, a highly purified preparation of follicle-stimulating hormone is administered to immature rats, followed by a single injection of human chorionic gonadotropin 2 days later. Owing to the different physical properties of fertilized rat eggs (e.g. a tougher zona pelucida and a more elastic pronuclear membrane) and some minor modifications to the surgical techniques, the microinjection procedure is more time-consuming. Nevertheless, the overall efficiency is almost comparable in both species (our own experience).

## 8.3 Transgenes to study gene regulation

The spatio-temporal regulation of gene expression during development and differentiation, and the subsequent coordinated regulation of cellular activities in response to physiological stimuli, are orchestrated primarily at the level of transcription. Differential gene expression is potentiated by the 5' upstream (promoter) region of genes and by enhancer elements located either more 5' to the gene, or in an intron, or 3' to the gene. A combination of tissue-specific, developmental-specific and general transcription factors bind to enhancer and promoter sequences, which are responsible for inducing gene expression (see Chapter 1).

In addition, protein–protein interactions between these bound factors in a three-dimensional array may have synergistic or antagonistic effects on transcription. It is the presence and combination of these factors which determines the developmental- and tissue-specific expression of a gene, and the signal transduction pathway to which the relevant gene will respond (for reviews, see Schöler *et al.*, 1988; Johnson and McKnight, 1989; Mitchell and Tjian, 1989).

Within an intact animal, transcription of the transgenic mRNA would then be determined by these regulatory sequences and should necessarily mirror the endogenous expression pattern. Unfortunately, this is not always the case. Inappropiate spatio-temporal expression patterns, or inadequate levels in anticipated tissue or even no expression have been obtained. This was caused by the inclusion of incomplete regulatory sequences for the control of gene expression or because it was attempted to express cDNAs in transgenic animals (for reviews, see Gordon, 1989; Rusconi, 1991; Wight and Wagner, 1994). It was shown that introns inserted into a transgenic construct could improve transcription and increase both the percentage of transgenic animals expressing the transgene and the expression level (Brinster *et al.*, 1988). Even heterologous introns inserted between promoters and cDNAs can improve transcription (Palmiter *et al.*, 1991); these authors concluded that, if possible, genomic coding sequences, or at least as much genomic sequence as possible, should be used for the generation of a transgenic construct.

Early experiments with transgenic mice clearly demonstrated the importance of *cis*-acting elements in or around genes in mediating the tissue specificity of expression (Brinster *et al.*, 1981; Hanahan, 1985; Ornitz *et al.*, 1985). By contrast to the sometimes relatively simple *cis*-acting elements (e.g. in the pancreatic insulin and elastase genes; Hanahan, 1985; Ornitz *et al.*, 1985), some elements are far more complex (e.g. β-globin). Transgenic mice carrying the human β-globin gene with 1.0 kb of 5'- and 3'-flanking sequence exhibited relatively weak but nevertheless erythroid-specific expression. Including DNA elements normally present as far as 50 kb 5' and 20 kb 3' to the β-globin gene resulted in high copy number-dependent, integration site-independent expression (Grosveld *et al.*, 1987; see Chapter 1). These distally located regulatory DNA elements' termed the *locus control region* (LCR), coincide with chromatin regions that exhibit hypersensitivity to DNase I digestion in erythroid, but not non-erythroid, cells. It has been postulated that the β-globin LCR acts primarily to open the chromatin domain. Van Assendelft *et al.* (1989) demonstrated that the β-globin LCR could activate homologous and heterologous promoters in a tissue-specific manner. However, it is unclear whether these regions, which have been identified in several other genes (Chapter 1), represent a specific type of positive control by exhibiting ubiquitous insulating properties or whether they are merely a particular localization of a general tissue-specific enhancer-like function (Rusconi, 1991).

The transgenic route provides the basis to define and characterize the entire region which *in vivo* is necessary to mirror the endogenous pattern of expression. To facilitate analysis of expression in transgenic animals, the promoter in question can be fused to a reporter gene not normally found in the mouse (e.g. chloramphenicol acetyltransferase, luciferase or β-galactosidase). Alternatively, a peptide that is easily identified immunologically, a hormone that has a pronounced physiological effect, or an oncogene that could induce tumour formation in the cell type in which it is expressed, might be used. In either case, the effect of systematic deletions of the putative regulatory sequences can be analysed *in vivo*. Based on this *in vivo* promoter mapping strategy, Püschel *et al.* (1990) defined the tissue- and developmental-specific *cis*-acting elements of the developmental control gene *Hox-A7* (*Hox-1.1*).

Traditionally, X-gal (5-bromo-4-chloro-3-indolyl-β-D-galactoside) has been used as a chromogenic substrate for the histochemical demonstration of β-galactosidase activity. We have improved this *in situ* staining method by replacing X-gal with Bluo-gal (5-bromoindolyl-β-*o*-galactopyranoside) which, after β-galactosidic cleavage, precipitates in the form of fine birefringent crystals. Microscopic examination under polarized light results in a greatly enhanced sensitivity of the staining method and allows for optimal morphological resolution (Aguzzi and Theuring, 1994, and discussion therein; Figure 8.2).

It is possible to study a gene of interest and its regulation within the intact animal under physiological or pathophysiological situations. Furthermore, we are entering a new era in which transgenic animals carrying promoter–reporter gene constructs are being screened in order to discover compounds or pharmacologically active drugs that interfere with transcription, either by elevating it or lowering it. Transcription factors are potential targets for therapeutic intervention. We decided to establish transgenic mice carrying human tissue plasminogen activator (t-PA) gene promoter–*lacZ* reporter gene constructs (Theuring *et al.*, 1994a). This transgenic approach will help to elucidate the genetic control elements and mechanisms involved in regulating human t-PA gene expression. Furthermore, these mice can be employed in a discovery programme for testing and screening compounds which potentially modulate fibrinolytic activity in man. In a first attempt to study the role of protein kinase C (PKC) in the regulation of t-PA gene expression *in vivo*, we initiated a study using 4-phorbol-12-myristate-13-acetate (PMA), a strong and stable PKC activator, for the modulation of the fibrinolytic activity in adult transgenic mice (Theuring *et al.*, 1994b). Our initial data indicate

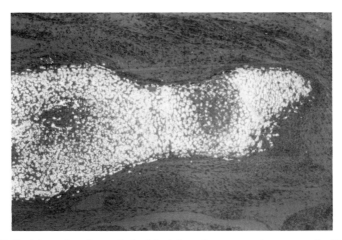

**Figure 8.2.** Serial section through the hip joint of a transgenic mouse. *LacZ* expression can be easily detected under polarized light.

that PMA treatment is able to induce a rapid time-dependent increase in both plasma t-PA antigen and tissue β-galactosidase activity. In addition, an increase in mRNA levels for both t-PA and *lacZ* was detected in some organs.

Performing promoter mapping and defining tissue- and developmental-specific promoter elements of a gene in transgenic animals is not restricted to eukaryotic genes. Viral regulatory elements can also be examined for tissue specificity. Encouraged by earlier reports dealing with the SV40 early promoter driving the large T oncogene and defining its tissue specificity (Brinster *et al.*, 1984; Palmiter *et al.*, 1985), we undertook a project aimed at the analysis of the tissue specificity of the Moloney murine sarcoma virus (MSV) enhancer in transgenic mice (Theuring *et al.*, 1990). Animals containing the intact 73/72- bp tandem repeat of the MSV enhancer exhibited lens-, pinealocyte- and pancreatic-specific expression of the oncogenic reporter gene SV40 T antigen, resulting in heritable lens cataracts (Götz *et al.*, 1991; Figure 8.3a), primitive neuroectodermal tumours (PNET; Figure 8.3b) and endocrine pancreatic tumours (Götz *et al.*, 1992, 1993). Analysis of the molecular phenotype and immunocytochemical properties of these PNETs revealed that these transgenic mice appear to provide a suitable animal model of human PNETs of the central nervous system (Korf *et al.*, 1990; Fung *et al.*, 1994).

By contrast, 5' deletion of the MSV enhancer to a residual 53 bp resulted in an altered expression pattern giving rise to a different spectrum of pathological changes. In particular, no pinealocyte-specific expression was obtained, whereas lens and pancreatic expression remained. Since both the entire MSV enhancer and the remaining 53 bp of the proximal enhancer element were found to be sufficient for directing oncogene expression to the lens of transgenic mice, we assumed that lens-specific transcription factors are able to interfere with the MSV regulatory elements and induce transgene

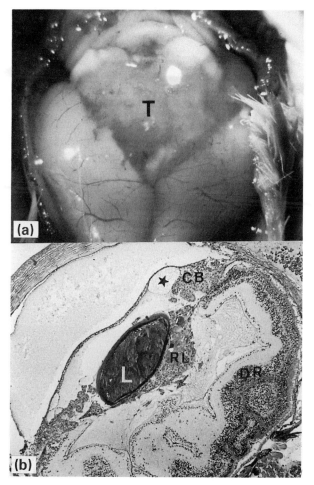

**Figure 8.3.** (a) Macroscopic view of the brain of a 74-day-old transgenic female mouse carrying an intact MSV–SV40T construct (line 123). The prominent tumour masses (T) originated from the pineal gland. (b) Horizontal section through the eye of a 3-month-old transgenic mouse (MSV–SV40T construct, line 125). A lens cataract and other severe malformation can be seen. L, cataractous lens with swollen fibres; RL, retrolental lentoid tissue, which is the result of a lens capsule defect; DR, dysplastic retina, CB, dysplastic ciliary body; *cysts of the iris.

expression (Theuring et al., 1992). This assumption is supported by the demonstration that the deleted proximal 53 bp enhancer element combined with the SV40 promoter is able to compete very efficiently with the g2-crystallin promoter for the lens-specific transcription factor, gF-1, whereas the SV40 promoter alone competes only very weakly (Q. Liu and M. Breitman, personal communication).

The transgenic system has provided new perspectives on the relative importance of *cis*- and *trans*-acting factors and sufficient information about the control of gene expression has been obtained to make this system useful for physiological studies. Regulation analysis allows us to:

(i) identify cells expressing the transgene;
(ii) identify *cis*-acting elements controlling expression in different cell types;
(iii) identify *trans*-acting enhancer/silencer factors;
(iv) study regulation *in vivo* under physiological and pathophysiological situations and after compound use.

It also has implications for somatic gene therapy. We are now able to direct, in a rational manner, the expression of a gene of interest in the whole animal, even if the transgene or parts of it are derived from evolutionarily distant species. Finally, knowledge of the appropriate regulatory regions controlling expression of certain genes will have an important impact on human gene therapy experiments, in which tightly controlled expression levels are important (e.g. β-globin; for review see, Dillon, 1993).

## 8.4 Transgenes to study gene function

A large proportion of the genes which will eventually be identified and cloned as a result of the Human Genome Project (Chapter 2) may be characterized by transgenic approaches (see also Chapter 9). It is the combination of *in vitro* and *in vivo* systems which will yield the most valuable information on newly identified genes and their functional analysis. The intact organism is characterized by cell–cell interactions, programmed cell death, and hormonal and neuronal input, to varying extents, either during development or adulthood. Since we are not able to reproduce these complex interactions within tissue culture systems, we are dependent on *in vivo* studies to understand complex systems such as the regulation of blood pressure and the function and influence of the proteins which make up these delicately regulated systems. Here is where the transgenic technology will have a major impact on our knowledge of these complex systems and about disease processes. Simply by adding new genetic material (i.e. by exerting a gene dosage effect), we may gain insight into the function of a gene of interest within the intact animal and its possible contribution to disease processes.

Transgenic animals have been instrumental in providing new insights into gene regulation and development, into the function of oncogenes, and into complex systems such as the immune system. Furthermore, the transgenic approach offers exciting possibilities for generating animal models for human diseases (i.e. in modelling cardiovascular disease, infectious and pulmonary diseases, tumorigenesis, immune and inflammatory diseases, developmental and reproductive disorders, and neurodegenerative disorders; Lathe and Mullins, 1993; Wagner and Theuring, 1993; see also Chapter 9).

Theoretically, transgenic methodology should provide the opportunity to target the expression of any cloned gene to any desired cell type or tissue. According to our current knowledge of differential gene expression, this would require combining the appropriate regulatory sequences (i.e. enhancer/silencer and promoter regions of a gene) known to be transcribed in the desired cell type, with the coding region for a protein of interest. The importance of regulatory sequences, intron-specific sequences and the existence of distant LCRs all favour the use of larger DNA fragments for the maintenance of normal gene expression. Therefore, the recent development of transgenic techniques using yeast artificial chromosomes (YACs; see Chapter 2) has yielded a complementary and alternative approach to the functional analysis of genes. YAC transgenes, which are believed to mimic the expression of the native gene, are expressed in a copy number-dependent, integration site-independent manner (Schedl *et al.*, 1993). Their potential use for functional transgenesis is even increased by inserting a pathogenic mutation into a YAC chromosome via homologous recombination, for example a mutation related to familial Alzheimer's disease (Duff *et al.*, 1994).

The different experimental strategies employing either gain-of-function or loss-of-function approaches might allow the generation of specific animal models for human genetic diseases.

Gain-of-function:
    Overexpression
    Ectopic expression
    Expression of oncogenes
    Expression of mutated transgenes, dominant

Loss-of-function:
    Homologous recombination
    Antisense
    Dominant negative mutations
    Expression of mutated transgenes
    Genetic ablation
    Antibodies

### *8.4.1 Functional analysis: gain-of-function*

*Overexpression.* Some ways of increasing the activity of gene products involve overexpression of proteins which are known or believed to be involved in the coordinated regulation of developmental processes or complex systems. By adding genetic information, either from mouse, rat or human sources, researchers have attempted to study the resulting phenotype achieved by this dosage effect. This approach has been adopted in different areas of biomedical research, for example in hypertension (Mullins *et al.*, 1990), arthritis (Keffer *et al.*, 1991), atherosclerosis (Rubin *et al.*, 1991), inflammatory disease (Hammer *et al.*, 1990; Tepper *et al.*, 1990) and neurodegenerative disorders (Quon *et al.*, 1991; for review, see Wagner and Theuring, 1993).

By introducing the apolipoprotein (a) gene into mice (this gene has either never been present in mice or has been lost during evolution), Lawn *et al.* (1992) generated an animal model for atherosclerosis. Apolipoprotein (a), which is missing from rodents and most other non-primate species (an exception is the European hedgehog), induced the formation of atherosclerosis in transgenic mice fed on high-fat diets. Both environmental and genetic factors play a role in the development of atherosclerosis, a multifactorial disorder of arterial hardening and fat deposition. The atherosclerotic lesions in the transgenic mice closely resembled the human situation in that they consisted of lipid-rich fibrous plaques with macrophage-like cells. Furthermore, it was shown that the overexpression of apolipoprotein AI could significantly protect susceptible mouse strains (e.g. C57bl) against diet-induced atherosclerotic disease (Rubin *et al.*, 1991). Overexpression of apolipoprotein CIII can result in severe hypertriglyceridaemia and it was suggested that this mechanism may represent one possible aetiology for this disorder in humans (Ito *et al.*, 1990). For a recent review on atherosclerosis and transgenic mice, see Rubin and Smith (1994).

*Ectopic expression.* Another method for assigning a function to genes that have been cloned examines the effects of their ectopic expression and their influence within the target tissue. The basic strategy utilizes hybrid genes in order to direct expression of the transgene to the desired cell type. This approach is best exemplified by the classical experiment of Palmiter *et al.* (1982) in which giant mice were generated; the use of a broad specificity promoter such as metallothionein fused to human growth hormone led to widespread growth hormone expression in the resulting transgenic mice. The consequent development of giant mice which grew significantly larger than their negative littermates (weight gain up to 1.8-fold) confirmed a role for growth hormone in controlling body growth. Ectopic expression of nerve growth factor under the insulin promoter led to hyperinnervation in the pancreatic islet cells by sympathetic, but not parasympathetic or sensory neurons (Edwards *et al.*, 1989), indicating a selectivity in the growth-promoting role of nerve growth factor.

To understand the function of murine homeobox genes, a transgenic analysis was performed. A gain-of-function mutation was induced by ectopic expression of the *Hox-A7* (*Hox-1.1*) gene under the control of a chicken β-actin promoter (Balling *et al.*, 1989; Kessel *et al.*, 1990). The normal spatial expression pattern of this developmental control gene was deregulated, resulting in craniofacial abnormalities and a posterior transformation of the cranial vertebrae in these mice, consistent with a homeotic function for this vertebrate homeobox gene.

*Expression of oncogenes.* Extensive analyses have been performed with proto-oncogenes, oncogenes, tumour suppressor genes, growth factors and their receptors, and other candidate genes fused to either homologous or heterologous 5' regulatory regions, thereby permitting targeted expression of relevant gene products to specific types of cells (for review, see Compere *et al.*, 1988; Hanahan, 1988; Adams and Cory, 1991; see also Chapter 9). The introduction of transgenic mice in the study of tumorigenesis has provided a wealth of information about the oncogenic potential of gene products, the kinetics of tumour development, cooperativity between oncogenes, synergy between oncogenes and tagged retroviruses, and has aided the molecular dissection of the multiple events leading from a normal cell to neoplastic transformation.

A remarkable proportion of these transgenic animals develop heritable tumours with characteristics reminiscent of human cancers and a number of general conclusions can be drawn from these studies:

(i) transgenic mice carrying oncogenes develop tumours in a predictable fashion, often with a very high incidence;
(ii) neoplastic transformation is not observed in all oncogene-expressing cells;
(iii) a latency period is observed before tumour formation; and
(iv) there is clonal tumour development (for reviews, see Christofori and Hanahan, 1994; Matzuk and Bradley, 1994; Merlino, 1994; Morgenbesser and DePinho, 1994; Ullrich *et al.*, 1994; Van Dyke, 1994; Webster and Muller, 1994; Zinkel and Fuchs, 1994).

Multiple genetic events are relevant for the stepwise progression to the tumour phenotype, and distinctive alterations have been recognized in transgenic mice, including tumour angiogenesis (Folkman *et al.*, 1989), chromosomal abnormalities (Lindgren *et al.*, 1989), cooperating oncogenes (Schackleford *et al.*, 1993) and deregulation of growth factors (Kandel *et al.*, 1991; Christofori *et al.*, 1994).

*Expression of mutated transgenes: dominant.* The high level expression of human Cu, Zn superoxide dismutase (SOD1) containing a substitution of glycine to alanine at position 93 caused motor neuron disease in transgenic mice (Gurney *et al.*, 1994). The mutant *SOD1* gene has been linked to an hereditary form of amyotrophic lateral sclerosis (ALS) and is found in about 20% of patients developing this invariably fatal condition. Mice that acquired the mutated *SOD1* gene developed weakness and neuronal degeneration and died by 5–6 months of age. The results demonstrated that dominant gain-of-function mutations in *SOD1* contribute to the pathogenesis of this inherited disease. The amino acid substitution had little effect on enzyme activity but instead conferred some new and harmful activity on SOD1 of hitherto unknown character which remains to be established. This transgenic model will help in understanding the pathological changes underlying ALS and might provide clues to possible future therapeutic intervention.

### 8.4.2 Functional analysis: loss-of-function

*Homologous recombination.* See Chapter 9.

*Antisense.* Alternative approaches to inactivate or at least partially reduce the activity of a desired gene involves the specific expression of either antisense or ribozyme molecules (see Chapter 11). For both approaches, very abundant expression of the RNA antagonist is needed before a reduction of the specific mRNA is achieved. Both approaches can be applied to transgenic animals, but thus far there have been reports only on the introduction of antisense constructs (Katsuki *et al.*, 1988; Pepin *et al.*, 1992; Erickson *et al.*, 1993; Matsumoto *et al.*, 1993).

*Dominant negative mutation.* This experimental strategy relies on the fact that many proteins are multimeric (i.e. they have multiple functional sites or subunits). The creation of a dominant negative mutation in one of the subunits of the protein will disrupt its entire function. Since the mutation is dominant and the phenotype manifested in the presence of functional wild-type protein, the consequences can be studied without having to disrupt each copy of the gene (Herskowitz, 1987). This methodological concept was proven in the following year by Stacey *et al.* (1988) who introduced a point mutation into the murine pro-α1(I) collagen gene identical to a lesion found in the human gene. Expression of the mutated gene in transgenic mice resulted in the dramatic dominant lethal phenotype of osteogenesis imperfecta (type II) which resembled clinically and biochemically the perinatal lethal phenotype of human patients; as little as 10% of mutant gene expression was sufficient to disrupt normal collagen function by distorting collagen fibrils. This study represented the first example of the generation of an animal model of a human genetic disease by introducing an identical mutation into the normal gene (encoding a multimeric protein) and inserting this as a dominant negative transgene into mice. By exerting a dominant phenotype, the consequences of the alterations could be tested in normal mice carrying the wild-type alleles (Stacey *et al.*, 1988).

A similar approach was employed by Krimpenfort *et al.* (1989) to study T-cell development and the requirements for allelic exclusion. A non-functional T-cell receptor (TCR) β-chain with a deletion in the major part of the variable region was introduced into transgenic mice. The results suggested that the deleted TCRβ was capable of inhibiting the endogenous β gene expression. Since no functional β-chains could be expressed on the surface of T cells, the mice were unable to express any αβ TCR at all. As a consequence, the mice were completely deficient in functional αβ T cells.

*Expression of mutated transgenes.* Sickle cell anaemia is a 'molecular disease' arising from a mutation ($\beta^s$) in codon 6 of the human β-globin (*HBB*) gene, causing a single amino acid substitution. One of the main constraints on research into the complicated and multifactorial pathology of this disease has been the lack of a good animal model. An animal model of sickle cell anaemia would not only allow a detailed understanding of the entire pathophysiological process, but would also permit the development of novel therapeutic approaches. Several groups have created and characterized transgenic mice expressing the human $\beta^s$ gene or variants of the sickle genes (for review, see Nagel, 1994). The use of the aforementioned LCR region resulted in high level, integration site-independent expression of the α- and/or β-globin gene. Although shortcomings of the transgenic animals (expressing mutant human haemoglobin molecules) in reproducing the human disease phenotype were observed, several important discoveries have been made with these models that would have been difficult without the transgenic approach (see Nagel, 1994, for discussion).

*Genetic ablation.* The basic strategy behind genetic ablation or toxigenics is the creation of an artificial gene in which a tissue-specific regulatory promoter element is fused to a coding region that either generates a toxic gene product or has the ability to metabolize a non-toxic substance to a toxic substance. Theoretically, targeted expression of this toxigene will result in cell ablation or at least partial tissue destruction, in aberrant elevation of signal molecules, or in inhibition of neurotransmission. It provides a method for generating mutant animals that lack, for example, specific cell types or even an entire cell lineage, and a novel approach to some aspects and problems of cell function and interaction, and cell lineage relationships (for reviews, see Evans, 1989; Breitman and Bernstein, 1992). Several strategies have been applied to the strategic elimination of cells. As prototype toxins for genetic ablation, the A chains of diphtheria toxin (*Corynebacterium diphtheriae*, *DT-A*) or ricin (*Ricinus communis*, *R-A*) were introduced. In these cases, protein synthesis is inhibited either through ribosylation of elongation factor 2 (*DT-A*) or through inactivation of 28S ribosomal RNA. This transgenic technology has been used to specifically ablate lens cells employing γ-crystallin–*DT-A* (Breitman *et al.*, 1987) and α-crystallin–*R-A* (Landel *et al.*, 1988) transgenes, to ablate acinar cells in the pancreas employing an elastase I–*DT-A* construct (Palmiter *et al.*, 1987), and to strategically ablate growth hormone-expressing cells in the pituitary (Behringer *et al.*, 1988). Subsequently, cholera toxin A1 (*Vibrio cholerae*; a non-cytotoxic and irreversible activator of $G_s$; Burton *et al.*, 1991) and tetanus toxin LC (*Clostridium tetani*; a highly potent neurotoxin; Eisel *et al.*, 1993) have been used to investigate cellular functions in transgenic mice.

Depending upon the transcriptional control elements used to target expression of the toxigenes, embryonic lethality might occur, thereby lowering the number of resulting transgenic animals and representing selection for certain threshold levels of toxic gene product. To overcome this and other problems which might be obtained with this strategy (e.g. promoter leakiness) an attenuated toxin gene would be an experimental advantage. Therefore, a mutant, attenuated form of the *DT-A* gene for ablating lens fibre cells was introduced (Breitman *et al.*, 1990; Breitman and Bernstein, 1992), which is about 30-fold less cytotoxic than the wild-type version as a consequence of a point mutation at residue 128.

An attractive and advantageous alternative would be an experimental system in which strategic cell death is inducible. This would provide a means to control the timing and duration of toxin action. Conditional ablation with the *tk* gene from herpes simplex virus (HSV) was achieved by exposing the mice to nucleoside analogues such as gancyclovir, acyclovir and related drugs (Borrelli *et al.*, 1988). The disadvantage of this approach is that cell ablation is effective only in actively dividing cells because the newly metabolized toxin (HSV-*tk* induces several phosphorylation steps, resulting finally in nucleoside triphosphates) acts by inhibition of DNA synthesis.

*Gene therapy.* Because *in vitro* studies have a limited role to play in developing somatic gene therapy protocols, many of the more important questions can only be elaborated *in vivo*. In view of the potential problems associated with somatic gene therapy, any such intervention will be required to be rigorously tested. As with any other new medical intervention, issues of safety and efficacy will be determining factors. For the correction of basic defects and the restoration of normal metabolism, either a repeated life-long treatment will be required or, perhaps eventually, a single lifetime treatment (Lathe and Mullins, 1993; Porteous and Dorin, 1993). In this regard, creating relevant animal models to test and develop effective and appropriate intervention and treatment is essential. Therapeutic development from early conception to pre-clinical trials is dependent on, and will benefit from, the availability of appropriate animal models and transgenic animals. In particular, mouse models for specific genetic diseases (and rat models, if the successful generation of rat ES cells can be achieved; see also Chapter 9) are likely to play a key role .

## 8.5 Conclusions

*Caenorhabditis* and *Drosophila* genetics have demonstrated convincingly that the study of mutants and their phenotypes is one of the most productive methods for the dissection of complex biological systems. Recent advances such as DNA microinjection and gene targeting by homologous recombination into the mammalian germline have made it possible that the same approach can be introduced into mammalian genetics.

The transgenic approach has contributed significantly to our understanding of gene regulation and function, and both mice and rats are being increasingly used as model systems for the analysis and understanding of common and complex polygenic disease conditions. Animal models have been created in which genes involved in these pathophysiological mechanisms have been deregulated, either by a gain-of-function or a loss-of-function approach. These precepts provide the basis for the application of transgenic animals to the study of the immune system, tumorigenesis and malignant diseases, the genetic alteration of the developmental programme of embryos, and the creation of animal models for human genetic diseases and gene therapy. Therefore, the use of transgenic technology will enable researchers to ascertain the *in vivo* function of these genes in disease processes and will allow research into ways of approaching specific therapies.

Whilst the generation of disease models represents a significant advance in the study of disease mechanisms, we have to keep in mind that there will never be a perfect animal model of a disease. These animals will contribute their own biochemical pathways, cellular characteristics, vascular system and gene products that contribute to the phenotype of a special pathophysiological situation. Nevertheless, the recognition and characterization of such possible biochemical and molecular differences between the experimental mammal and the human situation, and the understanding of the underlying mechanisms will enable us to better understand human pathology, its diagnosis and its therapy.

## Acknowledgements

Because I have restricted myself only to certain topics which should provide entry points into the transgenic technology, I apologize to those colleagues whose work I have not cited. I would like to thank H. Rohdewohld for discussion and J. Turner for critical reading of the manuscript.

## References

**Adams JM, Cory S.** (1991) Transgenic models of tumor development. *Science* **254**: 1161–1167.
**Aguzzi A, Theuring F.** (1994) An improved *in situ* β-galactosidase staining for histological analysis of transgenic mice. *Histochemistry* **102**: 477–481.
**Aguzzi A, Brandner S, Sure U, Rüedi D, Isenmann S.** (1994) Transgenic and knock-out mice: models of neurological disease. *Brain Pathol.* **4**: 3–20.
**Armstrong DT, Opavsky MA.** (1988) Superovulation of immature rats by continuous infusion of FSH. *Biol. Reprod.* **39**: 511–518.
**Balling R, Mutter G, Gruss P, Kessel M.** (1989) Craniofacial abnormalities induced by ectopic expression of the homeobox gene *Hox-1.1* in transgenic mice. *Cell* **58**: 337–347.
**Behringer RR, Mathews LS, Palmiter RD, Brinster RL.** (1988) Dwarf mice produced by genetic ablation of growth hormone-expressing cells. *Genes Dev.* **2**: 453–461.
**Borrelli E, Heyman R, Hsi M, Evans RM.** (1988) Targeting of an inducible toxic phenotype in animal cells. *Proc. Natl Acad. Sci. USA* **85**: 7572–7576.
**Bradley A, Evans M, Kaufman MH, Robertson E.** (1984) Formation of germ-line chimaeras from embryo-derived teratocarcinoma cell lines. *Nature* **309**: 255–256.
**Breitman ML, Bernstein A.** (1992) Engineering cellular deficits in transgenic mice by genetic ablation. In: *Transgenic Animal* (eds F Grosveld, G Kollias). Academic Press, New York, pp. 127–146.

**Breitman ML, Clapoff S, Rossant J, Tsui L-C, Glode LM, Maxwell IH, Bernstein A.** (1987) Genetic ablation: targeted expression of a toxin gene causes microphthalmia in transgenic mice. *Science* **238**: 1563–1565.

**Breitman ML, Rombola H, Maxwell IH, Klintworth GKK, Bernstein A.** (1990) Genetic ablation in transgenic mice with an attenuated diphtheria toxin A gene. *Mol. Cell. Biol.* **10**: 474–479.

**Brinster RL, Chen HY, Trumbauer M, Senear AW, Warren R, Palmiter RD.** (1981) Somatic expression of herpes thymidine kinase in mice following injection of a fusion gene into eggs. *Cell* **27**: 223–231.

**Brinster RL, Chen HY, Messing A, van Dyke T, Levine AJ, Palmiter R.** (1984) Transgenic mice harboring SV40 T-antigen genes develop characteristic brain tumors. *Cell* **37**: 367–379.

**Brinster RL, Chen HY, Trumbauer ME, Yagle MK, Palmiter RD.** (1985) Factors affecting the efficiency of introducing foreign DNA into mice by microinjecting eggs. *Proc. Natl Acad. Sci. USA* **82**: 4438–4442.

**Brinster RL, Allen JM, Behringer RR, Gelinas RE, Palmiter RD.** (1988) Introns increase transcription efficiency in transgenic mice. *Proc. Natl Acad. Sci. USA* **85**: 836–840.

**Burton FH, Hasel KW, Bloom FE, Sutcliffe JG.** (1991) Pituitary hyperplasia and gigantism in mice caused by a cholera toxin transgene. *Nature* **350**: 74–77.

**Capecchi MR.** (1980) High efficiency transformation by direct microinjection of DNA into cultured mammalian cells. *Cell* **22**: 479–488.

**Christofori G, Hanahan D.** (1994) Molecular dissection of multi-stage tumorigenesis in transgenic mice. *Sem. Cancer Biol.* **5**: 3–12

**Christofori G, Naik P, Hanahan D.** (1994) A second signal supplied by insulin-like growth factor II in oncogene-induced tumorigenesis. *Nature* **369**: 414–418.

**Compere SJ, Baldacci P, Jaenisch R.** (1988) Oncogenes in transgenic mice. *Biochim. Biophys. Acta* **948**: 129–149.

**Costantini F, Lacy E.** (1981) Introduction of a rabbit β-globin gene into the mouse germ line. *Nature* **294**: 92–94.

**Dillon N.** (1993) Regulating gene expression in gene therapy. *Trends Biotechnol.* **11**: 167–173.

**Duff K, McGuigan A, Huxley C, Schulz F, Hardy J.** (1994) Insertion of a pathogenic mutation into a yeast artificial chromosome containing the human amyloid precursor protein gene. *Gene Ther.* **1**: 70–75.

**Edwards RH, Rutter WJ, Hanahan D.** (1989) Directed expression of NGF to pancreatic β cells in transgenic mice leads to selective hyperinnervation of the islets. *Cell* **58**: 161–170.

**Eisel U, Reynolds K, Riddick M, Zimmer A, Niemann H, Zimmer A.** (1993) Tetanus toxin light chain expression in Sertoli cells of transgenic mice causes alterations of the actin cytoskeleton and disrupts spermatogenesis. *EMBO J.* **12**: 3365–3372.

**Erickson RP, Lai L-W, Grimes J.** (1993) Creating a conditional mutation of Wnt-1 by antisense transgenesis provides evidence that Wnt-1 is not essential for spermatogenesis. *Devel. Genet.* **14**: 274–281.

**Evans GA.** (1989) Dissecting mouse development with toxigenics. *Genes Dev.* **3**: 259-263.

**Evans MJ, Kaufman MH.** (1981) Establishment or culture of pluripotent cells from mouse embryos. *Nature* **292**: 154–156.

**Folkman J, Watson K, Ingber D, Hanahan D.** (1989) Induction of angiogenesis during the transition from hyperplasia to neoplasia. *Nature* **339**: 58–61.

**Fung K-M, Chikaraishi DM, Suri C, Theuring F, Messing A, Albert DM, Lee VM-Y, Trojanowski JQ.** (1994) Molecular phenotype of simian virus 40 large T antigen-induced primitive neuroectodermal tumors in four different lines of transgenic mice. *Lab. Invest.* **70**: 114–124.

**Gordon JW.** (1989) Transgenic animals. *Int. Rev. Cytol.* **115**: 171–229.

**Gordon JW.** (1993) Production of transgenic mice. *Methods Enzymol.* **225**: 747–771.

**Gordon JW, Ruddle FH.** (1981) Integration and stable germ line transmission of genes injected into mouse pronuclei. *Science* **214**: 1244–1246.

Gordon JW, Scangos GA, Plotkin DJ, Barbosa JA, Ruddle FH. (1980) Genetic transformation of mouse embryos by microinjection of purified DNA. *Proc. Natl Acad. Sci. USA* 77: 7380–7384.

Götz W, Theuring F, Favor J, Herken R. (1991) Eye pathology in transgenic mice carrying a MSV–SV40 large T-construct. *Exp. Eye Res.* 52: 41–49.

Götz W, Theuring F, Schachenmayr W, Korf H-W. (1992) Midline brain tumors in MSV–SV40-transgenic mice originate from the pineal organ. *Acta Neuropathol.* 83: 308–314.

Götz W, Schucht C, Roth J, Theuring F, Herken R. (1993) Endocrine pancreatic tumors in MSV–SV40 largeT transgenic mice. *Am. J. Pathol.* 142: 1493–1503.

Graessmann A, Graessmann M, Topp WC, Botchan M. (1979) Retransformation of a simian virus 40 cell line, which is resistant to viral and DNA infections, by microinjection of viral DNA. *J. Virol.* 32: 989–994.

Grosveld F, van Assendelft GB, Greaves DR, Kollias G. (1987) Position-independent, high-level expression of the human beta globin gene in transgenic mice. *Cell* 51: 975–985.

Gurney ME, Pu H, Chiu AY, Dal Canto MC, Polchow CY, Alexander DD, Caliendo J, Hentati A, Kwon YW, Deng H-X, Chen W, Zhai P, Sufit RL, Siddique T. (1994) Motor neuron degeneration in mice that express a human Cu, Zn superoxide dismutase mutation. *Science* 264: 1772–1775.

Hammer RE, Maika SD, Richardson JA, Tang J-P, Taurog JD. (1990) Spontaneous inflammatory disease in transgenic rats expressing HLA-B27 and human $\beta_2$m: an animal model of HLA-B27-associated human disorders. *Cell* 63: 1099–1112.

Hanahan D. (1985) Heritable formation of pancreatic β-cell tumors in transgenic mice expressing recombinant insulin/simian virus 40 oncogenes. *Nature* 315: 115–122.

Hanahan D. (1988) Dissecting multistep tumorigenesis in transgenic mice. *Annu. Rev. Genet.* 22: 479–519.

Harbers K, Jähner D, Jaenisch R. (1981) Microinjection of cloned retroviral genomes into mouse zygotes: integration and expression in the animal. *Nature* 293: 540–542.

Herskowitz I. (1987) Functional inactivation of genes by dominant negative mutations. *Nature* 329: 219–222.

Hogan B, Costantini F, Lacy E. (1986) *Manipulating the Mouse Embryo: a Laboratory Manual.* Cold Spring Harbor Laboratory Press, Cold Spring Harbor, NY.

Ito Y, Azrolan N, O'Connell A, Walsh A, Breslow JL. (1990) Hypertriglyceridemia as a result of human apo CIII gene expression in transgenic mice. *Science* 249: 790–793.

Jaenisch R. (1976) Germ line integration and Mendelian transmission of the exogenous Moloney leukemia virus. *Proc. Natl Acad. Sci. USA* 73: 1260–1264.

Jaenisch R, Mintz B. (1974) Simian virus 40 DNA sequences in DNA of healthy adult mice derived from preimplantation blastocysts injected with viral DNA. *Proc. Natl Acad. Sci. USA* 71: 1250–1254.

Johnson PF, McKnight SL. (1989) Eukaryotic transcriptional regulatory proteins. *Annu. Rev. Biochem.* 58: 799–839.

Kandel J, Bossy-Wetzel E, Radvanyi F, Klagsbrun M, Folkman J, Hanahan D. (1991) Neovascularization is associated with a switch to the export of bFGF in the multistep development of fibrosarcoma. *Cell* 66: 1095–1104.

Katsuki M, Sato M, Kimura M, Yokoyama M, Kobayashi K, Nomura T. (1988) Conversion of normal behavior to shiverer by myelin basic protein antisense cDNA in transgenic mice. *Science* 241: 593–595.

Keffer G, Probert L, Cazlaris H, Georgopoulos S, Kaslaris E, Kioussis D, Kollias G. (1991) Transgenic mice expressing human tumor necrosis factor: a predictive genetic model of arthritis. *EMBO J.* 13: 4025–4031.

Kessel M, Balling R, Gruss P. (1990) Variations of cervical vertebrae after expression of a *Hox*-1.1 transgene in mice. *Cell* 61: 301–308.

Korf H-W, Götz W, Herken R, Theuring F, Gruss P, Schachenmayr W. (1990) S-antigen and rod-opsin immunoreactions in midline brain neoplasms of transgenic mice: similarities to pineal cell tumors and certain medulloblastomas in man. *J. Neuropathol. Exp. Neurol.* **49**: 424–437.

**Krimpenfort P, Ossendorp F, Borst J, Melief C, Berns A.** (1989) T cell depletion in transgenic mice carrying a mutant gene for TCRβ. *Nature* **341**: 742–746.

**Landel CP, Zhao J, Bok D, Evans GA.** (1988) Lens-specific expression of recombinant ricin induces developmental defects in the eyes of transgenic mice. *Genes Dev.* **2**: 1168–1178.

**Lathe R, Mullins J.** (1993) Transgenic animals as models for human disease – report of an EC Study Group. *Transgen. Res.* **2**: 286–299.

**Lawn RM, Wade DP, Hammer RE, Chiesa G, Verstuyft JG, Rubin EM.** (1992) Atherogenesis in transgenic mice expressing human apolipoprotein (a). *Nature* **360**: 670–672.

**Lindgren V, Sippola-Thiele M, Skowronski J, Wetzel E, Howley PM, Hanahan D.** (1989) Specific chromosomal abnormalities characterize fibrosarcomas of bovine papillomavirus type I transgenic mice. *Proc. Natl Acad. Sci. USA* **86**: 5025–5029.

**Martin GR.** (1981) Isolation of a pluripotent cell line from early mouse embryos cultured in medium conditioned by teratocarcinoma stem cells. *Proc. Natl Acad. Sci. USA* **78**: 7634–7636.

**Matsumoto K, Kakidani H, Takahashi A, Nakagata N, Anzai M, Matsuzaki Y, Takahashi Y, Miyata K, Utsumi K, Iritani A.** (1993) Growth retardation in rats whose growth hormone gene expression was suppressed by antisense RNA transgene. *Mol. Reprod. Dev.* **36**: 53–58.

**Matzuk MM, Bradley A.** (1994) Identification and analysis of tumor suppressor genes using transgenic mouse models. *Sem. Cancer Biol.* **5**: 37–45.

**Merlino G.** (1994) Regulatory imbalance in cell proliferation and cell death during oncogenesis in transgenic mice. *Sem. Cancer Biol.* **5**: 13–20.

**Mitchell PJ, Tjian R.** (1989) Transcriptional regulation in mammalian cells by sequence-specific DNA binding proteins. *Science* **245**: 371–378.

**Morgenbesser SD, DePinho RA.** (1994) Use of transgenic mice to study *myc* family gene function in normal mammalian development and in cancer. *Sem. Cancer Biol.* **5**: 21–36.

**Mullins JJ, Peters J, Ganten D.** (1990) Fulminant hypertension in transgenic rats harbouring the mouse *Ren-2* gene. *Nature* **344**: 541–544.

**Nagel RL.** (1994) Lessons from transgenic mouse lines expressing sickle hemoglobin. *Proc. Soc. Exp. Biol. Med.* **205**: 274–281.

**Ornitz DM, Palmiter RD, Hammer RE, Brinster RL, Swift GH, MacDonald RJ.** (1985) Specific expression of an elastase–human growth hormone fusion gene in pancreatic acinar cells of transgenic mice. *Nature* **313**: 600–603.

**Palmiter RD, Brinster RL, Hammer RE, Trumbauer ME, Rosenfeld MG, Birnberg NC, Evans RM.** (1982) Dramatic growth of mice that develop from eggs microinjected with metallothionein–growth hormone fusion genes. *Nature* **300**: 611–615.

**Palmiter RD, Chen HY, Messing A, Brinster RL.** (1985) SV40 enhancer and large-T antigen are instrumental in development of choroid plexus tumours in transgenic mice. *Nature* **316**: 457–460.

**Palmiter RD, Behringer RR, Quaife CJ, Maxwell F, Maxwell IH, Brinster RL.** (1987) Cell lineage ablation in transgenic mice by cell-specific expression of a toxin gene. *Cell* **50**: 435–443.

**Palmiter RD, Sandgren EP, Avarbock MR, Allen DD.** (1991) Heterologous introns can enhance expression of transgenes in mice. *Proc. Natl Acad. Sci. USA* **88**: 478–482.

**Pepin M-C, Pothier F, Barden N.** (1992) Impaired type II glucocorticoid-receptor function in mice bearing antisense RNA transgene. *Nature* **355**: 725–728.

**Porteous DJ, Dorin JR.** (1993) How relevant are mouse models for human diseases to somatic gene therapy? *Trends Biotechnol.* **11**: 173–181.

**Püschel AW, Balling R, Gruss P.** (1990) Position-specific activity of the Hox1.1 promoter in transgenic mice. *Development* **108**: 435–442.

Quon D, Wang Y, Catalano R, Scardina JM, Murakami K, Cordell B. (1991) Formation of β-amyloid protein deposits in brains of transgenic mice. *Nature* **352**: 239–241.

Rohdewohld H, Weiher H, Reik W, Jaenisch R, Breindl M. (1987) Retrovirus integration and chromatin structure. Moloney murine leukemia proviral integration sites map near DNase I-hypersensitive sites. *J. Virol.* **61**: 336–343.

Rubin EM, Smith DJ. (1994) Atherosclerosis in mice: getting to the heart of a polygenic disorder. *Trends Genet.* **10**: 199–203.

Rubin EM, Krauss RM, Spangler EA, Verstuyft JG, Clift SM. (1991) Inhibition of early atherogenesis in transgenic mice by human apolipoprotein AI. *Nature* **353**: 265–267.

Rusconi S. (1991) Transgenic regulation in laboratory animals. *Experientia* **47**: 866–877.

Schackleford GM, MacArthur C, Kwan HC, Varmus HE. (1993) Mouse mammary tumor virus accelerates mammary carcinogenesis in Wnt-1 transgenic mice by insertional activation of int-2/FGF3 and hst/FGF4. *Proc. Natl Acad. Sci. USA* **90**: 740–744.

Schedl A, Montoliu L, Kelsey G, Schütz G. (1993) A yeast artificial chromosome covering the tyrosinase gene confers copy number dependent expression in transgenic mice. *Nature* **362**: 258–261.

Schöler HJ, Hatzopoulos AK, Schlokat U. (1988) Enhancers and *trans*-acting factors. In: *Architecture of Eukaryotic Genes* (ed. G Kahl). VCH, New York, pp. 89–120.

Stacey A, Bateman J, Choi T, Mascara T, Cole W, Jaenisch R. (1988) Perinatal lethal osteogenesis imperfecta in transgenic mice bearing an engineered mutant pro-α1(I) collagen gene. *Nature* **332**: 131–136.

Tepper RI, Levinson DA, Stanger BZ, Campos-Torres J, Abbas AK, Leder P. (1990) IL-4 induces allergic-like inflammatory disease and alters T cell development in transgenic mice. *Cell* **62**: 457–467.

Theuring F, Götz W, Balling R, Korf H-W, Schulze F, Herken R, Gruss P. (1990) Tumorigenesis and eye abnormalities in transgenic mice expressing MSV-SV40 large T-antigen. *Oncogene* **5**: 225–232.

Theuring F, Götz W, Korf H-W. (1992) Induced tumorigenesis in transgenic mice. In: *Molecular Approaches to the Study and Treatment of Human Diseases* (eds TO Yoshida, JM Wilson). Elsevier Science Publications, pp. 385–397.

Theuring F, Aguzzi A, Turner JD, Kropp C, Wohn K-D, Hoffmann S, Schleuning W-D. (1994a) Analysis of the human *tissue-type plasminogen activator* gene promoter activity during embryogenesis of transgenic mice and rats and its induction in the adult mouse brain. Submitted for publication.

Theuring F, Kooistra T, Schleuning W-D. (1994b) Transgenic mice carrying a human t-PA promoter – lacZ reporter gene: a model for the study of pharmacological modulation of fibrinolytic potential. *Fibrinolysis* **8** (Suppl. 2): 16–18.

Ullrich SJ, Zeng Z-Z, Jay G. (1994) Transgenic mouse models of human gastric and hepatic carcinomas. *Sem. Cancer Biol.* **5**: 61–68.

Van Assendelft GB, Hanscombe O, Grosveld F, Greaves DR. (1989) The beta-globin dominant control region activates homologous and heterologous promoters in a tissue-specific manner. *Cell* **56**: 969–977.

Van Dyke T. (1994) Analysis of viral–host protein interactions and tumorigenesis in transgenic mice. *Sem. Cancer Biol.* **5**: 47–60.

Wagner EF, Theuring F. (1993) *Transgenic Animals as Model Systems for Human Diseases*. E. Schering Foundation Workshop 6. Springer, Heidelberg.

Wagner EF, Stewart TA, Mintz B. (1981a) The human β-globin gene and a functional thymidine kinase gene in developing mice. *Proc. Natl Acad. Sci. USA* **78**: 5016–5020.

Wagner TE, Hoppe PC, Jollick JD, Scholl DR, Hodinka RL, Gault JB. (1981b) Microinjection of a rabbit β-globin gene in zygotes and its subsequent expression in adult mice and their offspring. *Proc. Natl Acad. Sci. USA* **78**: 6376–6380.

**Webster MA, Muller WJ.** (1994) Mammary tumorigenesis and metastasis in transgenic mice. *Sem. Cancer Biol.* **5**: 69–76.
**Wight DC, Wagner TE.** (1994) Transgenic mice: a decade of progress in technology and research. *Mutat. Res.* **307**: 429–440.
**Zinkel S, Fuchs E.** (1994) Skin cancer and transgenic mice. *Sem. Cancer Biol.* **5**: 77–90.

# 9

# Homologous recombination

Ahmed Mansouri

## 9.1 Introduction

The introduction of exogenous DNA into mammalian cells has proven to be a very useful approach to study the mechanisms controlling gene regulation. Normally, the introduced DNA molecule integrates randomly within the cellular genome. However, the specific targeting of exogenous modified DNA to a defined locus in the cellular genome is now possible by means of homologous recombination. Although in itself a rare event, gene-specific targeting does provide a valuable method not only for the study of gene function but also to correct genetic defects in somatic cells. In principle, the repair of genetic defects can be achieved by introducing one or more copies of the gene of interest into the genome by random integration. However, random integration is not controllable and may be deleterious to the target cells, either because of the possibility of inactivation of an important endogenous gene (insertion mutation) or, alternatively, as a result of the activation of a normally silent endogenous gene.

More than 10 years ago, it was demonstrated that mammalian somatic cells possess the basic machinery necessary for mediating homologous recombination between newly introduced DNA molecules and endogenous chromosomal loci (Folger *et al.*, 1982; Smithies *et al.*, 1985). Many studies have since been performed in order to determine the nature of the mechanism of homologous recombination in somatic cells. In practical terms, the technique of homologous recombination promises to provide new tools for the correction of gene defects in somatic cells, an important first step in the establishment of gene therapy strategies. Further, in conjunction with embryonic stem (ES) cells, the homologous recombination technique provides a new tool to study gene function through the creation of transgenic mice with designed mutations for any gene of interest and for the development of new animal models for inherited human diseases. The use of this technique has already led to a great increase in the availability of mouse mutants for different genes.

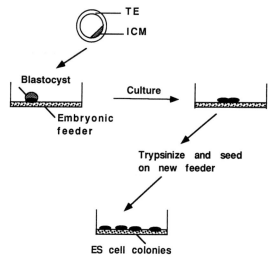

**Figure 9.1.** Establishment of embryonic stem cells from mouse blastocyst. Mouse blastocyst is cultured on mouse embryonic feeder cells (inactivated with mitomycin C). After a few days the inner cell mass (ICM) starts to grow. The growing ICM cells are mechanically disaggregated and trypsinized. Afterwards they are passaged on a new feeder layer. The passaging is repeated until a homogenous ES cell population is achieved. TE, trophectoderm.

## 9.2 Embryonic stem cells

ES cells are derived from the inner cell mass of the mouse blastocyst (Figure 9.1; Evans and Kaufman, 1981). The mouse blastocyst is cultured on primary embryonic fibroblasts and, after a few days, the blastocyst attaches to the feeder cells of fibroblasts and the inner cell mass starts to grow. Successive trypsinization of the growing inner cell mass cells and plating on new feeder cells yields a homogenous population of embryonic stem cells. These cells, when cultured on embryonic feeder cells (and/or in the presence of leukaemia inhibitory factor, LIF), remain undifferentiated and retain their pluripotency. This has been demonstrated by the introduction of the ES cells back into the mouse embryo via blastocyst injection. In chimeric animals, ES cells contribute to all tissues, including the germline, so that the ES genome is transmitted to the mouse germline (Bradley *et al.*, 1984). Moreover, ES cells can be manipulated *in vitro* by transfection with foreign exogenous DNA molecules, which can stably integrate into the cellular genome. After blastocyst injection of the ES cells which have integrated the foreign DNA, chimeric mice are able to transmit the newly introduced DNA into the germline to generate transgenic mice (Gossler *et al.*, 1986; Robertson *et al.*, 1986; Hooper *et al.*, 1987; Kuehn *et al.*, 1987). By contrast to transgenesis mediated by microinjection (see Chapter 8), where the integration, copy number and expression of the introduced gene cannot be controlled, in ES cells these factors may be analysed before proceeding to the generation of germline chimeras.

## 9.3 Principles of homologous recombination in mammalian cells

By contrast with yeast, random integration of the introduced DNA in mammalian cells is the dominant event whilst homologous recombination occurs at low frequency, depending upon the gene analysed. Many strategies have therefore been developed to enhance the homologous recombination frequency in mammalian cells, especially ES cells. Different vectors have also been designed to interrupt genes by homologous recombination and these will now be described.

### 9.3.1 Targeting vectors

In the first type of vector, the targeting construct is designed in such a way that, after linearization with a restriction enzyme to introduce DNA into the cells, it remains co-linear with the endogenous sequences of the gene to be targeted (Figure 9.2b). When a homologous recombination event occurs, the endogenous sequences are replaced by the targeting vector (Figure 9.2b). Sequences, which lie outside of the homology domains are excluded from the integration. The homology serves to define which sequences in the construct, identical to sequences in the endogenous gene, will participate in the recombination process (Figure 9.2a). This kind of vector is termed a *replacement vector* (Thomas and Capecchi, 1987).

The second type of vector is linearized so that the pairing of the targeting vector with the endogenous homologous sequences of the gene of interest results in the insertion of the whole targeting construct including plasmid sequences. This results in a partial duplication of the gene to be targeted. This kind of vector is therefore termed an *insertion vector* (Thomas and Capecchi, 1987).

Both types of vector exhibit similar targeting frequencies and their dependence on the extent of homology is identical (Thomas and Capecchi, 1987). Both types of vector have been successfully used in gene targeting experiments in ES cells (Doetschman *et al.*, 1987, 1988b; Thomas and Capecchi, 1987; Thompson *et al.*, 1989). However replacement vectors have been used in the majority of cases of genes which have been knocked out in ES cells. Replacement vectors are easy to construct and to handle. Insertion vectors, although more difficult to handle, have the advantage of generating point mutations or small deletions since the selectable marker can be inserted into an intron (Thomas and Capecchi, 1987). Insertion vectors have also been used successfully as the first step of knock-out experiments to design point mutations via the hit-and-run technique (Hasty *et al.*, 1991; Valancius and Smithies, 1991; see Section 9.3.4).

The exogenous DNA fragment or vector used for the targeting experiment contains two domains with homology to the gene to be targeted, one at the 5′

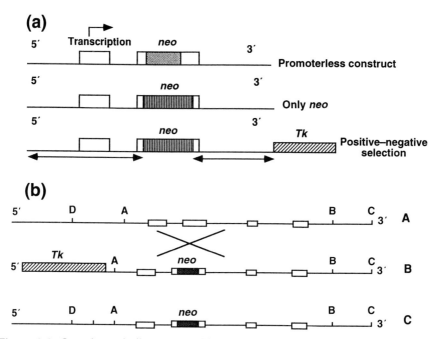

**Figure 9.2.** Open boxes indicate exons. The domains of homology are indicated schematically in the positive–negative selection (PNS) vector by the two arrows. *Tk*, herpes simplex virus–thymidine kinase gene. *neo*, neomycin gene. (a) Different types of targeting constructs. The promoterless construct contains the *neo* in-frame to the second exon. The construct containing only *neo* does not allow any enrichment. The PNS vector is designed for selection with G418 and gancyclovir and allows an enrichment factor between four- and 10-fold. (b) Principle of gene targeting using replacement vectors in combination with the PNS procedure. A, normal allele; B, targeting construct; C, genomic organization of the mutated allele after recombination. The letters on the constructs indicate restriction enzyme recognition sites.

end and the other at the 3′ end of the construct (Figure 9.2). Between the 5′ end and 3′ end homologous portions, a neomycin (*neo*) resistance gene is introduced as a selectable marker firstly to monitor the introduced DNA and secondly to interrupt the function of the gene to be analysed. The *neo* gene is therefore inserted into an exon or it may replace some of the coding sequence which must be deleted to abolish the function of the gene being targeted.

Electroporation of the targeting construct into ES cells and subsequent selection (normally 24–30 h after electroporation) with geneticin (G418) kills all the cells which have not integrated the targeting construct. After 8–10 days of selection, the resulting G418-resistant colonies are screened by polymerase chain reaction (PCR) or genomic Southern blot for homologous recombination events (Figure 9.3). The homologous recombinant clones are then injected into mouse blastocyst to generate germline chimeras, which transmit the introduced mutation into the mouse germline (Figure 9.3). Since the ES cells are normally isolated from 129Sv agouti strain, chimerism can be followed by coat colour of agouti against black (host strain C57Black/6 of the blastocyst).

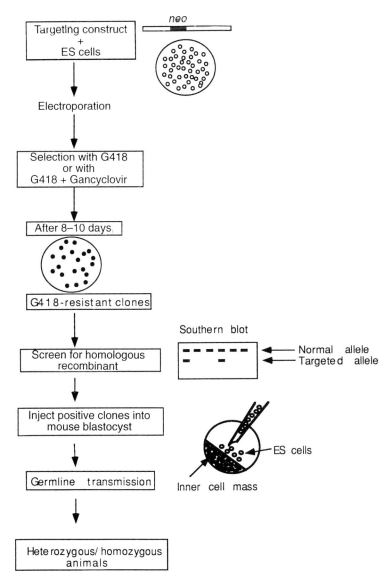

**Figure 9.3.** A typical targeting experiment is schematically shown. ES cells ($10^7$) are electroporated in the presence of 25 µg linearized construct and 24 h later selection is started (G418 or G418 + gancyclovir for PNS selection procedure). After 8–10 days, growing resistant colonies are picked and screened by PCR or Southern blot. Positive clones (indicated in the schematic Southern blot by the second band of the mutated allele) are then injected into mouse blastocysts and germline chimeras are generated which transmit the gene to the mouse germline.

### 9.3.2 Promoterless constructs

Apart from general factors found to influence the frequency of homologous recombination, several strategies have been developed to knock out genes by homologous recombination in ES cells. Generally two factors have been shown to affect targeting efficiency: the size of the genomic fragment included in the construct (Deng and Capecchi, 1992; Thomas *et al.*, 1992) and the origin of the DNA (Riele *et al.*, 1992). Good targeting efficiencies are usually obtained with constructs including more than 10 kb of homology. Increased targeting efficiencies have been observed when using isogenic DNA isolated from a 129 Sv genomic library, the mouse strain from which most ES cells are derived.

As mentioned above, the ES cells should be maintained in an undifferentiated state during the whole procedure to ensure germline transmission when injected into the mouse blastocyst. Genes which are expressed in undifferentiated ES cells can be disrupted using a *promoterless construct*. Basically, a promoterless *neo* gene is fused in-frame to the exon to be interrupted (Figure 9.2a). When introduced into the ES cells by electroporation, the *neo* gene is activated only if a random integration event places the sequence next to a promoter or when homologous recombination occurs under the target promoter (Schwartzberg *et al.*, 1989; Stanton *et al.*, 1990). The advantage of this strategy is the low number of clones to be screened and the high frequency of homologous recombination (the enrichment factor is about 100-fold).

### 9.3.3 Positive–negative selection procedure

When the gene to be targeted is not expressed in undifferentiated ES cells, the targeting frequency is very low. Capecchi and colleagues have developed a new strategy which provides enrichment for homologous recombinant events (Mansour *et al.*, 1988). The strategy is not dependent on gene function or expression and is especially suitable for those genes which are not expressed in ES cells. The only requirement is for information on the genomic organization and intron–exon boundaries of the gene to be targeted. As mentioned above, the *neo* gene is inserted into or replaces the coding region to be deleted. In addition, a thymidine kinase gene from herpes simplex virus (*Tk*) is placed outside of the targeting construct (Figure 9.2a, b). Two types of selection are now applied simultaneously: the first selection (positive selection) uses G418 to select for cells which have integrated the construct (random integrations + homologous recombinants) and the second (negative selection) uses gancyclovir to eliminate all cells which have integrated the construct but which have retained the *Tk* gene (random integrations). This positive–negative selection (PNS) procedure enriches for cell clones with the designed mutation. In this way, the PNS procedure provides an enrichment factor of between four- and 10-fold. Currently this method is used for most knockout experiments.

### 9.3.4 Hit-and-run and in–out targeting strategies

The generation of loss-of-function mutants by homologous recombination in ES cells uses the insertion of the neomycin gene or other selectable marker in the targeting construct (for review, see Capecchi, 1989). After targeting, the mutated allele contains the *neo* gene, generally associated with enhancer and promoter sequences. These sequences could affect gene regulation especially for those genes which are clustered, such as the *Hox* genes. Therefore Bradley and colleagues have developed an elegant technique to generate mutations without any marker being left in the mutated allele (Hasty *et al.*, 1991). Two steps are required for the experiment. In the first, an insertion vector containing the *neo* and *Tk* genes is used to introduce a subtle mutation in the homeobox of *Hox-B4* (*Hox-2.6*) using G418 as selection agent. In the second step, using gancyclovir to select for *Tk*, the inserted first mutation, which contains some duplication of the *Hox-B4* sequences, undergoes intramolecular recombination to release a mutated allele with the desired mutation in the homeobox and with the *neo* and *Tk* sequences removed (Hasty *et al.*, 1991). This method is termed 'hit-and-run targeting'. Smithies and colleagues have established a similar strategy termed as 'in–out targeting' (Valancius and Smithies, 1991) which uses hypoxanthine phosphoribosyltransferase (HPRT) as selection system and requires only ES cells which are HPRT negative.

## 9.4 Potential of homologous recombination in embryonic stem cells

### 9.4.1 Developmental biology

The study of the expression patterns of genes during development, essentially descriptive biology, is being steadily superceded by the new approach of homologous recombination in ES cells which provides a new tool to generate loss-of-function mutants to study gene function. The comparison of normal mice with those altered by inactivation of the gene of interest provides insights into the function of that gene during embryonic development (Yamada *et al.*, 1994). Gene families can be knocked out one by one, and single, double and eventually triple mutants can be analysed to elucidate the role of the different genes and to rule out redundancy effects. In the *MyoD* family, the inactivation of *MyoD*, *Myf-5* and myogenin revealed that although muscle is normally produced in mice lacking *MyoD* or *Myf-5* (Braun *et al.*, 1992; Rudnicki *et al.*, 1992), mice with null alleles for both genes produce no muscle and no myogenin is transcribed (Rudnicki *et al.*, 1993). This indicates that *Myf-5* and *MyoD* share a redundant function necessary for muscle generation. On the other hand, myogenin-deficient mice possess myoblasts but these cells are not able to differentiate into muscle, suggesting that this gene is required for differentiation to myotubes (Hasty *et al.*, 1993; Nabeshima *et al.*, 1993).

Among other gene families involved in embryonic development, the *Hox* family (Kessel and Gruss, 1990) and the *Pax* family (Gruss and Walther,

1992) are worthy of mention. These genes have been cloned by using *Drosophila* sequences to screen mouse libraries, making the assumption that developmental control genes are conserved between different species. The temporal and spatial expression patterns of these genes during mouse embryonic development suggest an important role during this process. Studies of loss-of-function mutants for some *Hox* genes (Chisaka and Capecchi, 1991; Lufkin *et al.*, 1991; Le Mouellic *et al.*, 1992; Ramírez-Solis *et al.*, 1993) support and confirm this idea and suggest that *Hox* gene expression is required to establish the anterior–posterior axis of the embryo (Kessel and Gruss, 1991). Finally, the inactivation of the *Wnt-1* (*Int-1*) gene results in the loss of the midbrain and some of the rostral metencephalon, suggesting that this gene is required for the normal development of the mouse brain (McMahon and Bradley, 1990; Thomas and Capecchi, 1990).

### 9.4.2 Animal models of human disease

In recent years, several spontaneous mouse mutants have been described which can be used as animal models for human genetic disorders (Darling and Abbott, 1992). However, the number of these mutants is limited and the search for other mutants would be extremely time consuming since more than 4000 human genetic traits/disorders are known. Therefore the homologous recombination technique remains the method of choice to generate animal models for known human genetic diseases. Although mouse models are not always suitable for all human disorders (many neurological and vascular systems are better studied in the rat), some existing mouse models are encouraging. In addition, ES cells from rat (Iannaccone *et al.*, 1994) and hamster (Doetschman *et al.*, 1988a) are already available and it is only a question of time before animals are generated from these species which may serve as models for human diseases.

A mouse model for Lesch–Nyhan syndrome was the first generated by the ES cell approach (Hooper *et al.*, 1987; Kuehn *et al.*, 1987). However, the total absence of HPRT activity in affected animals appears not to have any physiological effect. More interesting is the fact that HPRT-deficient male mice exhibit none of the neurological symptoms characteristic of Lesch–Nyhan syndrome (Hooper *et al.*, 1987; Kuehn *et al.*, 1987). The administration of high doses of amphetamine did cause a significant increase of stereotypic and locomotor behaviour in HPRT-deficient mice as compared to control animals (Jinnah *et al.*, 1991). Recently Wu and Melton (1993) observed Lesch–Nyhan syndrome symptoms, including the compulsive self-injurious behaviour, after administration of an adenine phosphoribosyltransferase (APRT) inhibitor to HPRT-deficient mice. Although these mice appear not to be a useful model for Lesch–Nyhan syndrome, the results reported by Wu and Melton (1993) are nevertheless encouraging.

Another animal model for a human disease has been generated for cystic fibrosis. Cystic fibrosis is an autosomal recessive disorder affecting about 1 in every 2500 newborn individuals in Caucasian populations. It is caused by a defect in chloride transport in epithelial cells. The gene encodes a protein termed cystic fibrosis transmembrane conductance regulator (CFTR). Several hundred different mutations of the *CFTR* gene are now known; the most frequent (δF508 in exon 10) constitutes approximately 70% of cystic fibrosis alleles in European populations (Collins, 1992; Tsui, 1992). Different groups have generated animal models for cystic fibrosis (Dorin *et al.*, 1992; Snouwaert *et al.*, 1992). Between them, these animal models exhibit most of the pathology observed in human patients. However, it is interesting to note that mice generated by a replacement vector had no obvious lung defect but did have a more severe intestinal defect which led to death of most animals in the first week after weaning (Snouwaert *et al.*, 1992). On the other hand, homozygous animals generated using the insertion vector were viable and exhibited the cystic fibrosis phenotype (including lung pathology) characteristic of human patients (Dorin *et al.*, 1992). It would appear that the mouse generated using the insertion vector is more suitable as an animal model since it survives and shows most of the symptoms characteristic of the human condition. However, the difference between the two knock-out animals (both mutations inserted in exon 10 of the *CFTR* gene) may not have been due to the type of targeting vector used. Rather, it is more likely that the mouse genetic background upon which the animals were crossed was responsible for the observed differences (Dorin *et al.*, 1992). The mice which exhibited early lethality were crossed on to inbred backgrounds, whereas the viable animals from the insertion vector were crossed with an outbred strain. Although many parameters remain to be carefully analysed, these mice are nevertheless useful as an animal model for cystic fibrosis and should prove invaluable in the establishment of a gene therapy programme which aims to correct the defect *in vivo*.

In Gaucher's disease, a lysosomal storage disorder, an animal model has been generated through the targeted disruption of the murine glucocerebrosidase gene using the ES cell approach (Tybulewicz *et al.*, 1992). In the field of cancer research, homologous recombination has permitted the generation of mice lacking the p53 protein. The mice are normal but develop cancer within 6 months of birth and thus serve as an animal model of carcinogenesis (Donehower *et al.*, 1992; Harvey *et al.*, 1993). The inactivation of the murine retinoblastoma gene (Clarke *et al.*, 1992; Jacks *et al.*, 1992; Lee *et al.*, 1992) surprisingly results in a different phenotype from that observed in humans; mice heterozygous for mutations in the *Rb* gene, which in humans is associated with retinoblastoma, do not develop tumours. Similarly, the inactivation of the Wilms' tumour (*WT1*) gene has a different phenotype in mice as compared to humans; further heterozygous animals do not exhibit any predisposition to develop tumours (Kreidberg *et al.*, 1993). These results may serve to emphasize differences in the pathogenesis of tumours in mice and human. However, long term and more careful analysis of the mutant mice may still reveal some similarities. These caveats notwithstanding, the development of animal models has been very useful for the analysis of gene function *in vivo*. For example, there could be a difference between animals manifesting germline and somatic homozygosity in the case of the retinoblastoma gene (Clarke et al., 1992).

## 9.5 Homologous recombination and gene therapy

The concept of introducing functional genes into those tissues where the expression of the corresponding gene is compromised has created a new research area, *gene therapy*. This has rapidly become a new discipline in molecular medicine. To introduce the DNA to the target cells or organs, different vectors have been developed. The most commonly used vectors are of retroviral or adenoviral origin. The first experiments have already been performed with a view to the development of therapies for acquired immunodeficiency syndrome (AIDS; Gilboa and Smith, 1994), neurological disorders (Friedmann, 1994), metabolic disorders (Kay and Woo, 1994) and cancer (Culver and Blaese, 1994). The introduction of the DNA to the target tissue or cell is random and there is as yet no report of the use of homologous recombination to correct a gene in the affected tissue. This is obvious, since in most cases random integration is the most probable event. It is therefore worthwhile to determine which parts of the homologous recombination mechanism could be involved in establishing homologous recombination as the dominant integration event after gene transfer. This notwithstanding, homologous recombination in ES cells is already the method of choice to produce animal models necessary for therapeutic trials, including gene therapy.

## 9.6 Future perspectives: *Cre/LoxP*-mediated gene targeting

Gene targeting often uses a selectable marker which can interfere with the analysis of the phenotype, especially if the disrupted gene belongs to a gene cluster like the *Hox* genes. As discussed above, this can be avoided by using the hit-and-run and in–out methods. There is, however, a newly developed technique, which makes it possible not only to remove the selectable marker but also, if a suitable promoter is available, to perform this in a tissue-specific manner.

This new technique is based on the *Cre/Lox* recombination system of bacteriophage P1 (Sauer, 1993). The *Cre* recombinase has been shown to perform recombination at *Lox* sites in bacteria and in eukaryotic cells (Sauer and Henderson, 1990). DNA, which is flanked by *Lox* sequences (for example, the selectable marker *neo*), is excised by the *Cre* recombinase.

The first experiments (Gu et al., 1993) using this system reveal that it can be used very efficiently for gene targeting in ES cells. Moreover, Rajewsky and his colleagues used this technique to knock out the β-polymerase gene specifically in T cells (Gu et al., 1994). This was achieved by expressing the *Cre* recombinase under a T cell-specific promoter. These results are very encouraging and hold out the promise that in the future it will be possible to design any mutation for any gene of interest. Furthermore, it may in principle be possible to knock out any gene in any tissue at any time of embryonic development. With the experiments cited above, the era of temporal gene knock-outs has started.

## Acknowledgements

I would like to thank Peter Gruss for his continuing support and encouragement, Jens Krull for technical assistance and David N. Cooper and Edward Stuart for reading the manuscript. This work was supported by the Max-Planck Society.

## References

Bradley A, Evans M, Kaufman MH, Robertson EJ. (1984) Formation of germline chimeras from embryo derived teratocarcinoma cell lines. *Nature* **309**: 255–256.

Braun T, Rudnicki MA, Arnold HH, Jaenisch R. (1992) Targeted inactivation of the muscle regulatory gene *Myf-5* results in abnormal rib development and perinatal death. *Cell* **71**: 369–382.

Capecchi MR. (1989) Altering the genome by homologous recombination. *Science* **244**: 1288–1292.

Chisaka O, Capecchi MR. (1991) Regionally restricted development defects resulting from targeted disruption of the mouse homeobox gene *Hox1.5*. *Nature* **350**: 473–479.

Clarke AR, Maandag ER, Van Roon M, Van der Lugt NMT, Van der Valk M, Hooper M, Berns A, Te Riele H. (1992) Requirement for a functional *Rb-1* gene in murine development. *Nature* **359**: 328–330.

Collins F. (1992) Cystic fibrosis: molecular biology and therapeutic implications. *Science* **256**: 774–779.

Culver KW, Blaese, RM. (1994) Gene therapy for cancer. *Trends Genet.* **10**: 174–178.

Darling SM, Abbott CM. (1992) Mouse models of human single gene disorders. I: non transgenic mice. *Bioessays* **14**: 359–366.

Deng C, Capecchi, MR. (1992) Reexamination of gene targeting frequency as a function of the extent of homology between the targeting vector and target locus. *Mol. Cell. Biol.* **12**: 3365–3371.

Doetschman TC, Gregg RG, Maeda N, Hooper ML, Melton DW, Thomson S, Smithies O. (1987) Targeted correction of mutant HPRT gene in mouse embryonic stem cells. *Nature* **330**: 576–578.

Doetschman TC, Williams P, Maeda, N. (1988a) Establishment of hamster blastocyst-derived embryonic stem cells. *Devel. Biol.* **127**: 224–227.

Doetschman TC, Maeda N, Smithies O. (1988b) Targeted mutation of the *HPRT* gene in embryonic stem cells. *Proc. Natl Acad. Sci. USA* **85**: 8583–8587.

Donehower LA, Harvey M, Slagle BL, McArthur MJ, Montgomery CA Jr, Butel JS, Bradley A. (1992) Mice deficient for p-53 are developmentally normal but susceptible to spontaneous tumours. *Nature* **356**: 215–221.

Dorin JR, Dickinson P, Alton EWFW, Smith SN, Geddes DM, Stevenson BJ, Kimber WL, Fleming S, Clarke AR, Hooper ML, Anderson L, Beddington RSP, Porteous DJ. (1992) Cystic fibrosis in the mouse by targeted insertional mutagenesis. *Nature* **359**: 211–215.

Evans MJ, Kaufman MH. (1981) Establishment in culture of pluripotent cells from mouse embryos. *Nature* **292**: 154–156.

Folger KR, Wong EA, Wahl G, Capecchi MR. (1982) Patterns of integration of DNA microinjected into cultured mammalian cells: evidence for homologous recombination between injected plasmid molecules. *Mol. Cell. Biol.* **2**: 1372–1378.

Friedmann T. (1994) Gene therapy for neurological disorders. *Trends Genet.* **10**: 210–214.

Gilboa E, Smith C. (1994) Gene therapy for infectious diseases: the AIDS model. *Trends Genet.* **10**: 139–144.

Gossler A, Doetschman TC, Korn R, Serfling E, Kemler R. (1986) Transgenesis by means of blastocyst-derived embryonic stem cell lines. *Proc. Natl Acad. Sci. USA* **83**: 9065–9069.

Gruss P, Walther, C. (1992) Pax in development. *Cell* **69**: 719–722.

Gu H, Zou Y-R, Rajewski K. (1993) Independent control of immunoglobulin switch recombination at individual switch regions: evidence through *Cre-loxP*-mediated gene targeting. *Cell* **73**: 1155–1164.

Gu H, Marth JD, Orban PC, Mossman H, Rajewski K. (1994) Deletion of a DNA polymerase β gene segment in T cells using cell type-specific gene targeting. *Science* **265**: 103–106.

Harvey M, McArthur MJ, Montgomery CA Jr, Butel JS, Bradley A, Donehover LA. (1993) Spontaneous and carcinogen induced tumorigenesis in p53-deficient mice. *Nature Genetics* **5**: 225–229.

Hasty P, Ramírez-Solis R, Krumlauf R, Bradley A. (1991) Introduction of a subtle mutation into the *Hox 2.6* locus in embryonic stem cells. *Nature* **350**: 243–246.

Hasty P, Bradley A, Morris JH, Edmondson DG, Venuti JM, Olson EN, Klein WH. (1993) Muscle deficiency and neonatal death in mice with a targeted mutation in the myogenin gene. *Nature* **364**: 501–506.

Hooper M, Hardy K, Handyside A, Hunter S, Monk M. (1987) HPRT-deficient (Lesch–Nyhan) mouse embryos derived from germline colonization by cultured cells. *Nature* **326**: 292–295.

Iannaccone PM, Taborn GU, Garton RL, Caplice MD, Brenin DR. (1994) Pluripotent embryonic stem cells from the rat are capable of producing chimeras. *Devel. Biol.* **163**: 288–292.

Jacks T, Fazeli A, Schmitt EM, Bronson RT, Goodel MA, Weinberg RA (1992) Effects of an *Rb* mutation in the mouse. *Nature* **359**: 295–300.

Jinnah HA, Gage FH, Friedman T. (1991) Amphetamine-induced behavioural phenotype in a hypoxanthine–guanine phosphoribosyltransferase deficient mouse model of Lesch–Nyhan syndrome. *Behav. Neurosci.* **105**: 1004–1012.

Kay MA, Woo SLC. (1994) Gene therapy for metabolic disorders. *Trends Genet.* **10**: 253–257.

Kessel M, Gruss P. (1990) Murine developmental control genes. *Science* **249**: 374–379.

Kessel M, Gruss P. (1991) Homeotic transformations of murine prevertebrae and concomitant alteration of the *Hox* codes induced by retinoic acid. *Cell* **67**: 89–104.

Kreidberg JA, Sariola H, Loring JM, Maeda M, Pelletier J, Housman D, Jaenisch R. (1993) WT-1 is required for early kidney development. *Cell* **74**: 679–691.

Kuehn MR, Bradley A, Robertson EJ, Evans MJ. (1987) A potential animal model for Lesch–Nyhan syndrome through introduction of HPRT mutations into mice. *Nature* **326**: 295–298.

Lee EY-P, Chang CY, Hu N, Wang Y-CJ, Lai C-C, Herrup K, Lee W-H, Bradley A. (1992) Mice deficient for Rb are nonviable and show defects in neurogenesis and haematopoiesis. *Nature* **359**: 288–294.

Le Mouellic H, Lallemand Y, Brûlet P. (1992) Homeosis in the mouse induced by a null mutation in the *Hox-3.1* gene. *Cell* **69**: 251–264.

Lufkin T, Dierich A, Le Meur M, Mark M, Chambon P. (1991) Disruption of the *Hox 1.6* homeobox gene results in defects in a region corresponding to its rostral domain of expression. *Cell* **66**: 1105–1119.

Mansour SL, Thomas KR, Capecchi MR. (1988) Disruption of the protooncogene *int-2* in mouse embryo-derived stem cells: a general strategy for targeting mutations to non-selectable genes. *Nature* **336**: 348–352.

McMahon AP, Bradley A. (1990). The *Wnt-1* (*int-1*) proto-oncogene is required for development of a large region of the mouse brain. *Cell* **62**: 1073–1085.

Nabeshima Y, Hanaoka K, Hayasaka M, Esumi E, Li S, Nonaka I, Nabeshima Y-I. (1993) Myogenin gene disruption results in perinatal lethality because of severe muscle defect. *Nature* **364**: 532-535.

Ramírez-Solis R, Zheng H, Whiting J, Krumlauf R, Bradley A. (1993) *Hoxb-4* (*Hox-2.6*) mutant mice show homeotic transformation of a cervical vertebra and defects in the closure of the sternal rudiments. *Cell* **73**: 279–294.

Riele HT, Maandag ER, Berns A. (1992) High efficiency gene targeting in embryonic stem cells through homologous recombination with isogenic DNA constructs. *Proc. Natl Acad. Sci. USA* **89**: 5128–5132.

Robertson E, Bradley A, Kuehn M, Evans M. (1986) Germline transmission of genes introduced into cultured pluripotent cells by retroviral vectors. *Nature* **323**: 445–447.

Rudnicki MA, Braun T, Hinuma S, Jaenisch R. (1992) Inactivation of MyoD in mice leads to upregulation of the myogenic HLH gene *MyF-5* and results in apparently normal muscle development. *Cell* **71**: 383–390.

Rudnicki MA, Schnegelsberg PNJ, Stead RH, Braun T, Arnold HH, Jaenisch R. (1993) *MyoD* or *MyF-5* is required for the formation of skeletal muscle. *Cell* **75**: 1351–1359.

Sauer B, Henderson N. (1990) Targeted insertion of exogenous DNA into the eukaryotic genome by the *Cre* recombinase. *New Biol.* **2**: 441–449.

Sauer B. (1993) Manipulation of transgenes by site-specific recombination: use of *Cre* recombinase. *Methods Enzymol.* **225**: 890–900.

Schwartzberg PL, Goff SP, Robertson EJ. (1989) Germline transmission of a c-*abl* mutation produced by targeted gene disruption in ES cells. *Science* **246**: 799–803.

Smithies O, Gregg RG, Boggs SS, Koralewski MA, Kucherlapati RS. (1985) Insertion of DNA sequences into human chromosomal β-globin locus by homologous recombination. *Nature* **317**: 1230–1234.

Snouwaert JN, Brigman KK, Latour AM, Malouf NN, Boucher RC, Smithies O, Koller BH. (1992) An animal model for cystic fibrosis made by gene targeting. *Science* **257**: 1083–1088.

Stanton BR, Reid SW, Parada LF. (1990) Germline transmission of an inactive N-*myc* allele generated by homologous recombination in embryonic stem cells. *Mol. Cell. Biol.* **10**: 6755–6758.

Thomas KR, Capecchi MR. (1987) Site-directed mutagenesis by gene targeting in mouse embryo-derived stem cells. *Cell* **51**: 503–512.

Thomas KR, Capecchi MR. (1990) Targeted disruption of the murine *int-1* proto-oncogene resulting in severe abnormalities in midbrain and cerebellar development. *Nature* **346**: 847–850.

Thomas KR, Deng C, Capecchi MR. (1992) High fidelity gene targeting in embryonic stem cells by using sequence replacement vectors. *Mol. Cell. Biol.* **12**: 2919–2923.

Thompson S, Clarke AR, Pow AM, Hooper ML, Melton DW. (1989) Germ line transmission and expression of a corrected *HPRT* gene targeting in embryonic stem cells. *Cell* **56**: 313–321.

Tsui L-C. (1992) The spectrum of cystic fibrosis mutations. *Trends Genet.* **8**: 392–398.

Tybulewicz VLJ, Tremblay ML, LaMarca ME, Willemsen R, Stubblefield BK, Winfield S, Zablocka B, Sidransky E, Martin BM, Huang SP, Mintzer KA, Westphal H, Mulligan RC, Ginns EI. (1992) Animal model of Gaucher's disease from targeted disruption of the mouse glucocerebrosidase gene. *Nature* **357**: 407–410.

Valancius V, Smithies O. (1991) Testing an in–out targeting procedure of making subtle genomic modifications in mouse embryonic stem cells. *Mol. Cell. Biol.* **11**: 1402–1408.

Wu CL, Melton DW. (1993) Production of a model for Lesch–Nyhan syndrome in hypoxanthine phosphoribosyltransferase deficient mice. *Nature Genetics* **3**: 235–240.

Yamada G, Sugimura K, Stuart ET. (1995) Gene targeting approaches in the study of cellular processes involved in growth or differentiation: advances in the analysis of oncogenes, tumour suppresser genes, cytokine/receptor system and developmental control genes. *Eur. J. Biochem.*, in press.

# 10

# Complementation analysis

Atusha Patel

## 10.1 Introduction

Somatic cell hybrids, derived from the fusion of two or more cells of the same or different species, provide a very useful strategy for the fusion of distinct characteristics present in different cells, in order to generate hybrid cells which can express both sets of characteristics. This principle has been exploited most efficiently in the generation of monoclonal antibodies, where an immortalized cell line (usually an Epstein–Barr virus-infected B-cell line) is fused with antibody-producing cells that have a limited lifespan, to generate immortalized hybrid cells which continue to produce the antibody for an indefinite period in culture (Köhler and Milstein, 1975). Somatic cell hybridization is also a particularly useful strategy for the study of the genetic control of specific phenotypic characteristics such as immortality, malignancy and differentiation. The success of this strategy is, however, dependent on the availability of appropriate cells with convenient genetic markers to enable the selection (or at least the identification) of the hybrid cell produced by the fusion of parent cells. In addition, the strategy is dependent on the availability of suitable means for examining the distinct characteristics which the hybrid cell has inherited from its parents. When these criteria are satisfied, somatic cell hybridization represents a particularly useful tool which allows the use of complementation analysis for the dissection of the individual steps involved in a biochemical pathway or process. Such studies have been used extensively in the study of specific DNA repair pathways using cells isolated from patients with DNA excision repair defects (Bootsma and Hoeijmakers, 1991; Hoeijmakers and Bootsma, 1992, 1994; Weeda *et al.*, 1993). Similar studies have also been employed in gene mapping studies, as well as in the analysis of genetic events involved in cellular immortalization, malignant transformation and differentiation. Complementation analysis based on somatic cell hybridization has also proved very useful for the identification of dominant and recessive genetic events in the control of a range of biological processes.

One specific application of somatic cell fusion, allowing the transfer of a subset of the chromosomes from one cell to another, has proved very useful in the assignment of specific genes to particular chromosomes. This technique, first developed by Fournier and Ruddle (1977), is termed *microcell fusion* and is distinct from conventional somatic cell hybridization in that only a fraction of the donor genome is introduced into recipient cells by the fusion process. Interspecies microcell hybrids containing a single human chromosome, or part of a chromosome, have proven invaluable in the construction of a physical map of the human genome (see Chapter 2) and the study of the chromosomal sites of integration of foreign genes.

## 10.2 Principles of somatic cell hybridization

Somatic cell hybrids can be generated by mixing the parental cell lines under conditions that promote cell fusion. Commonly used procedures employ polyethylene glycol (Köhler and Milstein, 1975), Sendai virus (see Okada, 1993) or an electric current (Finaz et al., 1984; Teissié and Rols, 1986). Polyethylene glycol (PEG)-induced cell fusion is by far the most convenient and frequently used procedure in the generation of somatic cell hybrids. However, for the other cell fusion procedures, the concentration of PEG and the precise culture conditions for different cell lines are often very different and must be optimized for a given pair of cell types. This can be achieved with relative ease using a range of concentrations of, and durations of exposure to, PEG.

Once cell fusion has been achieved, the generated heterokaryons must be isolated from the mixture containing not only the hybrid but also each of the two parental cell lines. A particularly convenient means of selecting hybrid cell lines is the elimination of the parent cells whilst allowing the hybrid cells to continue to grow. This can be achieved easily if the two cell lines used in the fusion possess selectable markers such as drug resistance, or are dependent on specific nutrients due to genetic defects in their biosynthetic pathways. The heterokaryotic hybrid cells, which have acquired combined drug resistance or have had their genetic defects complemented, can then be selected in media which contain the appropriate drugs, or are deficient in the appropriate nutrients (Choy et al., 1982). For the selection of viable hybrid cells, any counter-selectable recessive marker or selectable dominant marker can be used, provided that the two systems do not interfere. The best known method is the application of HAT (hypoxanthine–aminopterin–thymidine) selection to the fusion products of hypoxanthine guanine phosphoribosyltransferase-deficient (HPRT$^d$) or thymidine kinase-deficient (TK$^d$) cells (reviewed in Shay, 1982; Hooper, 1985).

HAT selection, which will be described in detail below, has been widely used for somatic cell hybridization experiments (Littlefield, 1964) and has encouraged the development of other selective media for the isolation of mutant cells (reviewed by Shay, 1982; Hooper, 1985). This selection relies on the existence of two sources for purine and pyrimidine nucleotides in mammalian

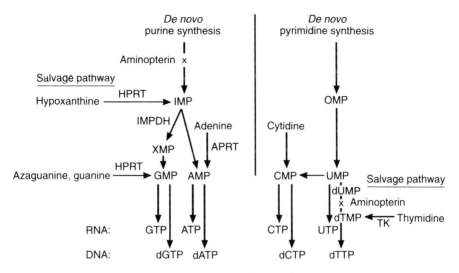

**Figure 10.1.** Schematic diagram showing pathways of nucleic acid biosynthesis.

cells, namely the *de novo* biosynthetic pathway and the salvage pathways. Under conditions which inhibit the *de novo* biosynthesis of nucleotides, mutants deficient in enzymes of the salvage pathways cannot survive. However, normal cells can survive such a selection because they are able to utilize hypoxanthine and thymidine to synthesize the nucleic acid precursors through the purine and pyrimidine salvage pathways (see Figure 10.1). Similarly, restoration of the salvage pathways in heterokaryons derived from the fusion of mutants with different lesions in their salvage pathways, allows the selection of such heterokaryons in HAT media. Before describing the HAT selection protocol, it is necessary to briefly describe nucleotide biosynthesis.

### 10.2.1 De novo *and salvage pathways of nucleotide synthesis*

In cultured cells, the *de novo* synthesis of nucleic acids can utilize simple compounds in the medium, such as formate, glycine and bicarbonate, provided that all the necessary enzymes are present. However, many cells in culture are found to be dependent on the salvage enzymes which have been extensively exploited for the selection of variant cells.

### 10.2.2 Purine nucleotide synthesis

The *de novo* synthesis of purine nucleotides is possible if 5-phosphoribosyl-1-pyrophosphate (PRPP), glutamine, glycine and aspartate are present in the medium in order to allow the production of inosine monophosphate (IMP). This process can be blocked by folic acid analogues, such as aminopterin and amethopterin, which block the synthesis of purine nucleotides (AMP and GMP). However, the enzyme HPRT in the salvage pathway can still phosphorylate hypoxanthine to make IMP. Therefore, if hypoxanthine is present, normal cells will be able to survive in aminopterin as shown in Figure 10.1 (reviewed by Choy et al., 1982; Morrow, 1982; Hooper, 1985).

The salvage pathway enzyme HPRT, which phosphorylates guanine to produce GMP (also made from the precursor IMP), mediates the phosphorylation of nucleotide analogues such as 8-azaguanine and thioguanine which can then be incorporated into DNA and RNA. Similarly, the enzyme adenine phosphoribosyltransferase (APRT), which directly converts adenine to AMP (AMP is also made from IMP), also mediates the phosphorylation of these analogues and their incorporation into RNA. Inosine monophosphate dehydrogenase (IMPDH) converts IMP to XMP which is then used to form GMP. The synthesis of these nucleoside analogues and their incorporation into DNA and RNA prevents cellular proliferation. Therefore, these enzymes allow the use of such nucleotide analogues for the isolation of mutant cell lines which are deficient in these enzymes. These mutants can in turn be used as suitable partners for somatic cell hybridization studies (reviewed by Choy et al., 1982; Morrow, 1982; Hooper, 1985).

### *10.2.3 The* HPRT *gene*

The *HPRT* gene is constitutively expressed and has been mapped to the X chromosome in both rodents (Franke and Taggart, 1980) and humans (Becker et al., 1979). The gene comprises nine exons, and the promoter resembles that of housekeeping genes in that it lacks CCAAT and TATA-like sequences but contains several copies of the sequence GGGCGG (Patel et al., 1986).

Partial deficiency of HPRT in human males causes gouty arthritis whereas total deficiency results in Lesch–Nyhan syndrome, a severe neurological disorder (Wilson et al., 1983; Kelley et al., 1986). The *HPRT* locus has been a primary focus of study on mutational mechanisms on account of our ability to select for and against gene expression in cultured animal cells (Hakala, 1957; Littlefield, 1964; Choy et al., 1982) and the predicted high number of independent mutations detectable at this locus (Haldane, 1935).

### *10.2.4 HPRT variants*

The *HPRT* gene is X-linked and therefore functionally hemizygous in all diploid mammalian cells. Coupled with the availability of efficient forward and reverse selective systems, this allows it to be exploited for genetic analysis and a large number of HPRT$^d$ variants have been isolated from a range of cell lines. Such variants arise at frequencies of around $10^{-6}$ in unmutagenized populations and at higher frequencies after treatment with a variety of mutagens. Some variants have no detectable HPRT activity whereas others exhibit residual activity. The majority of HPRT$^d$ variants undergo reversion at low, yet detectable, frequencies which can be increased by the use of mutagens. Some of the revertants possess an enzyme with altered kinetic properties or altered electrophoretic mobility as compared with the wild-type enzyme (reviewed by Shay, 1982; Hooper, 1985).

### 10.2.5 Pyrimidine nucleotide synthesis

A series of enzymatic reactions in the *de novo* synthesis of pyrimidine nucleotides result in the utilization of aspartate and carbamyl phosphate to generate UMP and CMP. The enzyme thymidylate synthetase, in conjunction with tetrahydrofolate reductase, reduces UMP to dUMP which is the immediate precursor to dTMP. This pathway is also blocked by aminopterin which completely inhibits DNA synthesis (reviewed by Choy *et al.*, 1982; Morrow, 1982; Hooper, 1985).

The salvage enzyme TK can catalyse the synthesis of dTMP from thymidine. Thus normal cells can continue to synthesize DNA even if thymidylate synthetase is blocked by aminopterin. TK is also responsible for the incorporation of the thymidine analogues iododeoxyuridine (IrdU) and bromodeoxyuridine (BrdU) which become incorporated into DNA, thereby making it more susceptible to ultraviolet (UV) light-induced damage. This property can be used to select for TK-deficient cell lines whereas aminopterin resistance can be used to select against TK-deficient mutants (reviewed by Choy *et al.*, 1982; Morrow, 1982; Hooper, 1985).

### 10.2.6 Metabolic cooperation

Interactions between cells can modify their phenotype in the absence of genetic alterations. When enzyme-positive and enzyme-negative cells are present in direct contact, toxic enzyme products (e.g. phosphorylated nucleotide analogues) can pass from the wild-type cells to the drug-resistant cells through intercellular gap junctions. This will result in the death of cells which are genetically drug resistant. This type of phenotypic interference in mutant selection and somatic cell hybridization studies is especially pronounced in normal diploid fibroblasts (DeMars and Held, 1972).

The avoidance of this transfer through intercellular gap junctions requires the variants to be selected at low densities (reviewed by Goldfarb *et al.*, 1974; Hooper, 1982, 1985). Similarly, wild-type cells can survive high concentrations of otherwise toxic compounds (e.g. hygromycin) when cultured in the presence of drug-resistant cells engineered to express the appropriate detoxifying enzyme (e.g. hygromycin phosphotransferase). This is due to the utilization of the drug by the resistant cells and the resulting reduction in the effective concentration of the drug in the culture media. This type of interference is highly dependent on culture conditions (e.g. frequency of culture media changes and drug replenishment) as well as both the absolute and the relative densities of the two cell populations. Also important are culture components, especially the source and batch of serum since serum can contain highly variable quantities of purines and pyrimidine, which may influence the selection of both mutants and their somatic cell hybrids (Shay, 1982).

## 10.2.7 The HAT selection system

Based on earlier studies of Hakala (1957) and Werkheiser (1961), a selective medium containing hypoxanthine, aminopterin and thymidine was designed to screen for rare revertants of HPRT$^d$ mutants that had acquired the ability to convert hypoxanthine to inosinic acid (Szybalski and Szybalski, 1962). This HAT selective medium was discovered to be extremely useful in the selection of somatic cell hybrids derived from parental variants defective in the salvage pathways of nucleotide biosynthesis (Littlefield, 1964, 1966).

Aminopterin blocks all reactions involving the enzyme dihydrofolate reductase, required for the synthesis of IMP, the precursor to purine nucleotides. Aminopterin also blocks the methylation of dUMP to form dTMP, therefore ensuring that cells become dependent on extraneous sources of purines and thymidine. Hypoxanthine can be used as an extraneous source of purines since it is phosphorylated by the enzyme HPRT to form IMP, which can in turn be converted to AMP and GMP. Thymidine can also be phosphorylated by TK to form TMP. Therefore, whereas normal cells can grow in HAT medium, cells which lack the enzymes HPRT or TK cannot, because both the *de novo* and the salvage pathways are blocked (reviewed by Choy *et al.*, 1982; Morrow, 1982; Hooper, 1985).

HPRT$^d$ or TK$^d$ cells can occur naturally, as in the human hereditary disease Lesch–Nyhan syndrome. Usually HPRT$^d$ or TK$^d$ cells exist as rare individual variants within a population and have to be selected out. Thus, HPRT$^d$ cells can be selected by growing the cells in 8-azaguanine or 6-thioguanine. The enzyme HPRT mediates the incorporation of the analogues into nucleic acid chains which can no longer be extended, thus preventing the growth of all cells except those which lack HPRT. TK$^d$ cells can be isolated by selection in BrdU which is phosphorylated by TK, resulting in its incorporation into DNA. Because BrdU-containing DNA is much more susceptible to UV-induced DNA damage, TK$^d$ mutants can be selected by their increased resistance to UV-induced toxicity (reviewed by Choy *et al.*, 1982; Morrow, 1982; Hooper, 1985).

HPRT$^d$ and TK$^d$ cells are unable to grow in HAT medium, but occasionally a back-mutation occurs and the cells revert. However, this only occurs with point mutations; cells with deletions, inversions or rearrangements cannot restore the normal coding sequence. The reversion frequency can be estimated by determining the number of colonies which arise when either HPRT$^d$ or TK$^d$ cells are grown in HAT. It is expressed as the ratio of colonies arising/number of cells inoculated. If the mutants are stable (non-leaky), the ratio should be less than $10^{-6}$.

## 10.2.8 Selection procedures for the isolation of hybrid cells

The HAT system was the first selective system developed to isolate somatic cell hybrids and has played an important part in the development of the field of somatic cell genetics (see Choy *et al.*, 1982).

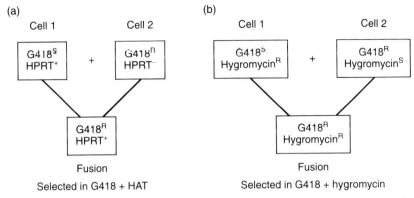

**Figure 10.2.** Schematic representation of selection procedures that may be used to isolate cell hybrids.

In Figure 10.2a, one of the fusion partners is a wild-type cell. This is sensitive to the cytotoxicity of the gentamycin analogue, G418, and has a proficient nucleotide salvage pathway (HPRT$^+$). These cells are fused to another cell line which is G418 resistant (G418$^R$) and deficient in HPRT activity (HPRT$^d$). Selection of the mixed population of the unfused partners and the fused somatic cell hybrids in the presence of both G418 and HAT medium results in the isolation of a cell hybrid that is both G418$^R$ and HPRT$^+$.

Hygromycin phosphotransferase or hygromycin B is an antibiotic which is toxic to prokaryotic and eukaryotic cells by inhibiting protein synthesis (Pettinger *et al.*, 1953; Blochlinger and Diggelmann, 1984). Transfection of a plasmid containing the hygromycin B phosphotransferase gene into eukaryotic cells has been shown to permit direct selection for hygromycin B resistance (Pettinger *et al.*, 1953; Blochlinger and Diggelmann, 1984). Figure 10.2b shows the product of a somatic cell hybridization between a cell line which is G418 sensitive (G418$^S$), but hygromycin B resistant, and another cell line which is, conversely, G418 resistant, but hygromycin sensitive. The cell hybrids can be selected in the presence of both G418 and hygromycin allowing the isolation of only those hybrids which have acquired both the neomycin phosphotransferase (*neo*) and hygromycin phosphotransferase (*hygro*) genes from each parent.

## 10.3 Identification of complementation groups and topological relationships

### 10.3.1 Extinction and activation

When a cell expressing a certain function is fused to a cell deficient in that process, the term extinction is used if the resulting hybrid does not express that function. However, if the ploidy of the expressing parent exceeds that of the non-expressing parent, there may be activation of the genome of the non-expressing parent. When re-expression occurs, it is unlikely to be due to chromosome(s) loss. However, re-expression may occur at a lower frequency in hybrids in which extinction is stable in the majority of cells in the population and, in such cases, is commonly associated with chromosome loss (reviewed in Hooper, 1985).

The extinction of a function is not necessarily caused by gene loss but is often due to the production of diffusible regulatory factors that inhibit gene expression. For example, hybrids formed between Friend erythroleukaemic cells and fibroblasts have been reported to be anchorage-dependent but also inducible for the expression of a wide spectrum of erythrocyte-specific differentiation products. In hybrids between melanoma and hepatoma cells, extinction is reciprocal, and re-expression of functions characteristic of the two parental cell types is mutually exclusive. By contrast, hybrids between Friend cells and lymphoma or myeloma cells have been reported to produce surface antigens characteristic of the lymphoma or myeloma parent and are also inducible for haemoglobin synthesis (Fougere and Weiss, 1978).

The re-expression of certain functions may occur in hybrid cells of the same type, and between de-differentiated variants and their parent cell lines. Extinction is seen in some cases but not in others, allowing conclusions to be drawn about the dominance of certain functions. In hybrids between the rat hepatoma line Fao and mouse hepatoma line BW1-J expressing a less extensive range of liver-specific functions, extinction was confined to normally neonatally activated functions. However, functions expressed earlier in the fetus were not suppressed. A brief extinction occurs in hybrids between a de-differentiated hepatoma variant and its differentiated parent (reviewed in Hooper, 1985).

### 10.3.2 Assignment of complementation groups in clinical diseases

One of the most extensively studied repair processes is the nucleotide excision repair (NER) pathway which is responsible for removing a broad spectrum of structurally unrelated DNA lesions. NER comprises at least five steps: (i) damage recognition, (ii) incision of the DNA strand, (iii) excision of the modified nucleotide, (iv) repair synthesis, and (v) DNA strand ligation.

Xeroderma pigmentosum (XP) is a rare autosomal recessive disease associated with subtle or major defects in DNA excision repair. Patients show skin hypersensitivity to sunlight (UV light). Other symptoms include abnormalities of skin pigmentation, as well as elevated risks of carcinoma and melanomas in the sun-exposed parts of the skin. Complementation analysis by cell fusion studies has demonstrated genetic heterogeneity within XP and has provided evidence for the existence of at least seven excision-deficient complementational groups: XPA–G and a group called XP-variant. The latter is apparently not affected in NER, but in post-replication repair, which is a poorly understood repair pathway (Bootsma and Hoeijmakers, 1991; Hoeijmakers and Bootsma, 1992, 1994; Weeda et al., 1993). Chediak–Higashi syndrome is an autosomal recessive disorder characterized by the presence of large intracellular granules, particularly lysosomes and melanosomes. Fusion of normal fibroblasts to Chediak fibroblasts complements the disorder, restoring normal lysosome size and distribution (Perou and Kaplan, 1993).

The genetic heterogeneity in peroxisome-deficient disorders, including Zellweger's cerebrohepatorenal syndrome, neonatal adrenoleukodystrophy and infantile Refsum disease, was examined by Yajima *et al.* (1992). After fusing the fibroblasts of patients, the formation of peroxisomes in the fused cells was examined. Eight complementation groups were identified, with no obvious relationship between the complementation groups and the clinical phenotype, indicating that the transport, intracellular processing and function of peroxisomal proteins were normalized in the complementary matching and that at least eight different genes are involved in the formation of normal peroxisomes and in the transport of peroxisomal enzymes.

Tsukamoto *et al.* (1990) made use of autoradiographic screening to isolate two Chinese hamster ovary (CHO) cell mutants deficient in peroxisomal dihydroxyacetone phosphate acyltransferase, a key enzyme for the biosynthesis of ether glycerolipids such as plasmalogens. The results suggested that assembly of the peroxisomes is defective in the mutants whereas the synthesis of peroxisomal proteins appeared to be normal. Cell fusion studies revealed that the two mutants were recessive to the wild-type CHO cells and belonged to different complementation groups, presumably containing different lesions in gene(s) encoding factor(s) required for peroxisomal assembly. In another study, McGuinness *et al.* (1990) carried out complementation studies using fused cell lines from patients with peroxisomal disorders and showed correction of defective plasmalogen synthesis and phytanic acid oxidation as well as an increase in the number of peroxisomes. At least six complementation groups have been reported.

### 10.3.3 Assignment of complementation groups in senescence

In order to examine the finite proliferative potential of normal cells, and to study the processes by which immortal cells have escaped growth cessation, Pereira-Smith and Smith (1988) fused a number of different normal human fibroblast cell lines with various immortal human cell lines. They found that, in all cases, the hybrids had a limited ability to proliferate in culture, suggesting that the finite proliferative capacity of normal human cells was dominant and that immortal cells had acquired recessive changes in their genetic programme, which allowed them to escape senescence. They assigned 30 immortal human cell lines to four complementation groups for the acquisition of an indefinite proliferative potential.

Duncan *et al.* (1993) fused three SV40-immortalized epithelial cell lines to cell lines representative of each of the four complementation groups. All three formed senescent hybrids with an SV40-immortalized cell line representative of group A, indicating that SV40 genes do not always cause immortalization via the same genetic mechanism. By contrast to other human cell lines, each of these three cell lines was assigned to more than one complementation group for immortalization. It would thus appear that the SV40-immortalized cell lines have lost the function of two or more putative senescence genes.

Similar results were obtained by Whitaker et al. (1992) who performed somatic cell hybridization analysis using an SV40-immortalized human bronchial epithelial cell line, BET-1A, to determine whether the genetic changes involved in the immortalization of normal human cells by SV40 genes are always identical. Fusion of BET-1A cells with an SV40-immortalized fibroblast cell line resulted in hybrids that senesced, indicating that these cell lines are in different complementation groups for immortalization.

Pendergrass et al. (1994) investigated the capacity of a murine cell line with a temperature-sensitive (ts) mutation in the DNA polymerase $\alpha$ (pol$\alpha$) locus and a series of ts non-pol$\alpha$ mutant cell lines from separate complementation groups to stimulate DNA synthesis in senescent fibroblast nuclei in heterokaryons. Their results suggest that there may be a functional deficiency in pol$\alpha$ of senescent human fibroblasts, and that this replication factor may be one of the rate-limiting factors involved in loss of the capacity to initiate DNA synthesis in senescent cells.

### 10.3.4 Assignment of complementation groups in biochemical pathways

Cell fusion studies have been useful in the dissection of the steps involved in a number of biochemical pathways. A mutation in the glycoprotein processing inhibitor-resistant (PIR) CHO cells was previously shown to result in a block at the $Man_5GlcNac_2$ stage of the dolichol–oligosaccharide biosynthetic pathway. These cells have normal mannose-P-dolichol synthase activity and are able to transfer the $Man_5GlcNac_2$ oligosaccharides to proteins. Cell fusion studies by Turka et al. (1992) demonstrated that the recessive PIR genotype could complement the mutation in cells which fail to synthesize mannose-P-dolichol. Thus the missing gene product is required by mannosyltransferase VI *in vivo* for proper utilization of either mannose-P-dolichol or $Man_5GlcNac_2$-P-P-dolichol.

### 10.3.5 Assignment of complementation groups in cytokine activity

A number of studies have investigated mechanisms involved in cytokine-mediated signalling between cells. The characterization of highly mutagenized human cell line 2fTGH led to the isolation of seven recessive interferon (IFN)-$\alpha$-unresponsive mutants which have been assigned to four complementation groups (U1–U4; John et al., 1991; McKendry et al., 1991). In another study, using immunoselection, Mao et al. (1993) isolated mutant HT1080 fibrosarcoma cell lines defective in the induction by IFN-$\gamma$ of the expression of human leukocyte antigen HLA-DRA. The mutants were all recessive and were divided into five complementation groups, with the mutants affected mainly in the expression of major histocompatibility class II and invariant-chain genes.

## 10.4 Identification of the dominant/recessive nature of genetic lesions

### 10.4.1 Chromosome segregation

One important application of somatic cell hybridization studies is analysis of the dominant or recessive nature of phenotypes of the two fusion partners. The chromosome complement of a hybrid cell is usually equal to the sum of the complements of one cell of each parental type with some chromosome loss. Chromosome loss is usually slight in intraspecies hybrids, but more extensive in interspecies hybrids where it is common for the chromosomes of one species to be selectively lost (reviewed in Hooper, 1985).

For an intraspecies cross, where chromosome loss is slight, it may be sufficient to examine a number of independently arising hybrids, since this lessens the chance that they will all have lost the same chromosomes. For interspecies crosses where chromosome loss is extensive, this is not sufficient and it is essential to have cytogenetic or other evidence for the presence of the chromosome carrying the gene in question. Again, several independent hybrids carrying the chromosome should be examined in case some contain translocations or microdeletions.

Chromosome loss in interspecies somatic cell hybrids is often unidirectional (i.e. a given hybrid cell tends to retain a complete set of chromosomes from one species and to segregate out the chromosomes of the other). In addition, it is possible to predict the direction of segregation from knowledge of the parental species and the cell types. For instance, in hybrids between cell lines of human and rodent origin, it is the human chromosomes that are selectively lost. Chromosome loss also depends on the nature of the cells used, particularly whether they are diploid and whether they are established cell lines. Hybrids generated by fusion of mouse primary culture cells and established human cell lines usually segregate out the mouse chromosomes (Ringertz and Savage, 1976).

One reason for the unidirectional loss of chromosomes is incompatibility between the mitochondria of one parent and the nucleus of the other. This is indicated by the rapid elimination of all the mitochondria of the parent whose chromosomes are preferentially lost. Segregation of chromosomes of the diploid parent, in hybrids between diploid cells and cells of established rodent cells, has provided a powerful method for the assignment of genes to specific chromosomes. This is achieved by the direct tracking of the nucleotide sequences of a given gene (Southern or *in situ* hybridization) and analysis of the chromosomes which have been retained in the hybrid cell. In addition, any gene which can be distinguished from the homologous gene of the fusion partner by the functional analysis of its product can also be assigned to a chromosome on the basis of coordinate segregation of the chromosomes and the gene product (Shows and Sakaguchi, 1982).

### 10.4.2 Dominant and recessive genetic changes involved in senescence

A number of studies based on the analysis of hybrids formed from the fusion of normal and immortal human cells have shown that the phenotype of cellular senescence is dominant, since the resultant hybrid cells have a finite life span. This demonstrates that immortality is a recessive trait resulting from the loss of function of a gene or genes that contribute to an active programme for cellular senescence (Pereira-Smith *et al.*, 1990; Smith *et al.*, 1992; Ryan *et al.*, 1994).

To determine whether programmed genetic events or random processes were responsible for cellular aging, Pereira-Smith and Ning (1992) used somatic cell hybridization to assign a large number of different immortal human cell lines to four complementation groups. In order to identify the chromosomes involved in regulating cell proliferation, single human chromosomes were introduced into immortal human cells representative of the different complementation groups. The introduction of chromosome 11 which is implicated in tumour suppression did not cause cellular senescence in three different immortal human cell lines tested (Ning *et al.*, 1991). However, introduction of a single normal human chromosome 4 into HeLa (cervical carcinoma) cells induced senescence in this immortal cell line. Hensler *et al.* (1994) found that the long arm of human chromosome 1 carries a senescence control gene or set of genes which are altered in the cell lines assigned to complementation group C. Similar studies by Ogata *et al.* (1993) have shown the presence of gene(s) on the long arm of chromosome 7 which affect the limited-division potential or senescence of normal human fibroblasts. These somatic cell hybridization studies support the hypothesis that senescence results from active genetic mechanisms. Further, Wadha *et al.* (1991) have shown that no random decline in genomic DNA methylation is detected during *in vitro* passaging of parental and hybrid cells representing the normal and conditional ageing populations, suggesting that it is unlikely that mortality of cells in culture is due to the random loss of epigenetic control imposed by 5-methylcytosine at CpG sites in the genome.

### 10.4.3 Dominant and recessive nature of viral genes

Somatic cell hybridization studies have also been useful in identifying mechanisms responsible for cell and species tropism of viruses (e.g. the restricted replication of HIV in murine cells; Trono and Baltimore, 1990). Similar studies have helped to identify a number of viral proteins, the function of which are crucial for viral replication (Freed *et al.*, 1992). Cell fusion studies have also proved to be fruitful in the investigation of the mechanisms by which viral genes cause transformation of a number of cell types (Kato and Sher, 1991; Rosl *et al.*, 1991; Renshaw *et al.*, 1992; Silverstein *et al.*, 1992; Chen *et al.*, 1993).

### 10.4.4 Dominant and recessive events in tumour progression

The suppression of tumour formation by somatic cell hybrids and microcell fusion products has helped to identify a group of genes that selectively suppress the growth of tumour cells but not normal cells. Such studies have shown, for instance, that the inheritance of a defective retinoblastoma gene (*RB1*) allele results in predisposition to the development of various cancers (Fung *et al.*, 1993). Knowles and Eydmann (1991) found that the loss of a suppressor function can contribute to urothelial transformation *in vitro*. As a step towards the identification of possible tumour suppressor genes in prostate cells, Peehl *et al.* (1990) created hybrids by fusing normal prostate cells with malignant HeLa cells. They found suppression of the malignant characteristics which demonstrated the dominance of the normal phenotype. Rodriguez-Alfageme *et al.* (1992) employed somatic cell hybridization studies to demonstrate the presence of a tumour suppressor gene on chromosome 5, the function of which is necessary for the regulated expression of c-*myc* in at least some colonic cells. Loss of this suppressor results in deregulated c-*myc* expression and is a necessary, but most likely insufficient, event for the expression of the tumorigenic phenotype in a subset of colon carcinomas (Tomita *et al.*, 1992).

### 10.4.5 Dominant nature of multi-drug resistance genes

A non-P-glycoprotein-mediated mechanism of multiple drug resistance, present in doxorubicin-selected sub-lines of the human non-small cell lung carcinoma has been identified by cell fusion studies. The somatic cell hybrids generated by fusion of these cells are cross-resistant to a number of drugs, possibly due to the presence of a single dominant mutation, which prevents the drug-induced inhibition of topoisomerase II (Eijdems *et al.*, 1992). Additionally, McClean *et al.* (1993) found that a multiple drug-resistant phenotype which develops following X-ray treatment of CHO cells *in vitro* was dominantly expressed, consistent with the hypothesis that this is a consequence of dominant genetic alterations resulting from exposure to X-irradiation.

### 10.4.6 Dominant and recessive events involved in immunological processes

Cell fusion studies have been useful in studying the expression of a number of antigens on the surface of hybrid cells for the identification of *trans*- or *cis*-acting diffusible regulators (Gocinski *et al.*, 1993), and the dissection of steps involved in the ultimate commitment of a cell to the B- or T-cell lineage by the analysis of hybrids between B- and T-cell lines (Kimata *et al.*, 1990).

### 10.4.7 Dominant and recessive developmentally regulated genes

The expression of the developmentally regulated genes has also been studied by somatic cell hybridization. Weintraub's group have demonstrated that transfer of human fibroblast chromosome 11 (containing the human *MyoD* gene) from primary cells into 10T1/2 mouse fibroblasts activates expression of the transferred human *MyoD* gene and converts these cells to myoblasts. These studies demonstrate that the *MyoD* locus is potentially functional in primary human fibroblasts, but is normally repressed in *trans* by a gene product which is active in human fibroblasts (Thayer and Weintraub, 1990). Zarrilli *et al.* (1993) have also employed somatic cell hybridization studies to show that the expression of the developmentally regulated insulin-like growth factor II (*IGF2*) gene is subject to dominant negative control.

## 10.5 Microcell fusion: principles and application to the chromosomal localization of genes

### 10.5.1 Introduction to microcell fusion

The technique for the production of microcells, initially developed by Ege and Ringertz (1974), was used to form viable microcell hybrids by Fournier and Ruddle (1977) for the transfer of one or a few intact chromosomes from one mammalian cell to another. Microcell hybridization is distinct from conventional somatic cell hybridization in that only a fraction of the donor genome is introduced into recipient cells. This procedure is also a fast method of generating stable hybrid cell lines and no selection against the donor cell line is needed because the microcells themselves are not viable in culture.

Interspecies microcell hybrids, containing a single human chromosome or part of a chromosome with suitable selectable markers, have facilitated the regional mapping and cloning of a number of genes. Microcell hybrids have also been invaluable in construction of physical maps of the human genome, and for the analysis of the chromosomal sites of integration of foreign genes. Furthermore, the transfer of parent donor chromosomes can be analysed by cytogenetic tests to identify hybrid cells containing translocated and rearranged chromosomes which can facilitate the cloning of these chromosomal fragments. Microcell hybridization has therefore provided a valuable tool in the investigation of development, differentiation, cell transformation and malignancy.

### 10.5.2 General principles for microcell-mediated transfer

The initial step of microcell-mediated chromosome transfer is to induce the putative donor cells to become micronucleate by prolonged mitotic arrest, which partitions the chromosomes into discrete nuclear subsets (Philips and Philips, 1969). Each micronucleus becomes surrounded by a rim of cytoplasm with an intact plasma membrane, serving as an efficient vector for the transfer of small numbers of chromosomes (Ege and Ringertz, 1974). The micronuclei can be physically isolated from the cells by centrifugation in the presence of cytochalasin B (Prescott *et al.*, 1977). Fusion of microcells with intact cells generates microcell heterokaryons which, under selection, may proliferate to produce microcell hybrid clones carrying only a single or a small number of chromosomes from the original donor cells. Fournier (1981) reported that sequential treatment of donor cells with colcemid and cytochalasin B yielded micronucleated cells that hybridized with the same efficiency as whole cells. With other modifications, the efficiency of generating microcell hybrids could be increased by 50- to 100-fold (Fournier, 1981).

### 10.5.3 Pinpointing chromosomes involved in specific disease processes

The transfer of abnormal chromosomes into normal cells has led to the identification of a number of genes involved in the pathogenesis of several diseases. Breast cancer development is associated with several genetic abnormalities. Loss of heterozygosity on the short arm of chromosome 11 has been observed in 30% of tumours, suggesting loss of one chromosome 11. Negrini *et al.* (1992) studied the transformed and tumorigenic phenotypes of MCF-7 cells following introduction of a normal human chromosome 11 via microcell fusion. In MCF-7/H11 cell hybrids containing a normal chromosome 11, the malignant phenotype had been suppressed and characteristics similar to those of the non-transformed parental cell line had been acquired.

Complementation of DNA excision repair defect in XP cells of group C (XP-C) has been achieved by the transfer of human chromosome 5 by microcell fusion from mouse–human hybrid cell lines, thus confirming the presence of a DNA repair gene on human chromosome 5 (Pal Kaur and Athwal, 1993). Microcell-mediated transfer of a single human chromosome 8 into severe combined immunodeficient (SCID) fibroblasts complemented the radiation sensitivity of these cells (Kurimasa *et al.*, 1994). Sub-clones from these microcell hybrids showed that the human gene able to restore normal resistance to radiation damage is located on the centromeric region of chromosome 8.

### 10.5.4 Identification of tumour suppressor genes

Putative tumour suppressor genes can be mapped to specific chromosomes by the introduction of individual chromosomes derived from normal cells by microcell fusion. Negrini *et al.* (1992) examined whether a highly malignant cell line HHUA can be suppressed by only one normal chromosome or by multiple chromosomes transferred by microcell fusion. The results indicated the presence of more than a single tumour suppressor gene for this human uterine endometrial carcinoma cell line, a finding which supports the involvement of multiple tumour suppressor genes in the control of the normal phenotype, and the multi-step nature of neoplastic transformation.

In order to identify chromosomes that carry putative tumour suppressor genes for the various phenotypes of Kirsten sarcoma virus-transformed NIH-3T3 (DT) cells, Yamada *et al.* (1990) transferred chromosome 1 into these cells by microcell fusion. Their results indicated that normal human chromosome 1 carries a putative tumour suppressor gene that affects various phenotypes of DT cells.

The loss and/or mutational inactivation of a gene or genes on the short arm of chromosome 3 (3p) may play a role in the development of human renal cell carcinoma (RCC). Thus the normal allele may carry suppressor activity for a tumour-associated phenotype(s). To test this hypothesis, Shimizu *et al.* (1990) introduced a single chromosome carrying 3p into a human RCC cell line by microcell fusion, and examined the tumorigenicity in nude mice and by analysis of *in vitro* growth properties. The introduction of 3p modulated at least the tumour growth, indicating the presence on 3p of a primitive tumour suppressor gene(s) for human RCC. In a similar study, Yoshida *et al.* (1994) introduced a chromosome 3 from normal human fibroblasts into the RCC cell line which exhibited loss of heterozygosity for 3p, resulting in *in vitro* growth suppression of the hybrid cells. However, during long-term cultivation, one of the clones that had lost the introduced chromosome 3 showed growth properties similar to the original RCC cells, indicating the presence of a dominant negative growth regulatory gene on chromosome 3p.

Functional loss of the retinoblastoma (*RB1*) gene has been implicated in the initiation or progression of several human tumour types including cancer of the eye, bone, bladder and prostate. To examine the growth regulatory potential of *RB1*, a neomycin resistance-marked chromosome 13 was transferred by microcell fusion into these tumour cell lines (Banerjee *et al.*, 1992). The extent of growth inhibition obtained in the different cell lines demonstrated the cell type-specific nature of *RB1*-mediated growth inhibition and tumour suppression. In the case of prostate cancer, the function of the *RB1* gene in tumour suppression appears to be independent of its growth regulatory function, since no growth inhibition in cell culture was noted in these cells, although there was significant tumour suppression.

Microcell fusion-mediated study of tumour suppressor genes on chromosome 11 has also indicated the presence of cell- and/or tumour- specific growth suppressors on this chromosome (Oshimura *et al.*, 1990).

### 10.5.5 Identification of genes involved in cellular senescence

The microcell fusion-mediated introduction of a single copy of human chromosome 1 into immortalized Syrian hamster fibroblasts induces the rapid senescence of these cells (Sugawara *et al.*, 1990). Microcell fusion has been used to demonstrate the presence of an X-linked senescence gene in both Chinese hamster embryo (CHE) cells and human fibroblasts (Wang *et al.*, 1992). These studies also demonstrated that inactivation of this gene was involved in the nickel-induced immortalization of CHE cells, but not the immortality of CHO or V9 cell lines (Wang *et al.*, 1992).

## References

**Banerjee A, Xu H, Hu S, Araujo D, Takahashi R, Stanbridge EJ, Benedict WF.** (1992) Changes in growth and tumorigenicity following reconstitution of retinoblastoma gene function in various human cancer cell types by microcell transfer of chromosome 13. *Cancer Res.* **52:** 6297–6304.

**Becker MA, Yen RCK, Itkin P, Goss SJ, Seegmiller JE, Bakay B.** (1979) Regional localization of the gene for human phosphoribosylphosphate synthetase on the X-chromosome. *Science* **203:** 1016–1019.

**Blochlinger K, Diggelmann H.** (1984) Hygromycin B phosphotransferase as a selectable marker for DNA transfer experiments with higher eucaryotic cells. *Mol. Cell. Biol.* **4:** 2929–2931.

Bootsma D, Hoeijmakers JHJ. (1991) The genetic basis of xeroderma pigmentosum. *Ann. Génét.* **34:** 143–150.

Chen T, Pecoraro G, Defendi V. (1993) Genetic analysis of *in vitro* progression of human papillomavirus-transfected human cervical cells. *Cancer Res.* **53:** 1167–1171

Choy WN, Gopalakrishnan TV, Littlefield JW. (1982) Techniques for using HAT selection in somatic cell genetics. In: *Techniques in Somatic Cell Genetics* (ed. JW Shay). Plenum Press, New York, pp. 11–21.

DeMars R, Held K. (1972) The spontaneous azaguanine-resistant mutants of diploid human fibroblasts. *Humangenetik* **16:** 87–110.

Duncan EL, Whitaker NJ, Moy EL, Reddel RR. (1993) Assignment of SV40-immortalized cells to more than one complementation group for immortalization. *Exp. Cell Res.* **205:** 337–344.

Ege T, Ringertz NR. (1974) Preparation of microcells by enucleation of micronucleate cells. *Exp. Cell Res.* **87:** 378–382.

Eijdems EWHM, Borst P, Jongsma APM, De Jong S, De Vries EGE, Van Groenigen M, Versantvoort CHM, Nieuwint AWM, Baas F. (1992) Genetic transfer of non-P-glycoprotein-mediated multidrug resistance (MDR) in somatic cell fusion: dissection of a compound MDR phenotype. *Proc. Natl Acad. Sci. USA* **89:** 3498–3502.

Finaz C, Lefevre A, Teissié J. (1984) Electrofusion – a new, highly efficient technique for generating somatic cell hybrids. *Exp. Cell Res.* **150:** 477–482.

Fougere C, Weiss MC. (1978) Phenotypic exclusion in mouse myeloma – rat hepatoma hybrid cells: pigment and albumin production are not re-expressed simultaneously. *Cell* **15:** 843–854.

Fournier REK. (1981) A general high-efficiency procedure for production of microcell hybrids. *Proc. Natl Acad. Sci. USA* **78:** 6349–6353.

Fournier REK, Ruddle FH. (1977) Microcell-mediated transfer of murine chromosomes into mouse, Chinese hamster and human somatic cells. *Proc. Natl Acad. Sci. USA* **74:** 319–323.

Franke U, Taggart RT. (1980) Comparative gene mapping; order of loci on the X-chromosome is different in mice and humans. *Proc. Natl Acad. Sci. USA* **77:** 3595–3599.

Freed EO, Delwart EL, Buchschacher GL, Panganiban AT. (1992) A mutation in the human immunodeficiency virus type 1 transmembrane glycoprotein gp41 dominantly interferes with fusion and infectivity. *Proc. Natl Acad. Sci. USA* **89:** 70–74.

Fung YT, T'Ang A, Murphree AL, Zhang F, Qiu W, Wang S, Shi X, Lee L, Driscoll B, Wu K. (1993) The Rb gene suppresses the growth of normal cells. *Oncogene* **8:** 2659–2672.

Gocinski BL, Sun D, Coleclough C. (1993) *Cis*-dominant regulation of CD4 and CD8 gene expression in rat/mouse T cell heterohybridomas. *J. Immunol.* **150:** 2243–2252.

Goldfarb PSG, Slack C, Subak-Sharpe J, Wright E. (1974) Metabolic cooperation between cells in tissue culture. In: *Symposia of the Society for Experimental Biology*. Cambridge University Press, London.

Hakala MT. (1957) Prevention of toxicity of amethopterin for sarcoma 180 cells in tissue culture. *Science* **126:** 255–256.

Haldane JBS. (1935) The rate of spontaneous mutation of a human gene. *J. Genet.* **31:** 317–326.

Hensler PJ, Annab LA, Barrett JC, Pereire-Smith OM. (1994) A gene involved in control of human cellular senescence on human chromosome 1q. *Mol. Cell. Biol.* **14:** 2291–2297.

Hoeijmakers JHJ, Bootsma D. (1992) DNA repair: two pieces of the puzzle. *Nature Genetics* **1:** 313–314.

Hoeijmakers JHJ, Bootsma D. (1994) Incisions for excision. *Nature* **371:** 654–655.

Hooper ML. (1982) Metabolic cooperation between mammalian cells in culture. *Biochim. Biophys. Acta* **651:** 85–103.

Hooper ML. (1985) *Mammalian Cell Genetics*. John Wiley & Sons, Chichester.

John J, McKendry R, Pellegrini S, Flavell D, Kerr IM, Stark GR. (1991) Isolation and characterization of a new mutant human cell line unresponsive to alpha and beta interferons. *Mol. Cell. Biol.* **11:** 4189–4195.

Kato J-Y, Sher CJ. (1991) A dominant suppressive mutation in a cellular gene restores the nontransformed phenotype to v-*fms*-transformed mink cells. *Oncogene* **6:** 687–693.

Kelley WN, Searle JG, Wilson JM. (1986) HGPRT-deficiency – the molecular basis of the clinical syndromes. *Verh. Deutsch. Ges. Inn. Med.* **92**: 465–9IS.

Kimata H, Berenson J, Kagan J, Saxon A. (1990) Functional human T cell–B cell hybridomas established from fusion of normal T cells and an EBV-transformed B cell line. *Clin. Immunol. Immunopathol.* **54**: 134–147.

Knowles MA, Eydmann ME. (1991) Loss of a tumor suppressor function during neoplastic progression of epithelial cells *in vitro. Int. J. Cancer* **47**: 726–731.

Köhler G, Milstein C. (1975) Continuous cultures of fused cells secreting antibody of predefined specificity. *Nature* **256**: 495–497.

Kurimasa A, Nagata Y, Shimizu M, Emi M, Nakamura Y, Oshimura M. (1994) A human gene that restores the DNA-repair defect in SCID mice is located on 8p11.1–q11.1. *Hum. Genet.* **93**: 21–26.

Littlefield JW. (1964) Selection of hybrids from matings of fibroblasts *in vitro* and their presumed recombinants. *Science* **145**: 709–710.

Littlefield JW. (1966) The use of drug-resistant markers to study the hybridization of mouse fibroblasts. *Exp. Cell Res.* **41**: 190–196.

Mao C, Davies D, Kerr IM, Stark GR. (1993) Mutant human cells defective in induction of major histocompatibility complex class II genes by interferon gamma. *Proc. Natl Acad. Sci. USA* **90**: 2880–2884.

McClean S, Hosking LK, Hill BT. (1993) Dominant expression of multiple drug resistance after *in vitro* X-irradiation exposure in intraspecific Chinese hamster ovary hybrid cells. *J. Natl Cancer Inst.* **85**: 48–53.

McGuinness MC, Moser HW, Watkins PA. (1990) Peroxisomal disorders: complementation analysis using beta-oxidation of very long chain fatty acids. *Biochem. Biophys. Res. Commun.* **172**: 364–369.

McKendry R, John J, Flavell D, Muller M, Kerr IM, Stark GR. (1991) High-frequency mutagenesis of human cells and characterization of a mutant unresponsive to both alpha and gamma interferons. *Proc. Natl Acad. Sci. USA* **88**: 11455–11459.

Morrow J. (1982) Selection of purine and pyrimidine nucleoside analog resistance in mammalian cells. In: *Techniques in Somatic Cell Genetics* (ed. JW Shay). Plenum Press, New York, pp. 1–9.

Negrini M, Castagnoli A, Sabbioni S, Recanatini E, Giovannini G, Possati L, Stanbridge EJ, Nenci I, Barbanti-Brodano G. (1992) Suppression of tumorigenesis by the breast cancer cell line MCF-7 following transfer of a normal human chromosome 11. *Oncogene* **7**: 2013–2018.

Ning Y, Shay JW, Lovell M, Taylor L, Ledbetter DH, Pereira-Smith OM. (1991) Tumor suppression by chromosome 11 is not due to cellular senescence. *Exp. Cell Res.* **192**: 220–226.

Ogata T, Ayusawa D, Namba M, Takahashi E, Oshimura M, Oishi M. (1993) Chromosome 7 suppresses indefinite division of nontumorigenic immortalized human fibroblast cell lines KMST-6 and SUSM-1. *Mol. Cell. Biol.* **13**: 6036–6043.

Okada Y. (1993) Sendai virus-induced cell fusion. *Meth. Enzymol.* **221**: 18–41.

Oshimura M, Kugoh H, Koi M, Shimizu M, Yamada H, Satoh H, Barrett JC. (1990) Transfer of a normal human chromosome 11 suppresses tumorigenicity of some but not all tumor cell lines. *J. Cell. Biochem.* **42**: 135–142.

Pal Kaur G, Athwal RS. (1993) Complememtation of DNA repair defect in xeroderma pigmentosum cells of group C by the transfer of human chromosome 5. *Somat. Cell Mol. Genet.* **19**: 83–93.

Patel PI, Framson PE, Caskey CT, Chinault AC. (1986) Fine structure of the human hypoxanthine phosphoribosyltransferase gene. *Mol. Cell. Biol.* **6**: 393–403.

Peehl DM, Wong ST, McNeal JE, Stamey TA. (1990) Analysis of somatic cell hybrids derived from normal human prostatic epithelial cells fused with HeLa cells. *Prostate* **17**: 123–136.

Pendergrass WR, Saulewicz AC, Hanaoka F, Norwood TH. (1994) Murine temperature-sensitive DNA polymerase alpha mutant displays a diminished capacity to stimulate DNA synthesis in senescent human fibroblast nuclei in heterkaryons at the nonpermissive condition. *J. Cell Physiol.* **158**: 270–276.

Pereira-Smith OM, Ning Y. (1992) Molecular genetic studies of cellular senescence. *Exp. Gerontol.* **27**: 519–522.

Pereira-Smith OM, Smith JR. (1988) Genetic analysis of indefinite division in human cells: identification of four complementation groups. *Proc. Natl Acad. Sci. USA* **85**: 6042–6046.

Pereira-Smith OM, Robetorye S, Ning Y, Orson FM. (1990) Hybrids from fusion of normal human T lymphocytes with normal immortal human cells exhibit limited life span. *J. Cell. Physiol.* **144**: 546–549.

Pereira-Smith OM, Stein GH, Robetorye S, Meyer-Demarest S. (1990) Immortal phenotype of the HeLa variant D98 is recessive in hybrids formed with normal human fibroblasts. *J. Cell. Physiol.* **143**: 222–225.

Perou CM, Kaplan J. (1993) Complementation analysis of Chediak–Higasi syndrome: the same gene may be responsible for the defect in all patients. *Somat. Cell Mol. Genet.* **19**: 459–468.

Pettinger RC, Wolfe RN, Hoehn MM, Marks PN, Dailey WA, McGuire JM. (1953) Hygromycin. 1. Preliminary studies on the production and biological activity of a new antibiotic. *Antibiot. Chemother.* **3**: 1286–1278.

Philips SG, Philips DM. (1969) Sites of nucleolus production in cultured Chinese hamster cells. *J. Cell Biol.* **40**: 248–268.

Prescott DM, Myerson D, Wallace J. (1977) Enucleation of mammalian cells with cytochalasin B. *Exp. Cell Res.* **71**: 480–492.

Renshaw MW, Kipreos ET, Albrecht MR, Wang JYJ. (1992) Oncogene v-*abl* tyrosine kinase can inhibit or stimulate growth, depending on the cell context. *EMBO J.* **11**: 3941–3951.

Ringertz NR, Savage RE. (1976) *Cell Hybrids*. Academic Press, New York.

Rodriguez-Alfageme C, Stanbridge EJ, Astrin SM. (1992) Suppression of deregulated c-*myc* expression in human colon carcinoma cells by chromosome 5 transfer. *Proc. Natl Acad. Sci. USA* **89**: 1482–1486.

Rosl F, Achtstatter T, Bauknecht T, Hutter K-J, Futterman G, Zur Hausen H. (1991) Extinction of the HPV18 upstream regulatory region in cervical carcinoma cells after fusion with non-tumorigenic human keratinocytes under non-selective conditions. *EMBO J.* **10**: 1337–1345.

Ryan PA, Maher VM, McCormick JJ. (1994) Failure of infinite life span human cells from different immortality complementation groups to yield finite life span hybrids. *J. Cell. Physiol.* **159**: 151–160.

Shay JW. (1982) *Techniques in Somatic Cell Genetics*. Plenum Press, New York.

Shimizu M, Yokota J, Mori N, Shuin T, Shinoda T, Terada M, Oshimura M. (1990) Introduction of normal chromosome 3p modulates the tumorigenicity of a human renal cell carcinoma cell line YCR. *Oncogene* **5**: 185–194.

Shows TB, Sakaguchi SL. (1982) *Adv. Hum. Genet.* **12**: 341–452.

Silverstein GH, Kohrman DC, Christensen JB, Brockman WM, Imperiale MJ. (1992) An SV40 transformation revertant due to a host mutation: isolation and complementation analysis. *Virology* **187**: 723–733.

Smith JR, Ning Y, Pereira-Smith OM. (1992) Why are transformed cells immortal? Is the process reversible? *Am. J. Clin. Nutr.* **55**: 1215S–1221S.

Sugawara O, Oshimura M, Koi M, Annab LA, Barret JC. (1990) Induction of cellular senescence in immortalized cells by human chromosome 1. *Science* **247**: 707–710.

Szybalski W, Szybalski E. (1962) Drug sensitivity as a genetic marker for human cell lines. *Univ. Mich. Med. Bull.* **28**: 277–293.

Teissié J, Rols MP. (1986) Fusion of mammalian cells in culture is obtained by creating contact between cells and after their electropermeabilisation. *Biochem. Biophys. Res. Commun.* **140**: 258–266.

Thayer M, Weintraub H. (1990) Activation and repression of myogenesis in somatic cell hybrids: evidence for *trans*-negative regulation of *MyoD* in primary fibroblasts. *Cell* **63**: 23–32.

**Tomita N, Jiang W, Hibshoosh H, Warburton D, Kahn SM, Weinstein B.** (1992) Isolation and characterization of a highly malignant variant of the SW480 human colon cancer cell line. *Cancer Res.* **52:** 6840–6847.

**Trono D, Baltimore D.** (1990) A human cell factor is essential for HIV-1 Rev action. *EMBO J.* **9:** 4155–4160.

**Tsukamoto T, Yokota S, Fujiki Y.** (1990) Isolation and characterizaton of Chinese hamster ovary cell mutants defective in assembly of peroxisomes. *J. Cell Biol.* **110:** 651–660.

**Turka LA, Linsley PS, Lin H, Brady W, Leiden JM, Wei RQ, Gibson ML, Zheng XG, Myrdal S, Gordon D.** (1992) T-cell activation by the CD28 ligand B7 is required for cardiac allograft rejection *in vivo*. *Proc. Natl Acad. Sci. USA* **89:** 11102–11105.

**Wadha R, Ikawa Y, Sugimoto Y.** (1991) Natural and conditional ageing of mouse fibroblasts: genetic vs. epigenetic control. *Biochem. Biophys. Res. Commun.* **178:** 269–275.

**Wang X-W, Lin X, Klein CB, Bhamra R, Lee Y-W, Costa M.** (1992) A conserved region in human and Chinese hamster X chromosomes can induce cellular senescence of nickel-transformed Chinese hamster cell lines. *Carcinogenesis* **13:** 555–561.

**Weeda G, Hoeijmakers JHJ, Bootsma D.** (1993) Genes controlling nucleotide excision repair in eukaryotic cells. *Bioessays* **15:** 249–258.

**Werkheiser WC.** (1961) Specific binding of 4-amino folic acid analogues by folic acid reductase. *J. Biol. Chem.* **236:** 888–893.

**Whitaker NJ, Kidston EL, Reddel RR.** (1992) Finite life span of hybrids formed by fusion of different simian virus 40-immortalized human cell lines. *J. Virol.* **66:** 1202–1206.

**Wilson JM, Young AB, Kelley WN.** (1983) Hypoxanthine–guanine phosphoribosyl transferase deficiency. The molecular basis of the clinical syndrome. *New Engl. J. Med.* **309:** 900–910.

**Yajima S, Suzuki Y, Shimozawa N, Yamaguchi S, Orii T, Fujiki Y, Osumi T, Hashimoto T, Moser HW.** (1992) Complementation study of peroxisome-deficient disorders by immunofluorescence staining and characterization of fused cells. *Hum. Genet.* **88:** 491–499.

**Yamada H, Horikawa I, Hashiba H, Oshimura M.** (1990) Normal human chromosome 1 carries suppressor activity for various phenotypes of a murine sarcoma virus-transformed NIH/3T3 cell line. *Jap. J. Cancer Res.* **81:** 1095–1100.

**Yoshida MA, Shimizu M, Ikeuchi T, Tonomura A, Yokota J, Oshimura M.** (1994) *In vitro* growth suppression and morphological change in a human renal cell carcinoma cell line by the introduction of normal chromosome 3 via microcell fusion. *Mol. Carcin.* **9:** 114–121.

**Zarrilli R, Casola S, Conti A, Bruni CB, Colantuoni V.** (1993) Extinction of insulin-like growth factor II gene expression in intratypic hybrids of rat liver cells. *Mol. Endocrinol.* **7:** 131–141.

# 11

# Antisense oligonucleotides: a survey of recent literature, possible mechanisms of action and therapeutic progress

Diana Pollock and Joop Gäken

## 11.1 Introduction

The use of oligonucleotides in antisense strategies was first proposed by Zamencik and Stephenson (1978). Over the past 15 years, there has been a proliferation of data derived from different antisense sequences employing different modifications and acting in different systems (Figure 11.1). Several systems have been explored, including *in vitro* (cell-free) systems (e.g. reticulocyte lysate and wheat germ extract), cell culture, closed systems with no endogenous RNA synthesis (e.g. *Xenopus* or murine oocytes) and *in vivo* systems (i.e. whole organisms). It is inappropriate to extrapolate from one type of system to another other than by inference. However, each study emphasizes, or sheds new light upon, a facet of antisense oligonucleotide action. In this chapter, we describe some examples of antisense action in different systems, discuss the implications of targeting, design, uptake and toxicity, describe the various modifications to oligonucleotide structure that have been made and their main effects, and relate these to antisense action. We shall examine problems encountered in antisense strategies, such as non-specific cleavage, and discuss ribozymes and triplex DNA and the introduction of antisense oligonucleotides into whole organisms as therapeutic agents.

# FUNCTIONAL ANALYSIS OF THE HUMAN GENOME

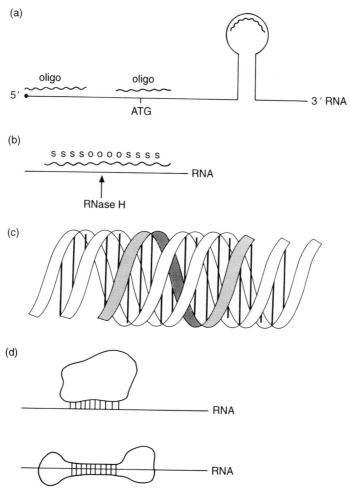

**Figure 11.1** (a) Diagrammatic representation of the hybridization of an oligonucleotide against a typical RNA strand. The RNA has a cap site at the 5' end, followed by the ATG codon which is the start of translation. Both of these sites are suitable targets for antisense oligonucleotides, possibly because they do not generally have secondary structure like other parts of the RNA. (b) The possible mode of action of a partially sulphurized oligonucleotide. The central phosphodiester portion of the oligonucleotide/RNA hybrid is given as a target for RNase H, which cleaves the RNA strand of the duplex. (c) Triple helix formation showing the binding of an oligonucleotide to the major groove of normal double-stranded DNA. (d) Diagrammatic representation of possible modes of action of circular oligonucleotides.

## 11.2 Some examples of antisense action in different systems

Most examples of antisense action involve the use of oligonucleotides as tools to reduce or remove the expression of a targeted mRNA. Crooke (1992) reviewed studies performed in cell culture targeting viral RNA, oncogenes and host cell genes with various types of oligonucleotide. Examples of antisense studies in cell culture include:

(i) antisense myogenin oligonucleotide used in L6A1 myoblast cells (Florini and Ewton, 1990);
(ii) antisense oligonucleotide against the c-*Ha-ras* oncogene used in NIH/3T3 fibroblast cells (Daaka and Wickstrom, 1990);
(iii) antisense oligonucleotide against herpes simplex virus 1 (HSV-1) in HeLa cells (Hoke *et al.*, 1991);
(iv) in archaebacteria (Stolt and Zillig, 1993);
(v) in plants (Picton *et al.*, 1993).

*Xenopus* and murine oocytes make attractive antisense systems because the maternal RNA pool may be depleted by antisense RNA action, and no mRNA synthesis occurs until later on in development. Examples of antisense experiments in these types of systems include:

(i) microinjection of antisense *veg*1 coding sequences into *Xenopus* oocytes (Shuttleworth and Coleman, 1988);
(ii) microinjection of c-*mos* antisense oligonucleotide into the maturing mouse egg (O'Keefe *et al.*, 1989);
(iii) microinjection of *An*2 and cyclin antisense oligonucleotides into *Xenopus* oocytes (Dagle *et al.*, 1990).

Examples of the use of antisense oligonucleotides *in vivo* are perhaps the most therapeutically relevant. Reports of the use of antisense oligonucleotides in whole organisms are much more limited but some examples include:

(i) administration of antisense c-*myb* during surgery in the rat (Simons *et al.*, 1992);
(ii) administration of antisense human immunodeficiency virus 1 (HIV-1) in rhesus monkeys (Iverson *et al.*, 1992; Spinolo *et al.*, 1992);
(iii) inhibition of restenosis in pigs (Shi *et al.*, 1994);
(iv) inhibition of leukaemic cell proliferation and tumour growth in mice (Kitajima *et al.*, 1992 a,b; Hijiya *et al.*, 1994; Skorski *et al.*, 1994).

## 11.3 Targeting and design

Daaka and Wickstrom (1990) described the synthesis of three antisense 15mers against human c-*Ha-ras* mRNA. The three oligonucleotides (Figure 11.1a) were targeted to the 5′ cap region, a predicted loop in the secondary structure of the 5′ untranslated region and the initiation codon region of c-*Ha-ras* mRNA. NIH/3T3 cells were treated with each of the three oligonucleotides for 24 h and the level of c-*Ha-ras* antigen analysed by radioimmunoprecipitation. The authors reported sequence-specific, dose-dependent inhibition of human c-*Ha-ras* protein levels by all three oligonucleotides. The most effective oligonucleotides were those targeted to the 5′ cap region, then those targeted to the initiation codon and, finally, those targeted to the loop in the 5′ untranslated region. This result was slightly unexpected, since secondary structure predictions had indicated that the initiation codon was in a more open part of the mRNA than the 5' cap site. Assuming computer simulations of RNA structure to be accurate, this indicates that the efficiency of targeting may not depend solely on secondary structure but also on other factors, probably including the ability of the oligonucleotide to enter the nucleus. These results are corroborated by those of Baker et al. (1992).

Ghosh et al. (1992) synthesized a series of phosphodiester (PO) and phosphorothioate (PS) oligonucleotides against rabbit β-globin, targeted to different parts of the RNA and found the 'openness' of the RNA structure to be the most important factor. In our own hands, antisense oligonucleotides have been used to target various regions of the murine poly(ADP-ribose) polymerase mRNA in L1210 cells (unpublished results). The levels of protein activity were later ascertained by performing an activity blot (Scovassi et al., 1984) in which total protein is electrophoresed, transferred to nitrocellulose or nylon filters, renatured, and incubated with radiolabelled [$^{32}$P]NAD, the substrate for the enzyme. The more active protein there is, the more radioactive product is formed and becomes bound to the membranes. If the production of active protein is interrupted and less is produced, there is a fall in the amount of radiolabelled substrate bound when compared to a control (unpublished results). These experiments showed that oligonucleotides targeted to the 5' end of the RNA, either the 5' cap site or the initiation codon, were the most effective inhibitors of RNA expression.

In 1991, Bacon and Wickstrom described a series of experiments that involved 'walking' along c-*myc* mRNA with antisense oligonucleotides to determine which target positions were the most effective. The primary target site for antisense inhibition was the 5' cap site, although the authors indicated that they did not know whether this was because the RNA was more 'open' at this point or was a function of target location. Clearly, regions upstream of the 5' cap site may only be targeted by antisense oligonucleotides in the nucleus, and regions downstream of this site may only be targeted in the cytoplasm. Perhaps the 5' cap site is available for targeting in both compartments. Also, it is possible that binding to the cap site prevents translation more efficiently than binding downstream of this site, as well as inducing RNase H cleavage of the target RNA (see later sections).

Non-specific activity of antisense oligonucleotides increases with increasing length and increasing G–C content. It would thus appear sensible to use the minimum length possible whilst still maintaining specificity; approximately 13–15 bases. Obviously, antisense oligonucleotides should be analysed for dimer formation and self complementarity since a high level of secondary structure in the oligonucleotide will reduce their efficiency of action, as it does in most experimental systems.

No hard and fast rules are possible because each mRNA will differ in terms of life-cycle, compartmentalization, secondary structure and associated factors. However, it seems likely that antisense oligonucleotides targeted to the 5' cap regions and the region spanning the initiation codon will be among the best for

RNA inhibition. It is also likely that different systems will have different profiles as far as uptake, degradation and activity of antisense oligonucleotides are concerned.

## 11.4 Uptake of antisense oligonucleotides

Cellular uptake and distribution of PS oligonucleotides in human cell culture were described in Stein *et al.* (1992a,b, 1993), Gao *et al.* (1992) and Akhtar *et al.* (1992). This last study also included the use of methylphosphonated (MeP) oligonucleotides. Uptake was described as involving pinocytosis and endocytosis, and the intracellular concentration was higher than that in the surrounding medium. Oligonucleotides were found in both the cytoplasmic and nuclear fractions. Some kind of trapping mechanism was implied by the biphasic efflux from the cell and a specific receptor implied for PS oligonucleotides by Akhtar *et al.* (1992).

These results corroborate those of Loke *et al.* (1989), who found that antisense oligonucleotides were taken up in a saturable, size-dependent manner compatible with receptor-mediated endocytosis. However, Stein and Cheng (1993) have suggested that oligonucleotide binding factors exist on the cell membrane, and oligonucleotides only enter the cell if the external concentration of oligonucleotides is 1–50 mM, which exceeds the apparent $K_d$s of such factors by several-fold. At these concentrations, pinocytosis is a likely mechanism of entry, although different cell lines are likely to display different uptake abilities and characteristics. Oligonucleotides also undergo exocytosis from a number of cell lines (Stein and Cheng, 1993) although the effect of this on antisense inhibition of gene expression is unknown. Uptake is increased if the oligonucleotides are conjugated to a lipophilic moiety such as poly-L-lysine or cholesterol. Transfer of oligonucleotides can be further

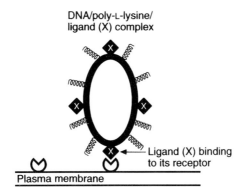

**Figure 11.2.** Possible mechanism of action of receptor-mediated uptake of oligonucleotides. Oligonucleotides can be bound to poly-L-lysine/ligand complexes. The ligand binds to its receptor on cell membranes, facilitating uptake of the complex. X can be transferrin, folic acid or asialoglycoprotein.

increased by coupling of ligands for cellular receptors to poly-L-lysine (see Figure 11.2). Transferrin, folic acid and asialoglycoprotein have all been used to direct the uptake of poly-L-lysine DNA complexes into cells from a variety of tissues (Wagner et al., 1990; Wu et al., 1991; Citro et al., 1994).

## 11.5 Toxicity of antisense oligonucleotides

A number of studies have reported a degree of toxicity of antisense oligonucleotides toward cells in culture (O'Keefe et al., 1989; Smith et al., 1990; Woolf et al., 1990) associated with a particular method of purification. Woolf et al. (1990) report that toxicity may be a sequence-related phenomenon that is abolished by DNase I treatment of the oligonucleotide. The implication is therefore that toxicity may result from a co-purification product. Alternatively, it may be a sequence-dependent effect with the oligonucleotide inhibiting the expression of an essential (and possibly unidentified) gene. A third possibility is that degradation of high concentrations of the oligonucleotide causes toxic concentrations of nucleotides within or around the cell. In our hands, no apparent toxicity was detected in L1210 (mouse) cells incubated in up to 200 mM PO or PS oligonucleotides.

## 11.6 Modifications to the structure of antisense oligonucleotides

All modifications to antisense oligonucleotides are primarily concerned with increasing their efficiency of action, and reducing anything which might hamper this action: Figure 11.3 depicts the structures of the two main types of modified oligonucleotide. Improvements in direct uptake into the cell can be achieved by conjugating the oligonucleotide to a lipid or cholesterol group (Shea et al., 1990); to poly-L-lysine (Lemaitre et al., 1987; Reinis et al., 1993) or by substituting the charged PO groups on the sugar–phosphate backbone with non-charged groups such as MeP (Giles and Tidd, 1992). All of these modifications, and others, improve uptake because they facilitate crossing of the plasma membrane.

Improvements in stability are likely to be very efficacious for oligonucleotide action because this ensures the maintenance of a constant antisense concentration in the system and allows action at a lower concentration (with all its considerable cost and feasibility implications). It is worth remembering that the degradation of antisense molecules will differ from system to system, so that general rules are impossible. Thus, for instance, Ryte et al. (1993) have described two human cell sub-lines, which differ radically in their degradation profile of antisense oligonucleotides.

Oligonucleotides are very susceptible to nucleolytic activity and their rapid degradation in various culture media and sera have been reported (Wickstrom, 1986; Daaka and Wickstrom, 1990; Hoke et al., 1991). Each of these studies uses 5' $^{32}$P-labelled oligonucleotides, and the presence of the labelled

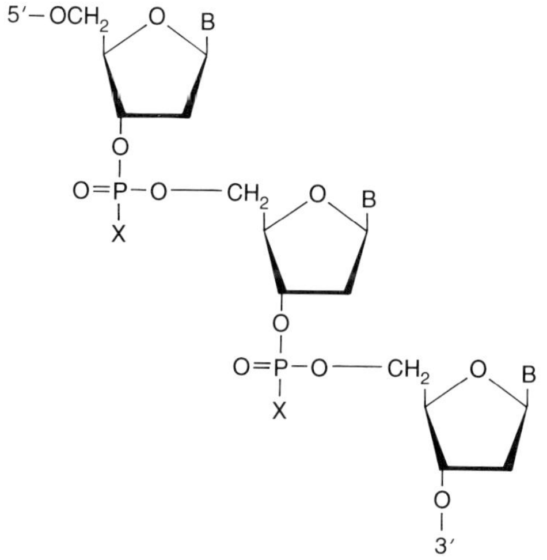

**Figure 11.3.** The structure of an oligonucleotide where X is oxygen (O) in normal phosphodiester bonding, but can be replaced with sulphur (S) in PS oligonucleotides, or a methyl group in MeP oligonucleotides. Other groups may also be substituted in the X position.

phosphate on smaller and smaller bands is taken as evidence of degradation. It is possible that some systems contain 5′ phosphatases which will enhance degradation and produce possible exaggeration of its extent, although the ladders produced during electrophoresis suggest degradation from the 3′ end. In our hands, PO and PS oligonucleotides showed little degradation in cell culture media (including heat-inactivated fetal calf serum) over periods ranging from several hours to 2 days. The oligonucleotides were studied by gel electrophoresis and ethidium bromide staining.

The point of nucleolytic attack is the phosphodiester bond, and modifications to this bond have been introduced to reduce degradation at this position. The most extensively characterized of these modifications are the MePs (Miller *et al.*, 1985) and the PSs (Stein *et al.*, 1988). MeP oligonucleotides have a methyl group instead of an oxygen on the phosphate of the PO link, whereas PS, or sulphurized oligonucleotides, have a sulphur instead of an oxygen on this phosphate. No direct comparison between MeP and PS oligonucleotides appears to have been made, although both types of modification decrease apparent degradation dramatically as compared to normal PO oligonucleotides. The resistance of PS oligonucleotides to degradation was extensively characterized by Stein *et al.* (1988) and Hoke *et al.* (1991). Stein *et al.* (1988) found the resistance of the PS oligonucleotide to nucleases P1, S1 and snake venom phosphodiesterase to be between 2 and 25 times greater than that of the PO oligonucleotide. A study of 5′ and 3′ PS-capped oligonucleotides revealed a resistance to degradation intermediate between the all-PS and all-PO oligonucleotides. In our hands, the PS-capped molecules exhibit a degradation resistance similar to all-PS oligonucleotides, except with DNase I, to which the partially sulphurized oligonucleotides show an increased susceptibility.

The synthesis of PS and MeP oligonucleotides, which can be performed using DNA synthesizers, delivers diastereoisomeric mixtures of $2^n$ isomers, where $n$ is the number of internucleotide PS or MeP junctions. All methods so far reported deliver mixtures of PS or MeP oligonucleotides that are not stereospecific. As demonstrated by W. Stec (personal communication), dinucleoside (3'/5') PS oligonucleotides prepared via phosphoroamidite/sulphurization methods do not consist of the mixture of Rp- and Sp-diastereoisomers in a strictly 1:1 ratio. In fact, the ratio of all-Rp isomer to all-Sp is 5:1. Therefore, evidence has been presented that the contribution of a single stereoisomer in a $2^n$ mixture is not predicted by the exponential function $2^{-n}$. Each component of a diastereoisomeric mixture has its own chirality; and this is the property responsible for individualized transport, affinity to proteins, stability against nucleases (for example, snake venom phosphodiesterase will digest one stereoisomer but not the other), and proclivity to target mRNA (Stec *et al.*, 1991; Lesser *et al.*, 1992; Koziolkiewicz and Stec, 1992). Degradation data concerning such modified oligonucleotides should therefore be interpreted with caution.

Hoke *et al.* (1991) demonstrated that all-PS, and 3' PS-capped oligonucleotides were superior in terms of nuclease resistance to all-PO, 5' PS-capped, and PO oligonucleotides with a central block of PS; thus corroborating and expanding the results of Stein *et al.* (1988). The degradation took place in cell culture media, with 10% heat-inactivated fetal calf serum, and confirmed earlier work that the predominant nucleases in fetal calf serum are 3' exonucleases. Similar results for MeP-protected oligonucleotides were reported by Tidd and Warenius (1989). Therefore MeP/PS-capping or MeP/PS-blocking throughout the oligonucleotide increases stability several-fold.

A recent study by Wagner *et al.* (1993) describes the synthesis and characterization of PS oligonucleotides containing C-5 propyne pyrimidines. These compounds are described as dose-dependent and very sensitive to mismatch. They cause gene-specific inhibition at nanomolar concentrations. However, uptake was only possible when the oligonucleotides were rendered more able to permeate cells by modification, or when they were microinjected into cells. These oligonucleotides may be promising therapeutic agents and may represent a new generation of antisense oligonucleotides for the study of gene function.

## 11.7 Possible mechanisms of action

### 11.7.1 Steric inhibition

Oligonucleotides are believed to inhibit RNA expression in a variety of ways. One such way, which is widely quoted, is steric inhibition. It is thought that the oligonucleotide binds to the RNA, and renders the normal processes of translocation, splicing and translation difficult. That the oligonucleotide binds to the target mRNA is not in doubt. Several studies have demonstrated sequence-specific binding (e.g. Stein *et al.*, 1988), but it is not possible to attribute the subsequent effect of inhibition of expression directly to steric interference. However, experiments in cell-free systems, where there are likely to be fewer uncontrolled factors, do imply that this is a probable mechanism of action.

### 11.7.2 RNase H-like cleavage of target RNA

Antisense oligonucleotides are thought to act via a eukaryotic RNase H-like activity (for human RNase H1, see Eder *et al.*, 1993) which cleaves the RNA portion of the RNA/DNA hybrid, after binding of the antisense oligonucleotide Figure 11.1b). Evidence for the occurrence of this type of cleavage is provided by the fact that labelled RNA molecules are cleaved at the sites of oligonucleotide hybridization (Shuttleworth and Coleman, 1988; Giles and Tidd, 1992a,b). *Escherichia coli* RNase H is a prokaryotic enzyme which cleaves the RNA strand of a DNA/RNA hybrid and is widely used in *in vitro* studies of RNase H-like activity. Its activity is related to the formation of the duplex and hence to the hybridization characteristics of the oligonucleotide and its target RNA. If an oligonucleotide hybridizes promiscuously with many RNA molecules, for example, many of these will be cleaved. If it is mismatched to its target, and duplex formation is weak, the induced RNase H cleavage of the target RNA will be inefficient.

Stein *et al.* (1988) studied the hybridization characteristics and *E. coli* RNase H susceptibility of duplexes of all-PS oligonucleotides and RNA, part-PS oligonucleotides and RNA, and all-PO oligonucleotides and RNA. The experiments were performed *in vitro* and with a prokaryotic enzyme. Therefore, only inference to the *in vivo* eukaryotic situation is appropriate. The duplex stability, as reflected by melting temperature ($T_m$), of each type of hybrid was studied by absorbance. The all-PS oligonucleotide was found to have a greatly reduced $T_m$ and the duplex was less stable than that of the all-PO oligonucleotide. The part-PS oligonucleotide formed a duplex, the stability of which was intermediate. Digestion of the duplexes by RNase H was also monitored by changes in absorbance; the all-PS oligonucleotide allowed greater RNase H digestion of the RNA than did the part-PS oligonucleotide, which in turn allowed greater susceptibility than the all-PO oligonucleotide. The authors concluded that the instability of the all-PS/RNA duplex contributed to its increased susceptibility to RNase H, possibly by allowing rapid release of the RNA fragments after digestion.

Similar findings were reported by Giles and Tidd (1992a,b) who studied RNase H activity with MeP/PO chimeric oligonucleotides. An *in vitro* experiment was set up with synthesized RNA molecules and *E. coli* RNase H, and oligonucleotides of various levels of MeP substitution were assayed for their stimulation of RNase H activity. Increasing levels of MeP substitution were found to decrease the $T_m$ of the RNA/DNA duplex and increase *E. coli* RNase H activity, corroborating the results of Stein *et al.* (1988) above, for modified oligonucleotides. These studies were re-addressed by Hoke *et al.* (1991) who used PS oligonucleotides directed against the initiation codon of

HSV UL13 mRNA in HSV-1-infected HeLa cells. The effectiveness of antisense action was measured by comparing virus titre in experimental and control wells. Six oligonucleotides were assessed in terms of antiviral activity, duplex DNA/RNA stability, *E. coli* RNase H susceptibility and nuclease resistance. The results of Hoke *et al.* (1991) are summarized in Table 11.1.

Table 11.1. Assessment of oligonucleotides with various levels of sulphurization in terms of antiviral activity, duplex DNA/RNA stability, *E. coli* RNase H susceptibility and nuclease resistance

| Oligonucleotide type | Nuclease resistance | Duplex stability | RNase H susceptibility | Anti-viral activity (%) |
|---|---|---|---|---|
| All-PS | High | Low | Good cleavage | 91 |
| 3' PS | ↓ | ↓ | Good cleavage | 45 |
| 3'-5' PS | ↓ | ↓ | Good cleavage | 23 |
| Central PS | ↓ | ↓ | Good cleavage | 29 |
| 5' PS | Low | High | Good cleavage | 19 |

In terms of nuclease resistance, and duplex stability, the results of Hoke *et al.* (1991) corroborate those of Stein *et al.* (1988). All the oligonucleotides gave good RNA cleavage with *E. coli* RNase H (although the eukaryotic enzyme is very likely different in profile to the prokaryotic enzyme). It can be seen from Table 11.1 that only the all-PS oligonucleotides produced very efficient inhibition of viral activity, which implies that all-PS oligonucleotides are the most effective antisense agents in this system. However, PS oligonucleotides suffer from a high level of non-specific antisense action, as will be discussed in Section 11.11, so a simple conclusion is not possible.

Dagle *et al.* (1990) compared two types of antisense oligonucleotide, targeted against *An2* and cyclin in *Xenopus* oocytes, and monitored RNA degradation using RNase protection assays. The constructed antisense sequences consisted of two MeP oligonucleotides, each with a central portion of either PO or PS. Direct comparison of PO and PS performance was not the main aim of the authors but some inferences may still be drawn. The oligonucleotide with a central PS portion directed RNA degradation at a much slower rate than if there was a central PO portion. This implied (in *Xenopus* oocytes at least) that PS bonds may be less accessible to RNase H activity than was implied by the work of Stein *et al.* (1988) and for MeP (Giles and Tidd, 1992). Corroboration the view of that PO/RNA duplexes are more susceptible to RNase H-directed cleavage of the RNA strand than are PS/RNA duplexes was provided by Agrawal *et al.* (1990) and Temsamani *et al.* (1991). In our own hands, a 3'-5' PS-capped oligonucleotide inhibits its target more effectively than a fully sulphurized one, and is more stable than its all-PO equivalent.

Condensation of the survey presented in Crooke (1992) reveals that several-fold more MeP oligonucleotide is required to inhibit RNA molecules than PS oligonucleotide (see Table 11.2). PS and MeP oligonucleotides both show increased uptake as compared to normal oligonucleotides, but PS

**Table 11.2.** Modified oligonucleotides, target sequences and concentration needed for activity

| Oligonucleotide modification | Target | Concentration range (μM) |
|---|---|---|
| MeP | Viruses | 20–100 |
| PS | Viruses | 0.5–20 |
| MeP | Oncogenes | Inactive |
| PS | Oncogenes | 10–1150 |
| MeP | Host genes | 100 |
| PS | Host genes | 0.01–2.5 |

oligonucleotides are effective at a much lower concentration than MeP, which suggests that their operation includes a different mechanism of action. Therefore, although MeP modification produces a stable antisense oligonucleotide which is readily taken up into the cell, sulphurization produces an oligonucleotide which seems more effective; perhaps by allowing the action of eukaryotic RNase H. Agrawal et al. (1990) also report site-specific RNase H action against the RNA strand of PO and PS oligonucleotide duplexes but not against the RNA strand of MeP oligonucleotide/RNA duplexes.

In summary, it is extremely difficult to draw reliable conclusions from work with modified oligonucleotides and RNase H data. The RNase H used in *in vitro* experiments is invariably *E. coli* RNase H, and so results do not necessarily apply to the *in vivo* eukaryotic situation. However, fully MeP or PS oligonucleotides are significantly more stable than their PO counterparts and can therefore be used in lower amounts, and may allow a reduction in non-specific toxicity. PS oligonucleotides may be more accessible to eukaryotic RNase H (Crooke, 1992) although not as accessible as PO oligonucleotides (Dagle et al., 1990). To allow this increased susceptibility, a central PO portion should be left in a 3'-5' PS-capped oligonucleotide. The length of the central sequence has not yet been optimized, although one study mentions a gap of as little as a single base.

### 11.7.3 Triplex DNA formation

DNA triplex formation occurs when a stretch of single-stranded nucleic acid fits into the major groove of a DNA double helix (Figure 11.1c; originally visualized by electron microscopy; for a recent study, see Cherny et al., 1993). The third strand is usually DNA, although it has been shown that single-stranded RNA forms an even more stable triplex (Escudé et al., 1992). It has been shown that Watson–Crick base pairs are able to hydrogen bond with an additional base, so forming a triplex (for a review, see Hélène, 1991; for mechanistic details, see Miller and Cushman, 1993). The third strand may run parallel or antiparallel to the helix, depending on its sequence; known as Hoogsteen or reverse-Hoogsteen binding respectively (Hoogsteen, 1959). Once in position, the third strand has been shown to prevent the action of certain DNA-modifying enzymes, such as methylases and restriction endonucleases. The triplex is also believed to inhibit the binding of transcription factors and RNA polymerases. A study which demonstrates inhibition of transcription of the sense strand is described in Duval-Valentin *et al.* (1992), which concerns expression of *E. coli* β-lactamase in a cell-free transcription system. Another study involved the targeting of oligonucleotide to the origin of replication of simian virus 40 DNA, in virally infected CV1 cells in culture (Birg *et al.*, 1990). Triplex formation was shown to occur and viral DNA replication was blocked. Several experiments indicate that oligonucleotide triplex formation may represent a successful antigene strategy, although most have not demonstrated this conclusively. For example, *E. coli* RNA polymerase was shown to possess reduced rates of transcription in the presence of three-stranded polynucleotides (Morgan and Wells, 1968), and repression of c-*myc* transcription in a HeLa cell nuclear extract was observed when Cooney *et al.* (1988) incubated the cells with a purine-rich 27mer that was designed to bind to duplex DNA about 115 bases upstream from the c-*myc* transcription start.

There are restrictions on the type of oligonucleotide that may be used for triplex binding owing to the limitations imposed on the triplex bond. Watson–Crick pairs may form a triplex only in homopurine or homopyrimidine stretches, with a third homopurine or homopyrimidine strand. The triplex may be stabilized by addition of an intercalating agent (Hélène, 1991; Grigoriev *et al.*, 1992, 1993) such as acridine, which is 'tethered' to the 5' end. Psoralen may also be used (Giovannangeli *et al.*, 1992a) or other ligands (Mergny *et al.*, 1992). In the β-lactamase gene of *E. coli* the triplex-forming oligonucleotide inhibited transcription at low temperatures, but at 37°C required a 5' intercalator to have any effect (personal communication, C. Hélène). It has been shown that the formation of triplex DNA is very slow, but once formed can be expected to last for several hours under physiological conditions and prevent transcription, or replication, of the gene during this time (Hélène, 1991). Duplex-binding oligonucleotides are subject to similar constraints as antisense oligonucleotides, in that stability and uptake are important. The stability of triplex-forming oligonucleotides has been enhanced by the use of oligo-α-deoxynucleotides instead of the natural oligo-β-deoxynucleotides. Oligonucleotides containing synthetic α-anomers are shown to bind to duplex DNA, and have a high resistance to nuclease digestion (Hélène, 1991). Another development is the production of oligonucleotides which bind single-stranded targets, and then fold back on themselves and the bound target to form triplex structures (Giovannangeli *et al.*, 1991). This is similar to a proposed mode of action for circular oligonucleotides (see Section 11.9). Uptake is a major factor for the success of this strategy because the oligonucleotides not only have to cross the plasma membrane but also enter the nucleus to inhibit eukaryotic host genes. The success of triplex binding as an antigene strategy will depend on the development of mechanisms which concentrate the oligonucleotides at their site of action.

An interesting review by Camarini-Otero and Hsieh (1993) postulates that triplex formation is a natural phenomenon and takes place in the pairing of homologous chromosomes, in recombination or 'crossing over', in chiasmatic pairing, ectopic recombination in yeast, and paramutation (where two non-linked alleles influence each other prior to segregation). Many data are consistent with the view that prokaryotic and eukaryotic recombinases form three-stranded molecules (Camarini-Otero and Hseih, 1993). When these are deproteinized, they are revealed as DNA triple helices, or triplexes, in which the third strand is in contact with an intact duplex. It is possible therefore that antisense strategies using triplex formation, like antisense oligonucleotides, represent an attempt to exploit the natural proclivities of genetic material (and like antisense oligonucleotides, will probably encounter endogenous, natural methods of control).

## 11.8 Double-stranded oligonucleotides

Double-stranded PS oligonucleotides have also been used to selectively bind DNA-binding proteins, so that the course of development could be studied without activation of the protein target element (Bielinska *et al.*, 1990).

## 11.9 Circular oligonucleotides

Circular DNA is very common in nature (e.g. forming the genomes of bacteria and viruses, plasmids, episomes, etc.). It has a high survival rate in that it is resistant to exonucleases and only susceptible to single-stranded endonucleases. Circular oligonucleotides have been synthesized by E. Kool (personal communication). They are found to bind in two different ways to single-stranded nucleic acid: in a so-called 'outside' formation (see Figure 11.1d), where the binding is reduced to 7–14 nucleotides due to topological constraints produced by the right-handed twist of the DNA; or in a triplex formation (Figure 11.1d), where the circular oligonucleotide forms a triplex and the stability of binding is increased compared to the linear. Circular oligonucleotide can form triplexes even when there are some non-complementary bases in the sequence. The 5' nucleotide loops appear to stabilize the structure. The binding specificity (Kool, 1991; Prakash and Kool, 1992; Wang and Kool, 1994) and stability of these structures has been looked at *in vitro* and has been found to match or better, linear oligonucleotides. It still remains to be seen, however, if circular oligonucleotides will be functional and useful in biological systems.

Another form of self-stabilization involves the formation of hairpin loop structures at the 3' end of the oligonucleotide (Tang *et al.*, 1993). These oligonucleotides also showed increased persistence with respect to their linear counterparts.

## 11.10 Ribozymes

Ribozymes are stretches of double-stranded RNA which have a catalytic, self-cleaving ability (for a recent review, see Vonahsen and Schroeder, 1993). They have been found in the protozoan *Tetrahymena thermophila* (Zaug et al., 1986) and also in plant viroids and virusoids, and in transcripts of newt satellite DNA (Xing and Whitton, 1992, and references therein). Ribozymes possess endoribonuclease activity, and most naturally occurring ribozymes have roles in RNA processing, such as intron removal by self-splicing and the cleavage of concatameric transcripts from replicating RNA viruses (Bennett and Cullimore, 1992). In nature, the ribozyme activity is intramolecular: the molecule splices itself. However, the catalytic activity of the ribozyme can be directed against other RNA molecules *in vitro*. A key study by Haseloff and Gerlach (1988) described the structure associated with cleavage and demonstrated that a single-stranded RNA molecule, of specific primary and secondary structure, would cleave other single-stranded RNA molecules in *trans* if the sequences on the ribozyme arms and the target RNA were complementary (Figure 11.4). The authors described the essential structure of the ribozyme core, and demonstrated that the target sequence depended on its complementarity with the ribozyme arms. The mechanism of cleavage is, as yet, unclear although it is likely to involve an oxygen in the ribose backbone (Pyle et al., 1992). Development of synthetic ribozymes meant that they could be used in antisense strategies to eliminate or reduce the levels of targeted RNA molecules. For example, synthetic ribozymes were constructed by inserting an oligonucleotide sequence into a plasmid coding for a ribozyme sequence (Haseloff and Gerlach, 1988). Transcription was controlled by a T7 promoter. Experiments *in vitro* demonstrated significant cleavage of target RNA molecules at various temperatures and in various conditions. The cleavage was found to be efficient and highly sequence specific.

Similar experiments were carried out by Kikuchi and Sasaki (1991) who used three synthetic ribozymes against *E. coli* prolipoprotein signal peptidase mRNA, at specific sites. Again, the ribozymes were transcribed from plasmids under the control of the T7 promoter. The specificity of cleavage was demonstrated, as well as the cyclical nature of repeated cleavage. Bennett and Cullimore (1992) showed that ribozymes could discriminate between closely related RNA molecules, even though the sites of cleavage shared up to 18 out of 20 bases. The substrate RNA has very few sequence requirements, other than to base-pair with the arms of the catalytic strand, and to contain a GUX motif, where X can be C, A or U (Figure 11.4). This comparative freedom of substrate structure widens the potential applications of ribozymes. Sarver *et al.* (1990) synthesized ribozymes against HIV-1 *gag* RNA, and showed precise cleavage in a cell-free system. These authors also performed human cell studies demonstrating the stable expression of ribozymes, reduction of target RNA and protein, and apparent lack of toxicity in cells expressing ribozymes over a period of months. In addition, it was shown that a protected ribozyme with a 5' GppppG cap structure cleaved its target with normal efficiency. Unmodified ribozymes are quickly degraded in culture, and the ability to modify them to resist such degradation greatly increases their potential. Recently, artificial ribozymes have been synthesized with modified (2'-*O*-allyl and 2'-*O*-methyl) ribonucleotides (Paolella *et al.*, 1992). Apart from a sensitive region of six nucleotides within the

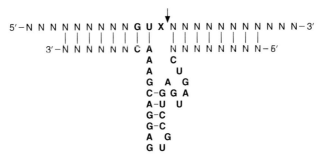

Figure 11.4. Structure of a hammerhead ribozyme. The site of cleavage is shown by an arrowhead. X can be C, A or U.

core, ribozymes could be fully substituted and still retain high activity compared to the native structures. These structures are very resistant to degradation in culture. The authors suggest that further modifications may also improve delivery into cells (e.g. increasing lipophilicity by decreasing the negative charge of the DNA backbone, using MeP or PS modifications, or conjugating the oligonucleotide to a lipophilic substance such as cholesterol).

The implementation of ribozymes for therapeutic use depends on the ability of the catalyst to function in the complex cellular milieu, which is not yet conclusively established. Bennett and Cullimore (1992) found that the presence of additional substrate or non-substrate RNA molecules reduced the efficiency of ribozyme action. This is possibly due to a rise in the level of non-specific cleavage.

In summary ribozymes are naturally occurring self-splicing double-stranded RNA regions. The catalytic activity can be mimicked by synthetically produced ribozymes, and can be specifically targeted to particular mRNAs by producing ribozymes with complementary sequences to these mRNAs. Ribozymes have been expressed in a number of *in vitro* systems and have been shown to cleave their target RNA specifically, although no work has been done on possible non-specific activity. That such activity exists seems likely, since ribozyme cleavage of target RNA is reduced by the addition of non-related RNA. Applications are presently limited to *in vitro* studies, transformed cell studies and systems where the ribozyme may be introduced by microinjection. Uptake and stability of the ribozyme (RNA) is very poor, although protected ribozymes have been synthesized recently which are more stable in serum. Implementation of ribozymes as therapeutic agents, in conjunction with gene therapy, may be a promising avenue to explore, although many questions, have yet to be answered.

## 11.11 Non-specific cleavage of host RNA

An intrinsic problem of using antisense oligonucleotides to ablate gene expression within cells is that the stringency of hybridization *in vivo* is not subject to control and may be sub-optimal. It is therefore possible that oligonucleotides could form partial hybrids with non-targeted RNA, causing cleavage, or otherwise inhibit their proper functioning. There would be a lowering of levels of

RNA other than that specified and the effects observed would not be specific to the RNA target. The number and extent of these effects would depend on the sequence. The observed occasional toxicity of antisense oligonucleotides may be related to this effect. The toxicity of all-PS oligonucleotides, in particular, may be related to promiscuity of hybridization of a particular sequence. This toxicity is found to increase as the number of PS bonds increases (Woolf et al., 1992).

Woolf et al. (1992) rigorously tested the specificity of an all-PO antisense oligonucleotide. The average mRNA pool within a higher cell is calculated at approximately $10^4$ different RNA molecules of about 2 kb each. Assuming random distribution of each of the four bases (which may not be the case), the shortest length of sequence likely to be unique is estimated to be 13 bases. Therefore it is argued that an antisense oligonucleotide of 13 or more bases must be specific for its target. The assumption is that only a perfect match between oligonucleotide and target mRNA will lead to hybridization and inhibition of the mRNA's function. This assumption may be wrong, and antisense oligonucleotides may cause RNase H-like mediated cleavage with shorter stretches of hybridization, and possibly promote other antisense effects, such as inhibition of translation. Woolf et al. (1992) designed oligonucleotides against *vg1* mRNA (a developmentally important maternal RNA) in *Xenopus* oocytes, which represent a closed system with no RNA synthesis until a relatively late stage after fertilization. The oocyte was used as a model to investigate the specificity of these oligonucleotides. Antisense inhibition was measured using various oligonucleotides including perfect and partial matches. The study shows that partial matches caused degradation of the target mRNA (although more slowly), demonstrating that it is probably impossible to obtain total specificity. The authors constructed three 25mers against fibronectin mRNA, of these, one was a perfect match, one matched in the central 17 of 25 bases; and one matched in the central 14 of 25 bases; all these oligonucleotides were found to cleave the target RNA with efficiencies of 79, 32 and 37%, respectively. This demonstrates that the mechanism of hybridization in this system, and with this sequence, involves a part, but not all, of the oligonucleotide sequence. A 13mer, with 10 central consecutive bases which matched the target, and oligonucleotides with internal mismatches also cleaved the target mRNA. Interestingly, unrelated control oligonucleotides cleaved the fibronectin RNA to a consistent level of 13%. This work indicated that flanking sequences may be mismatched, that an internal base may be mismatched and that only 10 complementary bases are required for target mRNA cleavage. Theoretically, longer oligonucleotides will cause more non-specific cleavage because they will hybridize non-specifically to more RNA molecules, although specific cleavage will probably always be greater than the non-specific cleavage. These results imply that the mechanisms and kinetics of hybridization play a very important role in antisense action. For any sequence, and any system, these kinetics are likely to be different. In cells other than *Xenopus* oocytes, there are two important factors which may actually modify the non-specific antisense response. The first is the continual replenishment of mRNA which may reduce the effects of a 13% removal; the second is the incubation temperature of other cells, which is likely to be higher than that of *Xenopus* oocytes, and hence hybridization is more stringent.

Giles and Tidd (1992a,b) studied non-specific cleavage of mRNA by antisense-directed RNase H. Digoxigenin-labelled RNA (c-*myc*, 1780 bp; p53, 1304 bp) were incubated *in vitro* with various oligonucleotides at an RNA:oligonucleotide ratio of 1:20. The RNA was electrophoresed, blotted, and the pattern of RNA cleavage was studied. Using a c-*myc* antisense 15mer, the authors found a high level of inappropriate cleavage and calculated that a match of 6 bp was all that was required for RNase H action. It should be borne in mind that these were *in vitro* experiments, carried out in conditions optimal for cleavage, and that the enzyme used was *E. coli* RNase H, not a mammalian or even a eukaryotic enzyme. The internal milieu of a eukaryotic cell would differ from these conditions, and, therefore, inference only is possible. Nevertheless, it is extremely interesting that 6 bp were all that were required for RNase H-mediated cleavage of the target RNA strand. Inappropriate cleavage was noticed with all of the antisense oligonucleotides used, even non-sense controls, depending on the number of 6 bp regions of complementarity to the RNA. The G–C contents of the central PO (cleaving) region also affected RNA degradation, in that more non-specific hybridization was seen with increasing G–C content. It is impossible to say whether this is due to increased promiscuity of the new cleaving sequence, or to an increase in duplex stability.

The above study did not include PS oligonucleotides, although it would have been very interesting to see the results. The PS bond reduces $T_m$, and decreases the strength of hybridization (as does the MeP bond), but allows RNase H action. It seems likely that a PS oligonucleotide would have a very high level of non-specific cleavage, a speculation that is backed indirectly by the observation by Woolf *et al.* (1992) that PS oligonucleotides occasionally demonstrate increased toxicity with increased number of PS bonds, and that this toxicity may be abolished by digestion with DNase I. The toxicity may be sequence related, possibly due to vital RNA destruction.

In conclusion, it is probably impossible to obtain total specificity with any antisense oligonucleotide and the observed effects are possibly due to the depletion of mRNA other than the target mRNA. If the inhibited phenotype can be rescued by addition of exogenous target RNA, the results would become clearer (as, for example, with the demonstration that *Oct*-3 is a required maternal factor for the first mouse embryonic division; Rosner *et al.*, 1991). Non-specific hybridization and cleavage can be reduced by reducing the length of the oligonucleotide. Performing the experiment more than once with oligonucleotides targeted to different parts of the same mRNA molecule will help to establish the relationship between ablation of that RNA and the observed effects. In any antisense oligonucleotide-based study, it is also necessary to include a control which consists of an oligonucleotide of similar composition, but of different sequence, to demonstrate the sequence specificity of the oligonucleotide of interest compared to this control oligonucleotide.

## 11.12 Therapeutic applications

Antisense oligonucleotides are potentially useful as therapeutic agents and it is hoped that they will help to rationalize drug design (for recent reviews, see Crooke, 1993; Cohen, 1993). In a review of the therapeutic potential of antisense oligonucleotides, Crooke (1992) summarizes the successful use of these agents against a range of targets, including viruses, oncogenes, host genes and others. The oligonucleotides are usually added to cell culture and reduction in mRNA expression measured within the culture. Crooke (1992) outlines key tasks, such as determining the pharmacological behaviour of antisense oligonucleotides and developing therapies for clinical application.

The design of antisense strategies to combat human disease is discussed by Miller and T'so (1987) in relation to chemotherapy; Jaroszewski *et al.* (1990) relating to therapy for multiple drug resistance; and by Calabretta (1991), Citro *et al.* (1992) and Bayever *et al.* (1992) regarding the inhibition of proto-oncogene expression in human cell culture, with a discussion of biological and therapeutic implications. All three papers describe some success in cell culture or animal studies against particular sequences and propose further studies.

A report detailing the use of antisense oligonucleotide *in vivo* was presented by Simons *et al.* (1992). Anti c-*myb* was locally delivered to rat carotid artery during surgery following balloon angioplasty (and therefore injury) to these vessels. It is known that this type of injury stimulates the mitogen-induced laying down of smooth muscle cells intimal to the artery, and the study aimed to determine whether anti-c-*myb* oligonucleotide would inhibit this induced response. It was shown that c-*myb* expression was indeed inhibited, and consequently mitogen-induced proliferation of vascular smooth muscle cells was also reduced. These studies represented one of the first reports on the use of antisense oligonucleotides to inhibit a normal gene product activity *in vivo*. The success of this study was attributed by the authors to:

(i) local delivery of high concentrations of oligonucleotide;
(ii) transient expression of the target gene;
(iii) low abundance of the target gene products.

Also important are the half-lives of the RNA and protein targets. Intimal arterial accumulation of smooth muscle cells is described as an occurrence responsible for the long-term failure of coronary and peripheral arterial by-pass grafts, and the development of restenosis following coronary artery angioplasty. Approaches using antisense oligonucleotides may represent a new class of therapeutic agents.

Oligonucleotides have been used in preliminary studies to pre-target toxic agents to cancer cells (W.H.A. Kuijpers, personal communication). An antigen on the cancer cell surface is targeted by a conjugated antibody/oligonucleotide, which binds to the cancer cell, leaving the oligonucleotide free (Kuijpers *et al.*, 1993). Optimal tumour uptake is allowed to take place, which usually takes 2–3 days. An oligonucleotide, antisense to the first one and bound to a radioisomer, is then incubated with the cells and the oligonucleotides hybridize. This targets the radioactivity directly to the cancer cells and, theoretically, will maintain it while the label is cleared quickly from the rest of the body. The theory is that less toxic agent can be used, and that the agent is targeted directly to the cancer cells. The research is preliminary, but exciting.

Oligonucleotides have also been used against influenza virus (Leiter et al., 1990; Agrawal and Sarin, 1991; Agrawal and Leiter, 1992; Lisiewicz et al., 1993) and also against the HIV viral genome (Agrawal and Tang, 1992; for a review, see Stein and Cheng, 1993). The oligonucleotide was tagged with fluorescein, and found to accumulate within cells, in a saturable manner. The mechanism of action of this antisense oligonucleotide was believed to involve mainly translation arrest due to binding and RNase H cleavage of target RNA. The oligonucleotide was generally digested from the 3' end, and various modifications were made by the authors to increase 3' stability. Among these was a short region of basepairing at the 3' end forming a small hairpin loop. As the length of this base-paired region increased, so did the oligonucleotide's stability to nucleases. The modified oligonucleotide hybridized to its target as efficiently as a linear molecule, and in vitro RNase H cleavage of the target was unaffected. A slight decrease in cellular uptake was observed due to the secondary structure, and there was no apparent improvement in anti-viral activity. These experiments nevertheless provide an example of antisense oligonucleotides being targeted against a viral gene, and antisense structure was optimized. Clinical trials have now been initiated using this oligonucleotide in human acquired immunodeficiency syndrome (AIDS) patients (S. Agrawal, personal communication). Targeting of the HIV viral genome with triplex-forming oligonucleotides is reported by Giovannangeli et al. (1992a).

The pharmacology of sulphurized antisense oligonucleotides in mice is described in Agrawal et al. (1991). In this study, an antisense PS oligonucleotide was internally labelled with $^{35}$S and administered to mice. The oligonucleotide was tracked as it was distributed throughout the mouse's body and it was found that it became quickly distributed throughout the tissues, a percentage being excreted in the urine. This excreted form showed increasing degradation with time. The oligonucleotide was found in all organs except brain, and was particularly concentrated in the kidney and liver. In the kidney and liver, degradation was more rapid than in other organs, approaching 50% of the total oligonucleotide dose by 48 h. An unexpected phenomenon was observed, in that bands of slower mobility appeared on the gels related to oligonucleotide extracted from these organs; in the kidney, one or two bands of slower mobility ran behind the oligonucleotide and, in the intestine, several discrete slower bands were seen. The authors postulated additions or modifications to the oligonucleotide chain, or possibly degradation of the labelled molecule and incorporation of the label into other structures, although the pattern of discrete bands appeared to support the first idea. Hybridization to target RNA molecules is not an explanation because the gels were run under denaturing conditions. Further studies with end-modified S-oligonucleotides suggest that stability in mice is increased if the oligonucleotide is 3'-capped, although overall distribution is unaffected (Temsamani et al., 1993). In our hands, sulphurized oligonucleotides incubated in serum produced two or more slow mobility bands when electrophoresed, and it is possible that the PS oligonucleotides form hybrids and concatamers, perhaps by virtue of disulphide bridges. Agrawal et al. (1991) report no ill effects of oligonucleotide administration, although the study was not long term (48 h only). Further animal studies are given in the 1992 volume of the *Proceedings of the American Association for Cancer Research* (vol. 33).

A study by Bigelow et al. (1992) describes the continuous subcutaneous infusion of an anti-HIV PS-27mer into mice. The authors describe the same general distribution and slow clearance from the tissues as Agrawal et al. (1991) and they suggest that subcutaneous administration may be a suitable method of delivery. Iverson et al. (1992) reported the systemic administration of anti-human p53 PS oligonucleotide to rhesus monkeys. Several pharmacological parameters affecting the dosage and distribution of oligonucleotides were measured, and no ill effects reported, although again this was not a long-term study. Spinolo et al. (1992) performed a similar series of experiments, using six rhesus monkeys and anti-p53. No treatment-related effects were discernible up to 118 days after administration.

Antisense oligonucleotides have therefore been shown to have wide distribution and slow clearance from animal tissues when introduced subcutaneously, intraperitoneally or intravenously. The slow clearance of PS oligonucleotides may be related to their ability to associate with certain proteins (e.g. albumin; Stein and Cheng, 1993). PS oligonucleotides have also been shown to be degraded *in vivo* and the presence of sulphurized mononucleosides is likely to have toxic effects, especially if incorporated into genomic DNA. A study describing the intraventricular administration in rats of antisense oligonucleotides is given by Whitesell et al. (1993) which reports the distribution of the antisense agent in the CNS and brain. Studies report no apparent illeffects in mice, rats or rhesus monkeys (although 'ill effects' would almost certainly be a sequence-related phenomenon). The strategy has been used *in vivo* in rats undergoing surgery and the antisense oligonucleotide used was shown to have its predicted beneficial effect. In spite of possible negative effects which require considerable investigation, the overall implication of these studies is that antisense strategies have considerable therapeutic potential.

## References

**Agrawal S, Leiter JME.** (1992) *Alternative Antiviral Approaches to Influenza Virus: Antisense RNA and DNA*. Wiley-Liss, New York.

**Agrawal S, Sarin PS.** (1991) Antisense oligonucleotides: gene regulation and chemotherapy of AIDS. *Adv. Drug Delivery Rev.* **6**: 251–270.

**Agrawal S, Tang J-Y.** (1992) GEM*91 An antisense oligonucleotide phosphorothioate as a therapeutic agent for AIDS. *Antisense Res. Dev.* **2**: 261–266.

**Agrawal S, Mayrand SH, Zamecnik PC, Pederson T.** (1990) Site-specific excision from RNA by RNase H and mixed-phosphate-backbone oligodeoxynucleotides. *Proc. Natl Acad. Sci. USA* **87**: 1401–1405.

**Agrawal S, Temsamani J, Tang JY.** (1991) Pharmacokinetics, biodistribution and stability of oligodeoxythioates in mice. *Proc. Natl Acad. Sci. USA* **88**: 7595–7599.

Akhtar S, Shoji Y, Juliano RL. (1992) Intracellular distribution of antisense oligonucleotides. *Proc. Am. Assoc. Cancer Res.* **33:** 495.

Bacon TA, Wickstrom E. (1991) Walking along human c-*myc* mRNA with antisense oligodeoxynucleotides: maximum efficacy at the 5' cap site. *Oncogene Res.* **6:** 13–19.

Baker BF, Miraglia L, Hagedorn CH, Bennett CF. (1992) The 5' cap of messenger RNA. A novel target for antisense technologies. *Proc. Am. Assoc. Cancer Res.* **33:** 399.

Bayever E, Haines KH, Iverson PL, Spinolo J, Kay HD, Smith L. (1992) Antisense p53 oligonucleotides as potential human anti-leukemic agents. *Proc. Am. Assoc. Cancer Res.* **33:** 523.

Bennett MJ, Cullimore JV. (1992) Selective cleavage of closely-related mRNAs by synthetic ribozymes. *Nucleic Acids Res.* **20:** 831–837.

Bielinska A, Shivdasani RA, Zhang L, Nabel GJ. (1990) Regulation of gene expression with double stranded phosphorothioate oligonucleotides. *Science* **250:** 997–1000.

Bigelow JC, Chrin LR, Mathews LA, McCormack JJ. (1992) Pharmacokinetics of an "antisense" phosphorothioate oligodeoxynucleotide after subcutaneous infusion in mice. *Proc. Am. Assoc. Cancer Res.* **33:** 532.

Birg F, Praseuth D, Zerial A, Thuong NT, Asseline U, Le Doan T, Hélène C. (1990) Inhibition of simian virus 40 DNA replication in CV-1 cells by an oligonucleotide covalently linked to an intercalating reagent. *Nucleic Acids Res.* **18:** 2901–2908.

Calabretta B. (1991) Inhibition of proto-oncogene expression by antisense oligodeoxynucleotides: biological and therapeutic implications. *Cancer Res.* **51:** 4505–4510.

Camerini-Otero RD, Hsieh P. (1993) Parallel DNA triplexes, homologous recombination and other homology-dependent DNA interactions. *Cell* **73:** 217–223.

Cherny DI, Malkov VA, Volodin AA, Frankkamenetskii MD. (1993). Electron microscopy visualization of oligonucleotide binding to duplex DNA via triplex formation. *J. Mol. Biol.* **230:** 379–383.

Citro G, Szczylik C, Ginobbi P, Zupi G, Callabretta B. (1994) Inhibition of leukaemia cell proliferation by folic acid–polylysine-mediated introduction of c-*myb* antisense oligodeoxynucleotides into HL-60 cells. *Br. J. Cancer* **69:** 463–467.

Cohen JS. (1993) Selective antigene therapy for cancer – principles and prospects. *Tohoku J. Exp. Med.* **168:** 351–359.

Cooney M, Czernuszewicz G, Postel EH, Flint SJ, Hogan ME. (1988) Site-specific oligonucleotide binding represses transcription of human c-*myc* gene *in vitro*. *Science* **241:** 456.

Crooke ST. (1992) Review. Therapeutic applications of oligonucleotides. *BioTechnology* **10:** 883–886.

Crooke ST. (1993) Progress toward oligonucleotide therapeutics – pharmacodynamic properties. *FASEB J.* **7:** 533–539.

Daaka Y, Wickstrom E. (1990) Target dependence of antisense oligodeoxynucleotide inhibition of c-*Ha-ras* p21 expression and focus formation in T24-transformed NIH3T3 cells. *Oncogene Res.* **5:** 267–275.

Dagle JM, Walder JA, Weeks DL. (1990) Targeted degradation of mRNA in *Xenopus* oocytes and embryos directed by modified oligonucleotides. Studies of *An2* and cyclin in embryogenesis. *Nucleic Acids Res.* **18:** 4751–4757.

Duval-Valentin G, Thuong NT, Hélène C. (1992). Specific inhibition of transcription by triple helix forming oligonucleotides. *Proc. Natl Acad. Sci. USA* **89:** 504–508.

Eder PS, Walder RY, Walder JA. (1993) Substrate specificity of human RNase H and its role in excision repair of ribose residues misincorporated in DNA. *Biochemie* **75:** 1–2.

Escudé C, Sun J-S, Rougée M, Garestier T, Hélène C. (1992) Stable triple helices are formed upon binding of RNA oligonucleotides and their 2'-O-methyl derivatives to double-helical DNA. *Mol. Biophys.* **315:** (Série III): 521–525.

Florini JR, Ewton DZ. (1990) Highly specific inhibition of IGF-I stimulated differentiation by an antisense oligodeoxyribonucleotide to myogenin mRNA. *J. Biol. Chem.* **265:** 13435–13437.

Gao WY, Storm C, Egan W, Cheng YC. (1992) Cellular pharmacology of phosphorothioate oligodeoxynucleotides in human cells. *Proc. Am. Assoc. Cancer Res.* **33:** 495.

Ghosh MK, Ghosh K, Jaroszewski J, Cohen JS. (1992) Improved antisense oligonucleotides. *Proc. Am. Assoc. Cancer Res.* 33: 555.

Giles RV, Tidd DM. (1992a) Increased specificity for antisense oligodeoxynucleotide targeting of RNA cleavage by RNase H using chimeric methylphosphonodiester/ phosphodiester structures. *Nucleic Acids Res.* 20: 763–770.

Giles RV, Tidd DM. (1992b) Enhanced RNase H activity with methylphosphonodiester/ phosphodiester chimeric antisense oligodeoxynucleotides. *Anti-Cancer Drug Design* 7: 37–48.

Giovannangeli C, Montenay-Garestier T, Rougée M, Chassignol M, Thuong NT, Hélène C. (1991) Single stranded DNA as a target for triple-helix formation. *J. Am. Chem. Soc.* 113: 7775–7777.

Giovannangeli C, Thuong NT, Hélène C. (1992) Oligodeoxynucleotide-directed photo-induced cross-linking of HIV proviral DNA via triple-helix formation. *Nucleic Acids Res.* 20: 4275–4281.

Grigoriev M, Praseuth D, Robin P, Hemar A, Saison-Behmoaras T, Dautry-Varsat A, Thuong NT, Hélène C, Harel-Bellan A. (1992) A triple helix-forming oligonucleotide–intercalator conjugate acts as a transcriptional repressor via inhibition of NF κB binding to interleukin 2 receptor-alpha regulatory sequence. *J. Biol. Chem* 267: 3389–3395.

Grigoriev M, Praseuth D, Guieysse AL, Robin P, Thuong NT, Hélène C, Harel-Bellan A. (1993) Inhibition of gene expression by triple helix-directed DNA cross-linking at specific sites. *Proc. Natl Acad. Sci. USA* 90: 3501–3505.

Haseloff J, Gerlach WL. (1988) Simple RNA enzymes with new and highly specific endoribonuclease activities. *Nature* 334: 585–591.

Hélène C. (1991) The anti-gene strategy: control of gene expression by triplex forming oligonucleotides. *Anti-Cancer Drug Design* 6: 569–584.

Hijiya N, Zhang J, Ratajczak MZ, Kant JA, Deriel K, Herlyn M, Zon G, Gewirtz AM. (1994) Biologic and therapeutic significance of *myb* expression in human melanoma. *Proc. Natl Acad. Sci. USA* 91: 4499–4503.

Hoke GD, Draper K, Freier SM, Gonzalez C, Driver VB, Zounes MC, Ecker DJ. (1991) Effects of phosphorothioate capping on antisense oligonucleotide stability, hybridisation and antiviral efficacy versus herpes simplex virus infection. *Nucleic Acids Res.* 19: 5743–5748.

Hoogsteen K. (1959) The structure of crystals containing a hydrogen-bonded complex of 1-methylthymine and 9-methyladenine. *Acta Crystallogr.* 12: 822.

Iverson P, Cornish K, Johansson S, Foy X, Bergot J, Fredani J, Smith L, Arneson M, Bayever E, Spinolo J. (1992) Systemic human p53 antisense oligonucleotide in Rhesus monkey. *Proc. Am. Assoc. Cancer Res.* 33: 522.

Jaroszewski JW, Kaplan O, Syi JL, Sehested M, Faustino PJ, Cohen JS. (1990) Concerning antisense inhibition of the multiple drug resistance gene. *Cancer Commun.* 2: 287–294.

Kikuchi Y, Sasaki N. (1991) Site specific cleavage of natural mRNA sequences by newly designed hairpin catalytic RNAs. *Nucleic Acids Res.* 19: 6751–6755.

Kitajima I, Shinohara T, Minor T, Bibbs L, Bilakovics J, Nerenberg M. (1992a) Human T-cell leukemia-virus type-I *tax* transformation is associated with increased uptake of oligodeoxynucleotides *in vitro* and *in vivo*. *J. Biol. Chem.* 267: 25881–25888.

Kitajima I, Shinohara T, Bilakovics J, Brown DA, Xu X, Nerenberg M. (1992b) Ablation of transplanted HTLV-I *tax*-transformed tumors in mice by antisense inhibition of NF-κB. *Science* 258: 1792–1795.

Kool ET. (1991) Molecular recognition by circular oligonucleotides – increasing the selectivity of DNA binding. *J. Am. Chem. Soc.* 113: 6265–6266.

Koziolkeiwicz M, Stec WJ. (1992) Application of phosphate-backbone-modified oligonucleotides in the studies on *Eco*RI endonuclease mechanism of action. *Biochemistry* 31: 9460–9466.

Kuijpers WHA, Bos ES, Kasperson FM, Veeneman GH, van Boeckel CAA. (1993) Specific recognition of antibody–oligonucleotide conjugates by radiolabeled antisense nucleotides: a novel approach for two-step radioimmunotherapy of cancer. *Bioconjug. Chem.* 4: 94–102.

**Leiter JME, Agrawal S, Palese P, Zamecnik PC.** (1990) Inhibition of influenza virus replication by phosphorothioate oligodeoxynucleotides. *Proc. Natl Acad. Sci. USA* **87**: 3430–3434.

**Lemaitre M, Bayard B, Lebleu B.** (1987) Specific antiviral activity of a poly(L)lysine conjugated oligodeoxynucleotide sequence complementary to vesicular stomatitis virus N progein mRNA initiation site. *Proc. Natl Acad. Sci. USA* **84**: 648–652.

**Lesser DR, Grajkowkski A, Kurpiewski MR, Koziolkiewicz M, Stec WJ, Jen-Jacobsen L.** (1992) Stereoselective interaction with chiral phosphorothioates at the central DNA kink of the EcoRI endonuclease–GAATTC complex. *J Biol. Chem.* **267**: 24810–24818.

**Lisziewicz J, Sun D, Metelev V, Zamecnik P, Gallo RC, Agrawal S.** (1993) Long-term treatment of human immunodeficiency virus-infected cells with antisense oligonucleotide phosphorothioates. *Proc. Natl Acad. Sci. USA* **90**: 3860–3864.

**Loke SL, Stein CA, Zhang XH, Mori K, Nakanishi M, Subasinghe C, Cohen JS, Neckers LM.** (1989) Characterization of oligonucleotide transport into living cells. *Proc. Natl Acad. Sci. USA* **86**: 3474–3478.

**Mergny JL, Duval-Valentin G, Nguygen CH, Perrouault L, Faucon B, Rougée M, Montenay-Garestier T, Bisagni E, Hélène C.** (1992) Triple helix-specific ligands. *Science* **256**: 1681–1684.

**Miller PS, Cushman C** (1993) Triplex formation by oligodeoxynucleotides involving the formation of XUA triads. *Biochemistry* **32**: 2999–3004.

**Miller PS, T'so POP.** (1987) A new approach to chemotherapy based on molecular biology and nucleic acid chemistry: matagen (masking tape for gene expression). *Anti-Cancer Drug Design* **2**: 117–128.

**Miller PS, Agris CH, Aurelian L, Blake KR, Murakami M, Reddy P, Spitz SA, T'so POP.** (1985) Control of ribonucleic acid function by oligonucleoside methylphosphonates. *Biochemie* **67**: 769–776.

**Morgan AR, Wells RD.** (1968) Specificity of the three-stranded complex formation between double-stranded DNA and single stranded RNA containing repeating nucleotide sequences. *J. Mol. Biol.* **37**: 63.

**O'Keefe SJ, Wolfes H, Kiessling AA, Cooper GM.** (1989). Microinjection of antisense c-*mos* oligonucleotides prevents meiosis II in the maturing mouse egg. *Proc. Natl Acad. Sci. USA* **86**: 7038–7042.

**Paolella G, Sproat BS, Lamond AI.** (1992) Nuclease resistant ribozymes with high catalytic activity. *EMBO J.* **11**: 1913–1919.

**Picton S, Barton SL, Bouzayen M, Hamilton AJ, Grierson D.** (1993) Altered fruit ripening and leaf senescence in tomatoes expressing an antisense ethylene-forming enzyme transgene. *Plant J.* **3**: 469–481.

**Prakash G, Kool ET.** (1992) Structural effects in the recognition of DNA by circular oligonucleotides. *J. Am. Chem. Soc.* **114**: 3523–3527.

**Pyle AM, Murphy F, Cech TR.** (1992) RNA substrate binding site in the catalytic core of the *Tetrahymena* ribozyme. *Nature* **324**: 123–128.

**Reinis M, Damkova M, Kores E.** (1993) Receptor-mediated transport of oligonucleotides into hepatic cells. *J. Virol. Methods* **42**: 99–106.

**Rosner MH, De Santo RJ, Armhiter H, Staudt LM.** (1991) Oct-3 is a maternal factor required for the first mouse embryonic division. *Cell* **64**: 1103-1110.

**Ryte A, Morelli S, Mazzei M, Alama A, Franco P, Canti GF, Nicolin A.** (1993) Oligonucleotide degradation contributes to resistance to antisense compounds. *Anti-Cancer Drugs* **4**: 197–200.

**Sarver N, Cantin EM, Chang PS, Zaia JA, Ladne PA, Stephens DA, Rossi JJ.** (1990) Ribozymes as potential anti-HIV-1 therapeutic agents. *Science* **247**: 1222–1225.

**Scovassi AI, Stefanini M, Bertazzoni U.** (1984) Catalytic activities of human poly(ADP-ribose) polymerase from normal and mutagenized cells detected after sodium dodecyl sulfate–polyacrylamide gel electrophoresis. *J. Biol. Chem.* **259**: 10973–10977.

**Shea RG, Marsters JC, Bischofberger N.** (1990) Synthesis, hybridization properties and antiviral activity of lipid–oligodeoxynucleotide conjugates. *Nucleic Acids Res.* **18**: 3777–3783.

Shi Y, Fard A, Galeo A, Hutchinson HG, Vermani P, Dodge GR, Hall DJ, Shaheen F, Zalewski A. (1994) Transcatheter delivery of c-*myc* antisense oligomers reduces neointimal formation in a porcine model of coronary-artery balloon injury. *Circulation* 90: 944–951

Shuttleworth J, Coleman A. (1988) Antisense oligonucleotide-directed cleavage of mRNA in *Xenopus* oocytes and eggs. *EMBO J.* 7: 427–434.

Simons M, Edelman ER, DeKeyser J-L, Langer R, Rosenberg R. (1992) Antisense c-*myb* oligonucleotides inhibit intimal arterial smooth muscle cell accumulation *in vivo*. *Nature* 359: 67–70.

Skorski T, Nieborowskaskorska M, Nicolaides NC, Szczylik C, Iversen P, Iozzo RV, Zon G, Calabretta B. (1994) Suppression of Philadelphia(1) leukemia-cell growth in mice by *bcr-abl* antisense oligodeoxynucleotide. *Proc. Natl Acad. Sci. USA* 91: 4504–4508.

Smith RC, Bennett WM, Dersch MA, Dworkin-Rastl E, Dworkin MB, Capco D. (1990) *Development* 110: 769–779.

Spinolo J, Bayever E, Iverson P, Johansson S, Cornish K, Pirrucello S, Smith L, Arneson M. (1992) Toxicity of human p53 antisense oligonucleotide infusions in *Rhesus macacca*. *Proc. Am. Assoc. Cancer Res.* 33: 523.

Stec WJ, Grajkowski A, Koziolkiewicz M, Uznanski B. (1991) Novel route to oligodeoxyribonucleoside phosphorothioates. Stereocontrolled synthesis of P-chiral oligodeoxyribonucleoside phosphorothioates. *Nucleic Acids Res.* 19: 5883–5888.

Stein CA, Cheng YC. (1993) Antisense oligonucleotides as therapeutic agents, is the bullet really magical? *Science* 261: 1004–1012.

Stein CA, Subasinghe C, Shinosuka K, Cohen JS. (1988) Physicochemical properties of phosphorothioate oligodeoxynucleotides. *Nucleic Acids Res.* 16: 3209–3221.

Stein CA, Tonkinson J, Zhang LM, Yakubov L, Fields S, Delohery T, Krishna S, Taub R, Gervasoni J. (1992a) Internalisation of oligonucleotides in HL60 cells may be PKC dependent. *Proc. Am. Assoc. Cancer Res.* 33: 555.

Stein CA, Tonkinson J, Zhang LM, Yakubov L, Fields S, Delohery T, Krishna S, Taub R, Gervasoni J. (1992b) Interaction of oligonucleotides with HL60 cells. *Proc. Am. Assoc. Cancer Res.* 33: 555.

Stein CA, Tonkinson JL, Zhang LM, Yakubov L, Gerasoni J, Taub R, Rotenberg SA. (1993) Dynamics of the internalization of phosphodiester oligonucleotides in HL60 cells. *Biochemistry* 32: 4855–4861.

Stolt P, Zillig W. (1993) Antisense RNA mediates transcriptional processing in an archaebacterium. *Mol. Microbiol.* 7: 875–882.

Tang J-Y, Temsamani J, Agrawal S. (1993) Self-stabilized antisense oligodeoxynucleotide phosphorothioates: properties and anti-HIV activity. *Nucleic Acids Res.* 21: 2729–2735.

Temsamani J, Agrawal S, Pederson T. (1991) Biotinylated antisense methylphosphonate oligodeoxynucleotides. *J. Biol. Chem.* 266: 468–472.

Temsamani J, Tang J-Y, Padmapriya A, Kubert M, Agrawal S. (1993) Pharmacokinetics, biodistribution and stability of capped oligodeoxynucleotide phosphorothioates in mice. *Antisense Res. Dev.* 3: 277–284.

Tidd DM, Warenius HM. (1989) Partial protection of oncogene, antisense oligodeoxynucleotides against serum nuclease degradation using terminal methylphosphonate groups. *Br. J. Cancer* 60: 343–350.

Vonahsen U, Schroeder R. (1993) RNA as a catalyst – natural and designed ribozymes. *Bioessays* 15: 299–307.

Wang S, Kool ET. (1994) Circular RNA oligonucleotides. Synthesis, nucleic acid binding properties, and a comparison with circular DNAs. *Nucleic Acid Res.* 22: 2326–2333.

Wagner E, Zenke M, Cotten M, Beug H, Birnstiel ML. (1990) Transferrin–polycation conjugates as carriers for DNA uptake into cells. *Proc. Natl Acad. Sci. USA* 87: 3410–3414.

Wagner RW, Matteucci MD, Lewis JG, Gutierrez AJ, Moulds C, Froehler BC. (1993) Antisense gene inhibition by oligonucleotides containing C-5 propyne pyrimidines. *Science* 260: 1510–1513.

Whitesell L, Geselowitz D, Chavany C, Fahmy B, Walbridge S, Alger JR, Neckers LM. (1993) Stability, clearance and disposition of intravenously administered oligodeoxynucleotides – implications for therapeutic application within the central nervous system. *Proc. Natl Acad. Sci. USA* **90:** 4665–4669.

**Wickstrom E.** (1986) Oligodeoxynucleotide stability in subcellular extracts and culture media. *J. Biochem. Biophys. Methods* **13:** 97–102.

**Woolf TM, Jennings CG, Rebagliati M, Melton DA.** (1990) The stability, toxicity and effectiveness of unmodified and phosphorothioate antisense oligodeoxynucleotides in *Xenopus* oocytes and embryos. *Nucleic Acids Res.* **18:** 1763–1769.

**Woolf TM, Melton DA, Jennings CGB.** (1992) Specificity of antisense oligonucleotides *in vivo*. *Proc. Natl Acad. Sci. USA* **89:** 7305–7309.

**Wu GY, Wilson JM, Shalaby F, Grossman M, Shafritz DA, Wu CH.** (1991) Receptor-mediated gene delivery *in vivo*. Partial correction of genetic analbuminemia in Nagase rats. *J. Biol. Chem.* **266:** 14338–14342.

**Xing Z, Whitton JL.** (1992) Ribozymes which cleave arenavirus RNAs: identification of susceptible target sites and inhibition by target site secondary structure. *J. Virol.* **66:** 1361–1369.

**Zamencik PC, Stephenson ML.** (1978) Inhibition of Rous sarcoma virus replication and cell transformation by a specific oligodeoxynucleotide. *Proc. Natl Acad. Sci. USA* **75:** 285–288.

**Zaug AJ, Been MD, Cech TR.** (1986) The *Tetrahymena* ribozyme acts like an RNA restriction endonuclease. *Nature* **324:** 429–433.

# Index

Adenosine phosphoribosyltransferase (APRT), 98, 226
Adenovirus(es)
    DNA methylation, *see* DNA methylation
    integration sites, 157–163
    major late promoter of, 175
    mechanism of integration, 163–167
    molecular biology of, 157
    oncogenic potential, 158, 161
    types, 160
Allele-specific oligonucleotides (ASO), 49
*Alu* sequence elements, 12–13
    chromosomal location, 13
    consensus sequence, 12
    distribution, 12
    evolutionary origin, 12
    methylation, 13
    nucleosome positioning, 13
    number, 12
    'source genes', 13
    spacing, 12
    structure, 12
    subfamilies, 13
    transcription, 12–13
*Alu*-PCR, *see* Polymerase chain reaction
Alzheimer's disease, 194
Amethopterin, 223
Aminopterin, 223, 226
Amyotrophic lateral sclerosis, 196
Angelman syndrome, 28
Animal models of human disease
    cancer, 215
    cystic fibrosis, 214–215
    Gaucher's disease, 215
    generation by production of transgenic animals, 194, 197
    Lesch–Nyhan syndrome, 214
    murine, 207, 214–215
Antibiotic resistance, *see* Plasmid(s)
Antibodies, monoclonal, 221
Antigene strategies, *see* DNA, triplex formation
Antisense oligonucleotides, *see* Oligonucleotides, antisense
Apolipoprotein (a), 195
Arthritis, 195
Atherosclerosis, 195

Beckwith–Wiedemann syndrome, 27

Boundary elements, 19
Branchpoint, 22

Cap site, *see* Transcriptional initiation
cDNA
    3' end isolation, 79
    5' end isolation, 79–81
    cloning, 58, 73–74
    expression in transgenic animals, 189
    identification by differential display, 58, 76, 81
    mapping, *see* Transcription map, human genome
    preparation, 72–73
    rapid amplification of cDNA ends (RACE), 78–79, 116
cDNA library, construction, 73–74
cDNA library, screening, 58, 70–71, 74–77
    antibodies, 74–75
    nucleic acid probes, 75
    PCR-based methods, 76–77
    phage display, 75
    subtracted nucleic acid probe, 75–76
Cell immortalization, 221, 229, 230
Centromere, 6
    function, 6
    protein binding, 6
    structure, 6
Chediak–Higashi syndrome, 228–230
Chi element, 12
Chiasmata, 4–5
    chromosomal localization, 4–5
    frequency, 4
    triplex formation in, 252
Chimeras, 187, 208, 210
Chromatin, 2–3
    condensation, 2, 5
    decondensation, 2–3, 17
    structure, 2–3
Chromosome(s)
    deletions, 49
    evolution, 44
    flow-sorted, 47, 53
    maps, 47, 48, 50, 54
    translocations, 49
    'walking', 51
Chromosome band(s)
    CpG islands and, 8
    DNA replication and, 7

'flavour', 4, 8
genes and, 8
Chromosome jumping/linking libraries, see Jumping libraries *and* Linking libraries
Cloning
  cDNA, see cDNA
  gene
    functional, 46
    positional, 46, 109
    positional candidate, 46
Comparative gene mapping, 44, 58–59
  genomic mismatch scanning, 59
  mouse–human, 58, 59
  representational difference analysis, 59
Complementation
  expression cloning, 73
Complementation analysis, 221–236
Complementation groups
  assignment in biochemical pathways, 230
  assignment in cytokine-mediated signalling, 230
  assignment in disease, 228–229
  identification, 227–230
  in senescence, 229–230
  xeroderma pigmentosum, 228
Contig mapping, 44, 47, 51, 52
COS cells, 135–136
Cosmids, 44, 51, 52, 118, 133, 141
CpG dinucleotide
  density and transcription, 24
  distribution, 23
  frequency, 23
  methylation, 23, 24, 170
  mutation, 23
  specific binding proteins, 24
CpG island(s)
  association with genes, 23, 53
  chromosome bands, 8
  identification of, 53, 58
  mapping, 53
  methylation and X-inactivation, 25, 26
CpG suppression, 23
Cytogenetic mapping, 49–51
  chromosomal deletions and, 48, 49
  fluorescence *in situ* hybridization (FISH), 50
  radiation hybrid mapping, 50
  somatic cell hybrid mapping, 49
Cytoplasmic polyadenylation element, 21

D-segments
  definition, 47
  isolated, 47
  mapped, 47
Databases, nucleic acid, 58, 118
Diagnosis, disease, 44
Differential display, see cDNA

DNA
  bends, 3, 174–175
  chromosomal integration of ingested, 156
  left-handed, see Z-DNA
  repetitive, see Repetitive DNA
DNA methylation, 23–28
  as a defence mechanism, 171
  cell senescence and, 232
  CpG islands, 23
  *de novo*, 24, 26, 123, 155–156, 167, 168, 169, 170, 171
  demethylation, 24, 26
  developmental changes, 26
  Epstein–Barr virus and, 168
  gene activity and, 24, 173–175
  hemimethylation, 24
  imprinting and, see Imprinting
  integrated adenovirus and, 167–170
  maintenance, 24
  mutation and, 23, 25
  patterns
    alterations in as a consequence of foreign DNA insertion, 167–171, 172–173
    during embryogenesis, 26
    inheritance, 24
  promoter(s) and, 24–25
  promoter
    adenoviral and, 174–175
    late frog virus 3 and, 170–171, 173–174
    topology and, 25, 174–175
  protein binding and, 25
  role in DNA replication and repair, 25
  sequence specificity, 23
  Sp1 sites and, 25–26
  tissue-specific patterns, 24, 26
  transcription and, 24–26
  triplet repeat amplification and, 171
  X-inactivation and, see X-inactivation
DNA methyltransferase(s), 23, 24, 25, 170, 171, 174, 175
DNA polymorphisms
  definition, 46
  frequency, 46
  restriction fragment length (RFLPs), 46
DNA repair
  defects, see Xeroderma pigmentosum
DNA replication
  chromosomal bands and, 7
  imprinted genes and, 28
  origins, see Origins of DNA replication
  timing and genes, 7
  units of, 7
DNA sequencing, 44, 53
DNA triplex formation
  role in normal cellular processes, 252–253
  use as an antigene strategy, 251–253
DNase footprinting, 20

# INDEX

DNase I sensitivity
  association with preferrred sites of
    retroviral integration, 187
  and relationship with gene activity, 4
  locus control regions and, 17–18
Dominance/recessiveness
  developmentally regulated genes and,
    233–234
  in immunological processes, 233
  in senescence, 232
  in tumour progression, 232–233
  multi-drug resistance genes, 233
  mutations, 109
  phenotypes recovered by gene trap
    mutagenesis, 120
Dominant selectable markers, *see* Markers
Dynabeads, 76, 81

Electroporation, *see* Transfection
Embryonic stem cells, 119–120, 208–209
  as a target for insertional mutagenesis,
    119–120
Enhancer(s), 15–16
  function, 15–16
  structure, 15–16
  *trans*-acting, identification of, 193
  trapping, 58, 110–111
Exon(s), 7
  amplification/trapping, 58, 117
  number, 7
  prediction, 57, 58
  sharing, 8
  size, 7
Expressed sequence tags (EST), 47, 118
Expression systems
  cell-free, 241, 252
  *E. coli*, 73, 82
  insect, 83
  *Xenopus* oocytes, 73, 82
  yeast, 83
Expression vectors
  eukaryotic, 83
  fusion, 82–83
  prokaryotic, 82–83
Extinction, *see* Somatic cell hybrids

Fingerprinting, restriction enzyme, 52, 54
Flow-sorted chromosomes, 47, 53
Fluorescence *in situ* hybridization, 50, 54,
  117, 158
Fusion proteins, expression of, 58, 82–83

Gel retardation analysis, 20
Gene(s)
  association with *Not*I linking clones, 53
  chromosomally mapped, 46
  cloned, 45–46
  cloning, *see* Cloning
  clusters, 9–11
  density, by chromosome, 8
  developmental expression, studies of,
    11, 14
  developmentally regulated, 11, 17–18
  disease, 44
  entrapment, 109–123
  evolution, 9–10
  families, *see* Multigene families
  functional organization, 9–11
  housekeeping, 8, 14, 226
  identification, 57–58
  isolation, 44
  linear order and regulation, 11
  multi-drug resistance, 233
  number, *see* Human Genome Project
  overlapping, 8
  promoter, *see* Gene promoter(s)
  recognition of transcribed, 57–58
  reporter, *see* Reporter genes
  senescence control, 236
  structure, 7
  targeting strategies, *see* Gene targeting
  therapy, *see* Gene therapy
  tissue-specific, 8
  transfer, 131–151
  untranslated regions, 22, 23, 135
  use in mapping, 45–46
  within genes, 8
Gene inactivation
  cystic fibrosis transmembrane
    conductance regulator (*CFTR*),
    214–215
  glucocerebrosidase, 215
  hypoxanthine phosphoribosyltransferase
    (*HPRT*), 214
  *p53*, 215
  β-polymerase, 216
  retinoblastoma (*RB*), 215–216
  Wilms' tumour (*WT1*), 215
Gene mapping
  microcell fusion in, 235–236
  somatic cell hybridization in, 49, 221
Gene promoter(s)
  CCAAT box, 14
  *cis*-acting motifs, 14–15
  cloning, 117
  constitutive elements, 14
  cytomegalovirus, 133
  enhancer(s), *see* Enhancer(s)
  GC box, 14
  identification of *cis*-acting elements
    using transgenic animals, 193
  inducible, 15, 137–138, 139
  initiator (Inr) element, 15
  *LacZ*, 74, 121
  multiple, 14
  murine mammary tumour virus

269

(MMTV), 133
negative regulatory elements, see
Repressors, Silencers and
Boundary elements
octamer motif, 15
polyoma, 138
responsive elements, 15
Rous sarcoma virus (RSV), 133
simian virus 40 (SV40), 133
TATAA box, 14
Gene targeting
Cre/LoxP, 216
hit-and-run, 213
in–out, 213
tissue-specific, 216
Gene therapy, 178, 193, 199, 216, 255
Gene transfer and transcript identification, 58
Gene trap vectors
cellular genes disrupted by, 113
integration, 114, 115
selectable marker genes in, 111
structure, 111, 112
use in cloning flanking sequences, 116, 120–121
use in cloning regulated sequences, 110–111, 121–123
use in mutagenesis, 118–121, 122
Genetic ablation, 198
Genetic linkage analysis, 44
Genetic mapping
human genome, 44, 45, 54–57
concept, 54–55
historical, 54–55
progress, 57
use of microsatellites in, 55, 57
use of RFLPs in, 55
mammals, 58
primitive organisms, 52
Genome, human
gene number, 1, 7, 46
mapping, see Genome mapping
proportion expressed as mRNA, 57
sequencing of, see Human Genome Project
size, 1, 44
Genome mapping
comparative, see Comparative mapping
cytogenetic, see Cytogenetic mapping
genetic, see Genetic mapping
human, 43–59
reasons for, 44
long range restriction, 48
physical, see Physical mapping
Genomic imprinting, see Imprinting, genomic
Genomic mismatch scanning, 59

Haemoglobin switching, 15, 18
HAT medium, 146, 222, 226
Hepatocyte nuclear factor 3 (HNF3), 3

Heterokaryons, see Somatic cell hybrids
Heterozygosity, loss of, 235, 236
High mobility group (HMG) proteins, 3
Histone(s)
displacement, 3
interactions with DNA, 3
interactions with transcription factors, 3
Hit-and-run targeting, 213
Homeobox (*Hox*) genes, 9, 11, 195–196, 213–214
Homologous recombination, see Recombination
Human Genome Project, 43, 44
Hybridization
*in situ*, 45
subtractive, 58
Hypertriglyceridaemia, 195
Hypoxanthine phosphoribosyltransferase (HPRT)
function, 226
gene, 226
insertional inactivation, 214
mutagenesis, retroviral, 98
variants, 226–227

Imprinting, genomic
control regions, 28
DNA methylation and, 26, 27–28, 167
gene loci, 27–28
genetic disorders and, 27–28
In–out targeting, 213
Inflammatory disease, 195
Ingested DNA
integration, 156
uptake, 175–177
*In situ* hybridization, see Hybridization, *in situ*
*In situ* polymerase chain reaction, see Polymerase chain reaction, *in situ*
Inter-*Alu* PCR probes, 48–49, 50, 52, 54
Intron(s)
number, 7, 21
size, 7, 21
Inverse PCR, 100–101, 116

Jumping libraries, 53
Junk DNA, 1

Kinetochore, 6
Knock-out, see Gene inactivation

Lesch–Nyhan syndrome, 214, 224, 226
Ligands, identification of, 148
LINE (L1) elements, 13–14
chromosomal location, 13
consensus sequence, 13
evolutionary origin, 13
number, 13

# INDEX

promoter, 14
retrotransposition, 13
size, 13
transcription, 13–14
use in fingerprinting clones, 54
use in identification of human chromosomes, 50
Linking libraries, 53–54
Locus control region(s), 2, 15, 17–19
  α-globin, 18
  β-globin, 4, 11, 15, 17–18, 190, 198
  chromatin conformation, 17
  function, 17–18
  importance in transgenic animals, 190
  insertional inactivation of, 18
  origins of DNA replication, 17
Long terminal repeats (LTRs), 88, 90

Markers
  allele-specific oligonucleotides, 49
  anonymous DNA fragments (D-segments), 47
  chromosome breakpoints, 45, 48
  fragile sites, 45
  genes, 45–46
  inter-*Alu* PCR, 48–49
  sequence-tagged sites (STSs), 47
Markers, dominant selectable, 132
  adenine phosphoribosyltransferase, 146
  blasticine S deaminase, 147
  bleomycin/phleomycin/zeomycin, 146
  dihydrofolatereductase, 146, 147
  guanine phosphoribosyltransferase, 135
  histidinol dehydrogenase, 147
  hygromycin phosphotransferase 133, 138, 146, 225, 227
  hypoxanthine phosphoribosyltransferase, 224
  neomycin phosphotransferase, 135, 146, 227
  puromycin, 146
  thymidine kinase, 140, 146, 147, 227
  xanthine guanine phosphoribosyl-transferase, 146
Microcell fusion
  application to chromosomal localization of genes, 235
  identification of cellular senescence genes, 236
  identification of tumour suppressor genes, 235–236
  principles, 234
  severe combined immune deficiency, 235
  to study tumorigenesis, 235–236
  xeroderma pigmentosum, 235
Microinjection, 73, 82, 187, 243

mRNA
  binding proteins, *see* RNA-binding proteins
  instability motifs, 22
  isolation of, 72
  isolation of total RNA, 71–72
  splicing, *see* mRNA splicing
  stability, *see* mRNA stability
  transcripts, enrichment for, 76
  3′ untranslated region, 22
  5′ untranslated region, 23
mRNA splicing, 21–22
  accuracy, 21
  alternative, 21
  stages, 21
mRNA stability, 22
  role of A–U rich elements, 22
Multi-drug resistance genes, *see* Dominance
Multigene families, 9
Muscle differentiation, 213
Mutagenesis, retroviral insertional, 87–103, 119
  cloning of provirus integration site, 95, 99, 100
  conditions required, 95
  identification of mutated genes, 88, 100–102
  in cultured cells, 118–119
  in mice, 119–121
  mechanisms, 96
  mutant selection procedures, 95
  mutation frequency, 95, 98
  use for identification of novel genes, 88
Mutation
  back-, 226
  temperature-sensitive, 230
MyoD
  gene transfer and differentiation, 213, 233–234

Neurodegenerative disorders, 195
Northern blotting, 58, 72, 73, 76
Nuclear localization signal(s), 113
Nucleosome, 2–3
  assembly, 3
  positioning, 3–4
  structure, 3
Nucleotide
  biosynthetic pathways, 223–224, 225
  salvage pathways, 224, 225

Oligonucleotides
  allele-specific, 49
  design for PCR, 77–78
  redundant primers, 78
  use in cDNA library screening, 75

271

Oligonucleotides, antisense, 241–260
   α-anomer-containing, 252
   binding factors, cellular, 245
   cellular uptake, 245–246
   circular, 252, 253
   conjugation of, 245–246
   delivery *in vivo*, 258–260
   design, 243–245
   diastereoisomers, 248
   double-stranded, 253
   duplex stability, 249
   fate *in vivo*, 258–260
   Hoogsteen binding and, 251
   in archaebacteria, 243
   in plants, 243
   inhibition of
      host cell genes, 252
      oncogenes, 243
         smooth muscle proliferation, 258
      viral genes, 243, 251
   mechanisms of action of, 248–253
   methylphosphonate-modified, 245, 246, 247, 249
   microinjection into mouse oocytes, 241, 243
   microinjection into *Xenopus* oocytes, 241, 243
   non-specific activity of, 244
   nuclease resistance of, 246, 247
   optimal length, 244
   phosphodiester, 244
   phosphorothioate-modified, 244, 250
   poly-L-lysine conjugation, 245, 246
   propyne pyrimidine-containing, 248
   RNase H susceptibility, 242, 244, 249–251, 256, 257
   specificity, 255–258
   stability, 246–248
   targeting, 243–245
   therapeutic applications, 258–260
   toxicity, 246, 251
Oncogenes
   cellular (c-*onc*), 87
   dominance/recessiveness, 232–233
   targeted by antisense oligonucleotides, 242, 243, 251
   viral (v-*onc*), 87
Oncogenesis, 177–178
Origins of DNA replication, 7
  consensus sequence, 7
  locus control regions and, 17
  plasmid, *see* Plasmid(s)
  viral, 134

Phage display, *see* cDNA library screening
Phagemid(s), 76
Physical mapping, human genome, 44, 45, 51–54
   chromosome jumping/linking libraries, 53–54
   chromosome maps, 54
   contig assembly, 52
   DNA sequencing, 54
   progress in, 54
   pulsed-field gel electrophoresis and CpG island mapping, 52–53
   yeast artificial chromosome (YAC), 51
Plasmid(s)
   antibiotic resistance genes, 132
   bovine papilloma virus-based, 139
   double insert, 136–137
   Epstein–Barr virus-based, 138–139
   multifunctional, 140
   origins of replication, 132–133, 134
   SV40-based, 134–135
Poly(A)
   addition, 20–21
   signal, 21, 135
   signal trapping, 58
Polyadenylation complex, 21
Polymerase chain reaction (PCR), 43, 77
   *Alu*-, 48, 49, 50, 52, 54
   amplification of very long DNA fragments, 77, 116
   in cDNA isolation, 76–81
   *in situ*, 51
   inverse PCR, *see* Inverse PCR
   primer design, 77–78
   reverse transcript, 58
Polymorphism, *see* DNA polymorphisms
Positive–negative selection (PNS) procedure, 212
Promoter(s), *see* Gene promoter(s)
Promoter-tagged sites isolation, 116–118
Promoterless constructs, 110, 212
Pseudoautosomal region, 5, 11
Pseudogene(s), 8–9, 118
   frequency, 9
   transcription, 8–9
   types, 8–9
Pulsed-field gel electrophoresis, 48, 52–53, 102
Purine biosynthesis, *see* Nucleotide biosynthetic pathways
Pyrimidine biosynthesis, *see* Nucleotide biosynthetic pathways
Pyrimidine tract, 22

Radiation hybrid mapping, 45, 48, 50
Rapid amplification of cDNA ends (RACE), *see* cDNA
Rare cutter restriction sites, 53
Recessiveness, *see* Dominance/recessiveness
Recombination,
   chiasmata, *see* Chiasmata

obligate within pseudoautosomal region, 5
sex difference, 4–5
transcription and, 5
triplex formation and, 252
Recombination hotspots
  chi element, 12
  repetitive DNA and, 5
  topoisomerase I cleavage sites, 5
Recombination, homologous, 207–219
  as a tool for identification of novel gene transcripts, 58
  efficiency, 212
  gene therapy, 216
  principles, 209–213
  study in a cell-free system, 165–167
Refsum disease, 229
Repetitive DNA
  alphoid, 12
  LINE (L1) elements, see LINE elements
  microsatellite, 12, 47
  minisatellite, 12
  Sau3A sequence, 12
  simple sequence, 12
  tandem repeats, 11–12
  telomeric, 12
Reporter genes, 15
  chloramphenicol acetyltransferase, 15, 147–148
  β-galactosidase, 147–148
  luciferase, 15, 148
  use in transgenic analysis, 190
Representational difference analysis, 59
Repressor(s), 16–17
Responsive elements, see Gene promoter(s)
Restenosis, 243
Restriction enzyme(s)
  maps, 52, 53
  rare cutters, 53
Restriction fragment length polymorphisms (RFLPs)
  frequency, 46
  use in mapping, 46, 55
Restriction mapping, 52–53
Retroposons, 14
Retrotranscripts, 14
Retroviruses
  amphotropic, 93
  cell surface receptors for, 90, 91–92
  cloning of integration site, 95, 99, 100, 102
  ecotropic, 92, 93
  endogenous, 14, 94, 95
  enhancer sequences, 88, 90, 96
  gene activation after integration into host cell genome, 96
  gene inactivation after integration into host cell genome, 96
  helper virus, 92, 93
  host range, 91–92
  infection, 95, 113
  infection efficiency, 93–94
  infection to create transgenic animals, 187
  insertional mutagenesis, see Mutagenesis, retroviral insertional
  integrase, 88
  integration into host cellular genome, 88, 96, 97
  life cycle, 88–90
  long terminal repeats (LTRs), 88, 90, 92, 96
  mutant selection procedures, 95, 99
  open reading frames (gag, pol, env), 88
  packaging cell lines, 93–95
  packaging (ψ) sequence, 88, 90
  preferred integration sites, 95, 115, 187
  provirus, 88
  receptor-mediated endocytosis, 90
  replication, 88, 90
  replication-competent, 93, 94, 99
  replication-defective, 93, 99
  reverse transcriptase activity, 90, 91
  role in identification of oncogenes, 87–88
  safety considerations associated with, 94
  self-inactivating vectors, 90
  slow transforming, 87
  super-infection, 99
  titres, 90, 93–94
Reverse transcriptase, 72
Ribozymes, 253–255
  mechanism of cleavage, 254
  naturally occurring, 254
  specificity of cleavage, 254, 255
  stability, 255
  structure, 254
  synthetic, 254
  uptake, 255
  use to reduce level of target mRNA, 254
RNA-binding proteins, 21
  cleavage poladenylation specificity factor (CPSF), 21
  cleavage stimulatory factor (CstF), 21
  translational regulation and, 23
RNase H, protection assay, 72
RNase H, cleavage of mRNA/DNA hybrids, 72–73, 91
Run-on assays, 102

Scaffold-associated regions (SARs), 4–6
  association with enhancers, 5–6
  association with topoisomerase II cleavage sites, 5
  DNA sequence characteristics, 5
  functional role, 5–6
  proteins binding, 5–6
  role as boundary elements, 4, 19
Selectable markers, see Markers, dominant selectable

Sendai virus, 222
Senescence, cellular, *see* Complementation groups, Dominance/recessiveness *and* DNA methylation
Sequence-tagged sites (STSs)
   expressed sequence tags (ESTs), 47, 118
   isolated, 47
   maps, 45
   microsatellite repeat, 47
   sources, 47
   use in mapping, 47, 48
Shuttle vector(s), 116–117, 120, 140
Sickle cell disease, 198
Silencer(s), 17, 18, 193
Small ribonucleoprotein particles (snRNPs), 22
Somatic cell hybrid analysis, 222–227
   chromosome loss in, 49, 231
   extinction of functions in, 49, 227–228
   generation of, 222
   heterokaryons in, 222
   metabolic cooperation in, 225
   principle, 49
   re-expression of functions in, 228
   selection procedures for, 222, 226–227
   use in gene mapping, 49
Southern blotting, 176, 210, 211, 231
Sp1 sites and *de novo* methylation, 24
Splice junctions, 21–22
   retroviral gene traps and, 111–113, 115
Splicing, *see* mRNA splicing
Spliceosome, 22
Subtractive hybridization, 58, 76
Superoxide dismutase, 196–197
Synteny, 9–11, 58

Telomerase, 6
Telomeres, 6
   DNA sequence characteristics, 6, 12
   DNA synthesis, 6
   function, 6
   proteins binding, 6
   recombination in vicinity of, 5
Toxigene, 198–199
*Trans*-acting factors, *see* Transcription factors
Transcription
   DNase sensitivity, 4
   initiation, 14–15
   levels, 118
   nucleosome assembly and, 3
   regulation, *see* Gene promoter
   repression, 3
   scaffold-associated regions and, 5–6
   termination, 20–21
   units of, 4
Transcription factors
   basal, 15
   DNA binding domains, 20

GATA1, 15, 18
HNF1, 15
identification of, 149–151
MyoD, 15
multimerization domains, 20
transcriptional activation domains, 20
Transcription map, human genome, 57
   rationale, 57
   requirements, 57
Transcriptional domains, 4
   boundaries, 19
Transcriptional initiation
   complex, 15, 20
   site(s), 14, 69
   sites and retroviral integration, 115
Transfection
   adenovirus-mediated, 144, 145
   biolistic methods, 145
   calcium phosphate coprecipitation, 141–142
   cotransfection, 136
   DEAE–dextran technique, 142
   definition, 131–132
   efficiency and DNA size, 141
   efficiency and DNA topology, 141
   electroporation, 143
   fate of introduced DNA, 131–132, 133
   liposome-mediated, 143
   microinjection, 145
   polybrene/DMSO, 145
   procedures, 140–141
   protoplast fusion, 145
   quality of DNA used in, 132–133, 141
   receptor-mediated, 144
   scrape loading, 145
   stable integration of DNA, 132, 133
   transferrin–poly-L-lysine, 144
   transient assays, 132, 133
Transgene(s), 185–200
   antisense, 197
   apolipoprotein (a), 195
   apolipoproteins AI/CIII, 195
   boundary elements and, 19
   collagen, 197
   dominant negative mutation, 197
   ectopic expression, 195–196
   expression, 189–193
   expression of mutated, 196–197, 198
   β–globin, 189, 190, 198
   genetic ablation and, 198
   growth hormone, 195
   homeobox, 195–196
   importance of *cis*-acting regulatory elements, 189–190
   importance of introns, 189
   influence of locus control regions, 18, 190
   inheritance, 188

integration, 187
methylation, 27
mosaicism, 187
nerve growth factor, 195
oncogenes and tumorigenesis, 196
overexpression, 194–195
position effects, 4
superoxide dismutase, 196–197
T-cell receptor, 197–198
tissue plasminogen activator, 190–191
yeast artificial chromosome, 194
Transgenic animals, 185–200
   as a tool to study gene function, 193–199
   as a tool to study gene regulation, 189–193
   atherosclerosis in, 195
   early experiments, 186, 187
   expression of cDNA in, 189
   gain-of-function, 194–197
   genetic ablation in, 198
   loss-of-function, 197–199
   methodology, 187–189
   motor neuron disease in, 196
   mice, 188
   osteogenesis imperfecta in, 197
   promoter mapping in, 189–193
   rats, 188
   studies of locus control regions in, 18–19
   tissue-specificity of viral promoters, 191
   tumours in, 196
   use of reporter genes in, 190
Translation, control of
   role of 5′ untranslated region, 23
Tumours
   adenovirus infection and, 158, 161

in transgenic animals, 196
inhibition of growth by antisense oligonucleotides, 243

Unstable trinucleotide repeats, 44
   DNA methylation and, 171
Vectors
   gene trap, 110–115
   insertion, 209
   lambda, 73, 74
   plasmid, 73
   replacement, 209
   retroviral, 87–93
   self-inactivating, 90
   shuttle, *see* Shuttle vector(s)
   targeting, 209–211

Xenopus oocytes, 73, 82, 241, 243
Xeroderma pigmentosum, 221, 228, 235
X-inactivation, 25–26
   centre, 28
   DNA methylation, 25–26
   genes which escape, 26
*XIST*, 28

Yeast artificial chromosomes (YACs), 44, 45, 51, 52, 54, 109, 118, 194

Z-DNA, 3, 5
   recombination and, 5
Zellweger's syndrome, 231
Zoo blots, 58

# ORDERING DETAILS

**Main address for orders**

BIOS Scientific Publishers Ltd
9 Newtec Place, Magdalen Road,
Oxford OX4 1RE, UK
Tel: +44 1865 726286
Fax: +44 1865 246823

**Australia and New Zealand**
DA Information Services
648 Whitehorse Road, Mitcham, Victoria 3132, Australia
Tel: (03) 873 4411
Fax: (03) 873 5679

**India**
Viva Books Private Ltd
4346/4C Ansari Road, New Delhi 110 002, India
Tel: 11 3283121
Fax: 11 3267224

**Singapore and South East Asia**
(Brunei, Hong Kong, Indonesia, Korea, Malaysia, the Philippines,
Singapore, Taiwan, and Thailand)
Toppan Company (S) PTE Ltd
38 Liu Fang Road, Jurong, Singapore 2262
Tel: (265) 6666
Fax: (261) 7875

**USA and Canada**
Books International Inc
PO Box 605, Herndon, VA 22070, USA
Tel: (703) 435 7064
Fax: (703) 689 0660

Payment can be made by cheque or credit card (Visa/Mastercard, quoting number and expiry date). Alternatively, a *pro forma* invoice can be sent.

Prepaid orders must include £2.50/US$5.00 to cover postage and packing for one item and £1.25/US$2.50 for each additional item.